Vector Analysis

K. A. Stroud
Formerly Principal Lecturer
Department of Mathematics
Coventry University

Dexter J. Booth
Principal Lecturer
School of Computing and Engineering
University of Huddersfield

INDUSTRIAL PRESS, INC.
NEW YORK

Library of Congress Cataloging-in-Publication Data
Stroud, K. A.
   Vector analysis / K.A. Stroud, Dexter Booth.
      p. cm.
   ISBN 0-8311-3208-6
      1. Vector analysis--Textbooks.   I. Booth, Dexter J. II. Title.

QA433.S77 2005
515'.63--dc22

2005043301

Published under license from Palgrave Publishers Ltd,
Houndmills, Basingstoke Hants RG21 6XS, United Kingdsom

Industrial Press Inc.
989 Avenue of the Americas
New York, NY 10018

Copyright © 2005 by Industrial Press Inc., New York,
Printed in the United States of America. All rights reserved.
This book, or parts thereof may not be reproduced, stored in a retrieval system, or transmitted in any form without the permission of the publisher.

2   3   4   5   6   7   8   9   10

Contents

Hints on using this book   vii
Useful background information   ix
Preface   xiii

Program 1   Partial differentiation   1

Learning outcomes   1
Partial differentiation   2
  Review exercise   14
Small increments   16
Can You? Checklist 1   22
Test exercise 1   22
Further problems 1   23

Program 2   Applications of partial differentiation   25

Learning outcomes   25
Introduction   26
Rate-of-change problems   29
Change of variables   37
Inverse functions   43
Review summary   51
Can You? Checklist 2   52
Test exercise 2   53
Further problems 2   54

Program 3   Polar coordinates   56

Learning outcomes   56
Introduction   57
Polar curves   59
Standard polar curves   61
Review summary   76
Can You? Checklist 3   77
Test exercise 3   77
Further problems 3   78

Program 4   Multiple integrals   80

Learning outcomes   80
Summation in two directions   81
Double integrals   84

Triple integrals	86
Applications	88
Review exercise	92
Alternative notation	93
Determination of volumes by multiple integrals	99
Review summary	102
Can You? Checklist 4	105
Test exercise 4	105
Further problems 4	106

Program 5 Differentials and line integrals — 108

Learning outcomes	108
Differentials	109
Exact differential	112
Integration of exact differentials	113
Area enclosed by a closed curve	115
Line integrals	119
Alternative form of a line integral	120
Properties of line integrals	123
Regions enclosed by closed curves	125
Line integrals round a closed curve	126
Line integral with respect to arc length	130
Parametric equations	131
Dependence of the line integral on the path of integration	132
Exact differentials in three independent variables	137
Green's theorem in the plane	138
Review summary	145
Can You? Checklist 5	147
Test exercise 5	148
Further problems 5	149

Program 6 Surface and volume integrals — 151

Learning outcomes	151
Double integrals	152
Surface integrals	157
Space coordinate systems	163
Cartesian coordinates	163
Cylindrical coordinates	164
Spherical coordinates	165
Element of volume in space in the three coordinate systems	167
Volume integrals	168
Alternative method	171
Change of variables in multiple integrals	177
Curvilinear coordinates	179
Transformation in three dimensions	187
Review summary	189

Can You? Checklist 6	191
Test exercise 6	192
Further problems 6	192

Program 7 Vectors 195

Learning outcomes	195
Introduction: scalar and vector quantities	196
Vector representation	197
Components of a given vector	201
Components of a vector in terms of unit vectors	205
Vectors in space	207
Direction cosines	209
Scalar product of two vectors	210
Vector product of two vectors	212
Angle between two vectors	215
Direction ratios	218
Triple products	218
Properties of scalar triple products	219
Coplanar vectors	220
Vector triple products of three vectors	222
Review summary	225
Can You? Checklist 7	227
Test exercise 7	228
Further problems 7	228

Program 8 Vector differentiation 230

Learning outcomes	230
Differentiation of vectors	231
Differentiation of sums and products of vectors	235
Unit tangent vectors	236
Partial differentiation of vectors	238
Integration of vector functions	239
Scalar and vector fields	241
Grad (gradient of a scalar field)	242
Directional derivatives	245
Unit normal vectors	248
Grad of sums and products of scalars	250
Div (divergence of a vector function)	251
Curl (curl of a vector function)	252
Summary of grad, div and curl	254
Multiple operations	255
Review summary	258
Can You? Checklist 8	260
Test exercise 8	261
Further problems 8	261

Program 9   Vector integration — 263

Learning outcomes	263
Line integrals	264
Scalar field	264
Vector field	267
Volume integrals	271
Surface integrals	275
Scalar fields	276
Vector fields	279
Conservative vector fields	284
Divergence theorem (Gauss' theorem)	289
Stokes' theorem	295
Direction of unit normal vectors to a surface S	298
Green's theorem	304
Review summary	307
Can You? Checklist 9	309
Test exercise 9	310
Further problems 9	311

Program 10   Curvilinear coordinates — 314

Learning outcomes	314
Curvilinear coordinates	315
Orthogonal curvilinear coordinates	319
Orthogonal coordinate systems in space	320
Scale factors	324
Scale factors for coordinate systems	325
General curvilinear coordinate system (u, v, w)	327
Transformation equations	328
Element of arc ds and element of volume dV in orthogonal curvilinear coordinates	329
Grad, div and curl in orthogonal curvilinear coordinates	330
Particular orthogonal systems	333
Review summary	335
Can You? Checklist 10	337
Test exercise 10	338
Further problems 10	339

Answers	340
Index	345

Hints on using this book

This book contains lessons called *Programs*. Each Program has been written in such a way as to make learning more effective and more interesting. It is like having a personal tutor because you proceed at your own rate of learning and any difficulties you may have are cleared before you have the chance to practise incorrect ideas or techniques.

You will find that each Program is divided into numbered sections called *frames*. When you start a Program, begin at Frame 1. Read each frame carefully and carry out any instructions or exercise that you are asked to do. In almost every frame, you are required to make a response of some kind, testing your understanding of the information in the frame, and you can immediately compare your answer with the correct answer given in the next frame. To obtain the greatest benefit, you are strongly advised to cover up the following frame until you have made your response. When a series of dots occurs, you are expected to supply the missing word, phrase, number or mathematical expression. At every stage you will be guided along the right path. There is no need to hurry: read the frames carefully and follow the directions exactly. In this way, you must learn.

Each Program opens with a list of **Learning outcomes** which specify exactly what you will learn by studying the contents of the Program. The Program ends with a matching checklist of **Can You?** questions that enables you to rate your success in having achieved the **Learning outcomes**. If you feel sufficiently confident then tackle the short **Test exercise** which follows. This is set directly on what you have learned in the Program: the questions are straightforward and contain no tricks. To provide you with the necessary practice, a set of **Further problems** is also included: do as many of these problems as you can. Remember, that in mathematics, as in many other situations, practice makes perfect – or more nearly so.

Useful background information

Symbols used in the text

$=$	is equal to	$\rightarrow$	tends to
$\approx$	is approximately equal to	$\neq$	is not equal to
$>$	is greater than	$\equiv$	is identical to
$\geq$	is greater than or equal to	$<$	is less than
$n!$	factorial $n = 1 \times 2 \times 3 \times \ldots \times n$	$\leq$	is less than or equal to
$\|k\|$	modulus of k, i.e. size of k irrespective of sign	∞	infinity
$\sum$	summation	$\underset{n \to \infty}{\text{Lim}}$	limiting value as $n \to \infty$

Useful mathematical information

1 Algebraic identities

$$(a+b)^2 = a^2 + 2ab + b^2 \qquad (a+b)^3 = a^3 + 3a^2b + 3ab^2 + b^3$$
$$(a-b)^2 = a^2 - 2ab + b^2 \qquad (a-b)^3 = a^3 - 3a^2b + 3ab^2 - b^3$$
$$(a+b)^4 = a^4 + 4a^3b + 6a^2b^2 + 4ab^3 + b^4$$
$$(a-b)^4 = a^4 - 4a^3b + 6a^2b^2 - 4ab^3 + b^4$$
$$a^2 - b^2 = (a-b)(a+b) \qquad a^3 - b^3 = (a-b)(a^2 + ab + b^2)$$
$$a^3 + b^3 = (a+b)(a^2 - ab + b^2)$$

2 Trigonometrical identities

(a) $\sin^2 \theta + \cos^2 \theta = 1$; $\sec^2 \theta = 1 + \tan^2 \theta$; $\operatorname{cosec}^2 \theta = 1 + \cot^2 \theta$

(b) $\sin(A + B) = \sin A \cos B + \cos A \sin B$

$\sin(A - B) = \sin A \cos B - \cos A \sin B$

$\cos(A + B) = \cos A \cos B - \sin A \sin B$

$\cos(A - B) = \cos A \cos B + \sin A \sin B$

$\tan(A + B) = \dfrac{\tan A + \tan B}{1 - \tan A \tan B}$

$\tan(A - B) = \dfrac{\tan A - \tan B}{1 + \tan A \tan B}$

(c) Let $A = B = \theta$ $\therefore$ $\sin 2\theta = 2 \sin \theta \cos \theta$

$$\cos 2\theta = \cos^2 \theta - \sin^2 \theta = 1 - 2 \sin^2 \theta = 2 \cos^2 \theta - 1$$

$$\tan 2\theta = \dfrac{2 \tan \theta}{1 - \tan^2 \theta}$$

(d) Let $\theta = \dfrac{\phi}{2}$ $\therefore$ $\sin\phi = 2\sin\dfrac{\phi}{2}\cos\dfrac{\phi}{2}$

$$\cos\phi = \cos^2\dfrac{\phi}{2} - \sin^2\dfrac{\phi}{2} = 1 - 2\sin^2\dfrac{\phi}{2} = 2\cos^2\dfrac{\phi}{2} - 1$$

$$\tan\phi = \dfrac{2\tan\dfrac{\phi}{2}}{1 - 2\tan^2\dfrac{\phi}{2}}$$

(e) $\sin C + \sin D = 2\sin\dfrac{C+D}{2}\cos\dfrac{C-D}{2}$

$\sin C - \sin D = 2\cos\dfrac{C+D}{2}\sin\dfrac{C-D}{2}$

$\cos C + \cos D = 2\cos\dfrac{C+D}{2}\cos\dfrac{C-D}{2}$

$\cos D - \cos C = 2\sin\dfrac{C+D}{2}\sin\dfrac{C-D}{2}$

(f) $2\sin A\cos B = \sin(A+B) + \sin(A-B)$
$2\cos A\sin B = \sin(A+B) - \sin(A-B)$
$2\cos A\cos B = \cos(A+B) + \cos(A-B)$
$2\sin A\sin B = \cos(A-B) - \cos(A+B)$

(g) Negative angles: $\sin(-\theta) = -\sin\theta$
$\cos(-\theta) = \cos\theta$
$\tan(-\theta) = -\tan\theta$

(h) Angles having the same trigonometrical ratios:
  (i) Same sine: θ and $(180° - \theta)$
  (ii) Same cosine: θ and $(360° - \theta)$, i.e. $(-\theta)$
  (iii) Same tangent: θ and $(180° + \theta)$

(i) $a\sin\theta + b\cos\theta = A\sin(\theta + \alpha)$
$a\sin\theta - b\cos\theta = A\sin(\theta - \alpha)$
$a\cos\theta + b\sin\theta = A\cos(\theta - \alpha)$
$a\cos\theta - b\sin\theta = A\cos(\theta + \alpha)$

where $\begin{cases} A = \sqrt{a^2 + b^2} \\ \alpha = \tan^{-1}\dfrac{b}{a} \quad (0° < \alpha < 90°) \end{cases}$

3 Standard curves
(a) *Straight line*

Slope, $m = \dfrac{dy}{dx} = \dfrac{y_2 - y_1}{x_2 - x_1}$

Angle between two lines, $\tan\theta = \dfrac{m_2 - m_1}{1 + m_1 m_2}$

For parallel lines, $m_2 = m_1$
For perpendicular lines, $m_1 m_2 = -1$

Equation of a straight line (slope $= m$)
(i) Intercept c on real y-axis: $y = mx + c$
(ii) Passing through (x_1, y_1): $y - y_1 = m(x - x_1)$
(iii) Joining (x_1, y_1) and (x_2, y_2): $\dfrac{y - y_1}{y_2 - y_1} = \dfrac{x - x_1}{x_2 - x_1}$

(b) *Circle*

Centre at origin, radius r: $\quad x^2 + y^2 = r^2$

Centre (h, k), radius r: $\quad (x - h)^2 + (y - k)^2 = r^2$

General equation: $\quad x^2 + y^2 + 2gx + 2fy + c = 0$

with centre $(-g, -f)$: radius $= \sqrt{g^2 + f^2 - c}$

Parametric equations: $x = r\cos\theta,\ y = r\sin\theta$

(c) *Parabola*

Vertex at origin, focus $(a, 0)$: $\quad y^2 = 4ax$

Parametric equations: $\quad x = at^2,\ y = 2at$

(d) *Ellipse*

Centre at origin, foci $\left(\pm\sqrt{a^2 + b^2}, 0\right)$: $\dfrac{x^2}{a^2} + \dfrac{y^2}{b^2} = 1$

where $a =$ semi-major axis, $b =$ semi-minor axis

Parametric equations: $x = a\cos\theta,\ y = b\sin\theta$

(e) *Hyperbola*

Centre at origin, foci $\left(\pm\sqrt{a^2 + b^2}, 0\right)$: $\dfrac{x^2}{a^2} - \dfrac{y^2}{b^2} = 1$

Parametric equations: $x = a\sec\theta,\ y = b\tan\theta$

Rectangular hyperbola:

Centre at origin, vertex $\pm\left(\dfrac{a}{\sqrt{2}}, \dfrac{a}{\sqrt{2}}\right)$: $xy = \dfrac{a^2}{2} = c^2$

i.e. $xy = c^2 \qquad\qquad$ where $c = \dfrac{a}{\sqrt{2}}$

Parametric equations: $x = ct,\ y = c/t$

4 Laws of mathematics

(a) *Associative laws* – for addition and multiplication

$a + (b + c) = (a + b) + c$

$a(bc) = (ab)c$

(b) *Commutative laws* – for addition and multiplication

$a + b = b + a$

$ab = ba$

(c) *Distributive laws* – for multiplication and division

$a(b + c) = ab + ac$

$\dfrac{b + c}{a} = \dfrac{b}{a} + \dfrac{c}{a}$ (provided $a \neq 0$)

Preface

It is now over 35 years since Ken Stroud first developed his approach to personalized learning with his classic text *Engineering Mathematics*, now in its fifth edition. That unique and hugely successful programmed learning style is exemplified in this text and I am delighted to have been asked to contribute to it. I have endeavored to retain the very essence of his style that has contributed to so many students' mathematical abilities over the years, particularly the time-tested Stroud format with its close attention to technique development throughout.

The first two Programs deal with the elements and application of partial differentiation and present a standard approach with a thorough coverage. The next Program introduces polar coordinates to acclimatize the student to the possibility of different coordinate systems. Program 4 extends the students' awareness of the integration of functions of a single variable to integrals of two and three variables. Program 5 then discusses the concept of a differential and applies this to line integrals, culminating in Green's theorem in the plane. Program 6 looks at surface and volume integrals including the change of variables; the Program ends with an introduction to curvilinear coordinates. The next three Programs deal with the essence of the vector calculus, covering the algebra of vectors, vector fields, their calculus, and the integral theorems. The final Program deals with curvilinear coordinates as applied to all that was learned in the previous Programs dealing with the vector calculus.

To give the student as much assistance as possible in organizing their study there are specific **Learning outcomes** at the beginning and **Can You?** checklists at the end of each Program. In this way, the learning experience is made more explicit and the student is given greater confidence in what has been learnt. Test exercises and Further problems follow, in which the student can consolidate their newly-found knowledge.

This is the fourth opportunity that I have had to work on the Stroud books, having made additions to both the *Engineering Mathematics* and *Advanced Engineering Mathematics* texts. It is as ever a challenge and an honor to be able to work with Ken Stroud's material. Ken had an understanding of his students and their learning and thinking processes which was second to none, and this is reflected in every page of this book. As always my thanks go to the Stroud family for their continuing support for and encouragement of new projects and ideas which are allowing Ken's work an ever wider public.

Huddersfield Dexter J Booth
February 2005

Program 1

Partial differentiation

Frames 1 to 40

Learning outcomes

When you have completed this Program you will be able to:
- Find the first partial derivatives of a function of two real variables
- Find second-order partial derivatives of a function of two real variables
- Calculate errors using partial differentiation

Partial differentiation

1

The volume V of a cylinder of radius r and height h is given by

$$V = \pi r^2 h$$

i.e. V depends on two quantities, the values of r and h.

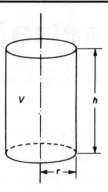

If we keep r constant and increase the height h, the volume V will increase. In these circumstances, we can consider the derivative of V with respect to h – but only if r is kept constant.

i.e. $\left[\dfrac{dV}{dh}\right]_{r \text{ constant}}$ is written $\dfrac{\partial V}{\partial x}$

Notice the new type of 'delta'. We already know the meaning of $\dfrac{\delta y}{\delta x}$ and $\dfrac{dy}{dx}$. Now we have a new one, $\dfrac{\partial V}{\partial h}$. $\dfrac{\partial V}{\partial h}$ is called the *partial derivative* of V with respect to h and implies that for our present purpose, the value of r is considered as being kept ....

2

constant

$V = \pi r^2 h$. To find $\dfrac{\partial V}{\partial h}$, we differentiate the given expression, taking all symbols except V and h as being constant $\therefore \dfrac{\partial V}{\partial h} = \pi r^2 \cdot 1 = \pi r^2$

Of course, we could have considered h as being kept constant, in which case, a change in r would also produce a change in V. We can therefore talk about $\dfrac{\partial V}{\partial r}$ which simply means that we now differentiate $V = \pi r^2 h$ with respect to r, taking all symbols except V and r as being constant for the time being.

$\therefore \dfrac{\partial V}{\partial r} = \pi 2 r h = 2\pi r h$

In the statement $V = \pi r^2 h$, V is expressed as a function of two variables, r and h. It therefore has two partial derivatives, one with respect to .... and one with respect to ....

3

One with respect to *r*; one with respect to *h*

Another example:

Let us consider the area of the curved surface of the cylinder $A = 2\pi rh$

A is a function of r and h, so we can find $\dfrac{\partial A}{\partial r}$ and $\dfrac{\partial A}{\partial h}$

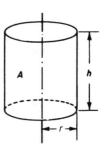

To find $\dfrac{\partial A}{\partial r}$ we differentiate the expression for A with respect to r, keeping all other symbols constant.

To find $\dfrac{\partial A}{\partial h}$ we differentiate the expression for A with respect to h, keeping all other symbols constant.

So, if $A = 2\pi rh$, then $\dfrac{\partial A}{\partial r} = \ldots\ldots\ldots\ldots$ and $\dfrac{\partial A}{\partial h} = \ldots\ldots\ldots\ldots$

4

$$\dfrac{\partial A}{\partial r} = 2\pi h \text{ and } \dfrac{\partial A}{\partial h} = 2\pi r$$

Of course, we are not restricted to the mensuration of the cylinder. The same will happen with any function which is a function of two independent variables. For example, consider $z = x^2 y^3$.

Here z is a function of x and y. We can therefore find $\dfrac{\partial z}{\partial x}$ and $\dfrac{\partial z}{\partial y}$.

(a) To find $\dfrac{\partial z}{\partial x}$, differentiate with respect to x, regarding y as a constant.

$$\therefore \dfrac{\partial z}{\partial x} = 2xy^3$$

(b) To find $\dfrac{\partial z}{\partial y}$, differentiate with respect to y, regarding x as a constant.

$$\dfrac{\partial z}{\partial y} = x^2 3y^2 = 3x^2 y^2$$

Partial differentiation is easy! For we regard every independent variable, except the one with respect to which we are differentiating, as being for the time being ....

5

$$\boxed{\text{constant}}$$

Here are some examples. 'With respect to' is abbreviated to w.r.t.

Example 1

$u = x^2 + xy + y^2$

(a) To find $\dfrac{\partial u}{\partial x}$, we regard y as being constant.

Partial diff w.r.t. x of $x^2 = 2x$
Partial diff w.r.t. x of $xy = y$ (y is a constant factor)
Partial diff w.r.t. x of $y^2 = 0$ (y^2 is a constant term)

$$\dfrac{\partial u}{\partial x} = 2x + y$$

(b) To find $\dfrac{\partial u}{\partial y}$, we regard x as being constant.

Partial diff w.r.t. y of $x^2 = 0$ (x^2 is a constant term)
Partial diff w.r.t. y of $xy = x$ (x is a constant factor)
Partial diff w.r.t. y of $y^2 = 2y$

$$\dfrac{\partial u}{\partial y} = x + 2y$$

Another example in Frame 6

6

Example 2

$z = x^3 + y^3 - 2x^2 y$

$$\dfrac{\partial z}{\partial x} = 3x^2 + 0 - 4xy = 3x^2 - 4xy$$

$$\dfrac{\partial z}{\partial y} = 0 + 3y^2 - 2x^2 = 3y^2 - 2x^2$$

And it is all just as easy as that.

Example 3

$z = (2x - y)(x + 3y)$

This is a product, and the usual product rule applies except that we keep y constant when finding $\dfrac{\partial z}{\partial x}$, and x constant when finding $\dfrac{\partial z}{\partial y}$.

$$\dfrac{\partial z}{\partial x} = (2x - y)(1 + 0) + (x + 3y)(2 - 0) = 2x - y + 2x + 6y = 4x + 5y$$

$$\dfrac{\partial z}{\partial y} = (2x - y)(0 + 3) + (x + 3y)(0 - 1) = 6x - 3y - x - 3y = 5x - 6y$$

▶

Partial differentiation

Here is one for you to do.

If $z = (4x - 2y)(3x + 5y)$, find $\dfrac{\partial z}{\partial x}$ and $\dfrac{\partial z}{\partial y}$

Find the results and then move on to Frame 7

7

$$\boxed{\dfrac{\partial z}{\partial x} = 24x + 14y; \qquad \dfrac{\partial z}{\partial y} = 14x - 20y}$$

Because $z = (4x - 2y)(3x + 5y)$, i.e. product

$$\therefore \dfrac{\partial z}{\partial x} = (4x - 2y)(3 + 0) + (3x + 5y)(4 - 0)$$

$$= 12x - 6y + 12x + 20y = 24x + 14y$$

$$\dfrac{\partial z}{\partial y} = (4x - 2y)(0 + 5) + (3x + 5y)(0 - 2)$$

$$= 20x - 10y - 6x - 10y = 14x - 20y$$

There we are. Now what about this one?

Example 4

If $z = \dfrac{2x - y}{x + y}$, find $\dfrac{\partial z}{\partial x}$ and $\dfrac{\partial z}{\partial y}$

Applying the quotient rule, we have:

$$\dfrac{\partial z}{\partial x} = \dfrac{(x + y)(2 - 0) - (2x - y)(1 + 0)}{(x + y)^2} = \dfrac{3y}{(x + y)^2}$$

and $\dfrac{\partial z}{\partial y} = \dfrac{(x + y)(0 - 1) - (2x - y)(0 + 1)}{(x + y)^2} = \dfrac{-3x}{(x + y)^2}$

That was not difficult. Now you do this one:

If $z = \dfrac{5x + y}{x - 2y}$, find $\dfrac{\partial z}{\partial x}$ and $\dfrac{\partial z}{\partial y}$

When you have finished, on to the next frame

Vector Analysis

8

$$\boxed{\frac{\partial z}{\partial x} = \frac{-11y}{(x-2y)^2}; \quad \frac{\partial z}{\partial y} = \frac{11x}{(x-2y)^2}}$$

Here is the working:

(a) To find $\dfrac{\partial z}{\partial x}$, we regard y as being constant.

$$\therefore \frac{\partial z}{\partial x} = \frac{(x-2y)(5+0) - (5x+y)(1-0)}{(x-2y)^2}$$

$$= \frac{5x - 10y - 5x - y}{(x-2y)^2} = \frac{-11y}{(x-2y)^2}$$

(b) To find $\dfrac{\partial z}{\partial y}$, we regard x as being constant.

$$\therefore \frac{\partial z}{\partial y} = \frac{(x-2y)(0+1) - (5x+y)(0-2)}{(x-2y)^2}$$

$$= \frac{x - 2y + 10x + 2y}{(x-2y)^2} = \frac{11y}{(x-2y)^2}$$

In practice, we do not write down the zeros that occur in the working, but this is how we think.

Let us do one more example, so move on to the next frame

9

Example 5

If $z = \sin(3x + 2y)$ find $\dfrac{\partial z}{\partial x}$ and $\dfrac{\partial z}{\partial y}$

Here we have what is clearly a 'function of a function'. So we apply the usual procedure, except to remember that when we are finding:

(a) $\dfrac{\partial z}{\partial x}$, we treat y as constant, and

(b) $\dfrac{\partial z}{\partial y}$, we treat x as constant.

Here goes then.

$$\frac{\partial z}{\partial x} = \cos(3x+2y) \times \frac{\partial}{\partial x}(3x+2y) = \cos(3x+2y) \times 3 = 3\cos(3x+2y)$$

$$\frac{\partial z}{\partial y} = \cos(3x+2y) \times \frac{\partial}{\partial y}(3x+2y) = \cos(3x+2y) \times 2 = 2\cos(3x+2y)$$

There it is. So in partial differentiation, we can apply all the ordinary rules of normal differentiation, except that we regard the independent variables other than the one we are using, as being for the time being ....

Partial differentiation

10

[constant]

Fine. Now here is a short exercise for you to do by way of revision.

In each of the following cases, find $\dfrac{\partial z}{\partial x}$ and $\dfrac{\partial z}{\partial y}$:

1. $z = 4x^2 + 3xy + 5y^2$
2. $z = (3x + 2y)(4x - 5y)$
3. $z = \tan(3x + 4y)$
4. $z = \dfrac{\sin(3x + 2y)}{xy}$

Finish them all, then move on to Frame 11 for the results

Here are the answers:

11

1. $z = 4x^2 + 3xy + 5y^2$ $\dfrac{\partial z}{\partial x} = 8x + 3y$ $\dfrac{\partial z}{\partial y} = 3x + 10y$

2. $z = (3x + 2y)(4x - 5y)$ $\dfrac{\partial z}{\partial x} = 24x - 7y$ $\dfrac{\partial z}{\partial y} = -7x - 20y$

3. $z = \tan(3x + 4y)$ $\dfrac{\partial z}{\partial x} = 3\sec^2(3x + 4y)$ $\dfrac{\partial z}{\partial y} = 4\sec^2(3x + 4y)$

4. $z = \dfrac{\sin(3x + 2y)}{xy}$

$\dfrac{\partial z}{\partial x} = \dfrac{3x\cos(3x + 2y) - \sin(3x + 2y)}{x^2 y}$ $\dfrac{\partial z}{\partial y} = \dfrac{2y\cos(3x + 2y) - \sin(3x + 2y)}{xy^2}$

If you have got *all* the answers correct, turn straight on to Frame 15. If you have not got all these answers, or are at all uncertain, move to Frame 12.

Let us work through these examples in detail.

12

1. $z = 4x^2 + 3xy + 5y^2$

   To find $\dfrac{\partial z}{\partial x}$, regard y as a constant:

   $\therefore \dfrac{\partial z}{\partial x} = 8x + 3y + 0$,   i.e. $8x + 3y$ $\therefore \dfrac{\partial z}{\partial x} = 8x + 3y$

   Similarly, regarding x as constant:

   $\dfrac{\partial z}{\partial y} = 0 + 3x + 10y$,   i.e. $3x + 10y$ $\therefore \dfrac{\partial z}{\partial y} = 3x + 10y$

2. $z = (3x + 2y)(4x - 5y)$ Product rule

   $\dfrac{\partial z}{\partial x} = (3x + 2y)(4) + (4x - 5y)(3)$

   $\phantom{\dfrac{\partial z}{\partial x}} = 12x + 8y + 12x - 15y = 24x - 7y$

   $\dfrac{\partial z}{\partial y} = (3x + 2y)(-5) + (4x - 5y)(2)$

   $\phantom{\dfrac{\partial z}{\partial y}} = -15x - 10y + 8x - 10y = -7x - 20y$

Move on for the solutions to **3** *and* **4**

13

3. $z = \tan(3x + 4y)$

$$\frac{\partial z}{\partial x} = \sec^2(3x + 4y)(3) = 3\sec^2(3x + 4y)$$

$$\frac{\partial z}{\partial y} = \sec^2(3x + 4y)(4) = 4\sec^2(3x + 4y)$$

4. $z = \dfrac{\sin(3x + 2y)}{xy}$

$$\frac{\partial z}{\partial x} = \frac{xy\cos(3x + 2y)(3) - \sin(3x + 2y)(y)}{x^2 y^2}$$

$$= \frac{3x\cos(3x + 2y) - \sin(3x + 2y)}{x^2 y}$$

Now have another go at finding $\dfrac{\partial z}{\partial y}$ in the same way.

Then check it with Frame 14

14

Here it is:

$$z = \frac{\sin(3x + 2y)}{xy}$$

$$\therefore \frac{\partial z}{\partial y} = \frac{xy\cos(3x + 2y) \cdot (2) - \sin(3x + 2y) \cdot (x)}{x^2 y^2}$$

$$= \frac{2y\cos(3x + 2y) - \sin(3x + 2y)}{xy^2}$$

That should have cleared up any troubles. This business of partial differentiation is perfectly straightforward. All you have to remember is that for the time being, all the independent variables except the one you are using are kept constant – and behave like constant factors or constant terms according to their positions.

On you go now to Frame 15 and continue the Program

15

Right. Now let us move on a step.

Consider $z = 3x^2 + 4xy - 5y^2$

Then $\dfrac{\partial z}{\partial x} = 6x + 4y$ and $\dfrac{\partial z}{\partial y} = 4x - 10y$

The expression $\dfrac{\partial z}{\partial x} = 6x + 4y$ is itself a function of x and y. We could therefore find its partial derivatives with respect to x or to y.

Partial differentiation

(a) If we differentiate it partially w.r.t. x, we get:

$$\frac{\partial}{\partial x}\left\{\frac{\partial z}{\partial x}\right\} \text{ and this is written } \frac{\partial^2 z}{\partial x^2} \text{ (much like an ordinary second derivative, but with the partial } \partial)$$

$$\therefore \frac{\partial^2 z}{\partial x^2} = \frac{\partial}{\partial x}(6x + 4y) = 6$$

This is called the second partial derivative of z with respect to x.

(b) If we differentiate partially w.r.t. y, we get:

$$\frac{\partial}{\partial y}\left\{\frac{\partial z}{\partial x}\right\} \text{ and this is written } \frac{\partial^2 z}{\partial y \cdot \partial x}$$

Note that the operation now being performed is given by the left-hand of the two symbols in the denominator.

$$\frac{\partial^2 z}{\partial y \cdot \partial x} = \frac{\partial}{\partial y}\left\{\frac{\partial z}{\partial x}\right\} = \frac{\partial}{\partial y}\{6x + 4y\} = 4$$

So we have this:

$$z = 3x^2 + 4xy - 5y^2$$

$$\frac{\partial z}{\partial x} = 6x + 4y \qquad \frac{\partial z}{\partial y} = 4x - 10y$$

$$\frac{\partial^2 z}{\partial x^2} = 6$$

$$\frac{\partial^2 z}{\partial y \cdot \partial x} = 4$$

Of course, we could carry out similar steps with the expression for $\frac{\partial z}{\partial y}$ on the right. This would give us:

$$\frac{\partial^2 z}{\partial y^2} = -10$$

$$\frac{\partial^2 z}{\partial x \cdot \partial y} = 4$$

Note that $\frac{\partial^2 z}{\partial y \cdot \partial x}$ means $\frac{\partial}{\partial y}\left\{\frac{\partial z}{\partial x}\right\}$ so $\frac{\partial^2 z}{\partial x \cdot \partial y}$ means ....

16

17

$$\boxed{\frac{\partial^2 z}{\partial x \cdot \partial y} \text{ means } \frac{\partial}{\partial x}\left\{\frac{\partial z}{\partial y}\right\}}$$

Collecting our previous results together then, we have:

$z = 3x^2 + 4xy - 5y^2$

$\dfrac{\partial z}{\partial x} = 6x + 4y$ $\qquad$ $\dfrac{\partial z}{\partial y} = 4x - 10y$

$\dfrac{\partial^2 z}{\partial x^2} = 6$ $\qquad$ $\dfrac{\partial^2 z}{\partial y^2} = -10$

$\dfrac{\partial^2 z}{\partial y \cdot \partial x} = 4$ $\qquad$ $\dfrac{\partial^2 z}{\partial x \cdot \partial y} = 4$

We see in this case, that $\dfrac{\partial^2 z}{\partial y \cdot \partial x} = \dfrac{\partial^2 z}{\partial x \cdot \partial y}$. There are then, *two* first derivatives and *four* second derivatives, though the last two seem to have the same value. Here is one for you to do.

If $z = 5x^3 + 3x^2 y + 4y^3$, find $\dfrac{\partial z}{\partial x}, \dfrac{\partial z}{\partial y}, \dfrac{\partial^2 z}{\partial x^2}, \dfrac{\partial^2 z}{\partial y^2}, \dfrac{\partial^2 z}{\partial x \cdot \partial y}$ and $\dfrac{\partial^2 z}{\partial y \cdot \partial x}$

When you have completed all that, move to Frame 18

18

Here are the results:

$z = 5x^3 + 3x^2 y + 4y^3$

$\dfrac{\partial z}{\partial x} = 15x^2 + 6xy$ $\qquad$ $\dfrac{\partial z}{\partial y} = 3x^2 + 12y^2$

$\dfrac{\partial^2 z}{\partial x^2} = 30x + 6y$ $\qquad$ $\dfrac{\partial^2 z}{\partial y^2} = 24y$

$\dfrac{\partial^2 z}{\partial y \cdot \partial x} = 6x$ $\qquad$ $\dfrac{\partial^2 z}{\partial x \cdot \partial y} = 6x$

Again in this example also, we see that $\dfrac{\partial^2 z}{\partial y \cdot \partial x} = \dfrac{\partial^2 z}{\partial x \cdot \partial y}$. Now do this one.

It looks more complicated, but it is done in just the same way. Do not rush at it; take your time and all will be well. Here it is. Find all the first and second partial derivatives of $z = x \cos y - y \cos x$.

Then to Frame 19

Partial differentiation

Check your results with these.

$z = x\cos y - y\cos x$

When differentiating w.r.t. x, y is constant (and therefore $\cos y$ also).
When differentiating w.r.t. y, x is constant (and therefore $\cos x$ also).

So we get:

$$\frac{\partial z}{\partial x} = \cos y + y \cdot \sin x \qquad \frac{\partial z}{\partial y} = -x \cdot \sin y - \cos x$$

$$\frac{\partial^2 z}{\partial x^2} = y \cdot \cos x \qquad \frac{\partial^2 z}{\partial y^2} = -x \cdot \cos y$$

$$\frac{\partial^2 z}{\partial y \cdot \partial x} = -\sin y + \sin x \qquad \frac{\partial^2 z}{\partial x \cdot \partial y} = -\sin y + \sin x$$

And again, $\dfrac{\partial^2 z}{\partial y \cdot \partial x} = \dfrac{\partial^2 z}{\partial x \cdot \partial y}$

In fact this will always be so for the functions you are likely to meet, so that there are really *three* different second partial derivatives (and not four). In practice, if you have found $\dfrac{\partial^2 z}{\partial y \cdot \partial x}$ it is a useful check to find $\dfrac{\partial^2 z}{\partial x \cdot \partial y}$ separately. They should give the same result, of course.

What about this one?

If $V = \ln(x^2 + y^2)$, prove that $\dfrac{\partial^2 V}{\partial x^2} + \dfrac{\partial^2 V}{\partial y^2} = 0$

This merely entails finding the two second partial derivatives and substituting them in the left-hand side of the statement. So here goes:

$V = \ln(x^2 + y^2)$

$$\frac{\partial V}{\partial x} = \frac{1}{(x^2 + y^2)} 2x$$

$$= \frac{2x}{x^2 + y^2}$$

$$\frac{\partial^2 V}{\partial x^2} = \frac{(x^2 + y^2)2 - 2x.2x}{(x^2 + y^2)^2}$$

$$= \frac{2x^2 + 2y^2 - 4x^2}{(x^2 + y^2)^2} = \frac{2y^2 - 2x^2}{(x^2 + y^2)^2} \qquad (a)$$

Now you find $\dfrac{\partial^2 V}{\partial y^2}$ in the same way and hence prove the given identity.

When you are ready, move on to Frame 21

21

We had found that $\dfrac{\partial^2 V}{\partial x^2} = \dfrac{2y^2 - 2x^2}{(x^2 + y^2)^2}$

So making a fresh start from $V = \ln(x^2 + y^2)$, we get:

$$\dfrac{\partial V}{\partial y} = \dfrac{1}{(x^2 + y^2)} \cdot 2y = \dfrac{2y}{x^2 + y^2}$$

$$\dfrac{\partial^2 V}{\partial y^2} = \dfrac{(x^2 + y^2)2 - 2y \cdot 2y}{(x^2 + y^2)^2}$$

$$= \dfrac{2x^2 + 2y^2 - 4y^2}{(x^2 + y^2)^2} = \dfrac{2x^2 - 2y^2}{(x^2 + y^2)^2} \quad \text{(b)}$$

Substituting now the two results in the identity, gives:

$$\dfrac{\partial^2 V}{\partial x^2} + \dfrac{\partial^2 V}{\partial y^2} = \dfrac{2y^2 - 2x^2}{(x^2 + y^2)^2} + \dfrac{2x^2 - 2y^2}{(x^2 + y^2)^2}$$

$$= \dfrac{2y^2 - 2x^2 + 2x^2 - 2y^2}{(x^2 + y^2)^2} = 0$$

Now on to Frame 22

22

Here is another kind of example that you should see.

Example 1

If $V = f(x^2 + y^2)$, show that $x\dfrac{\partial V}{\partial y} - y\dfrac{\partial V}{\partial x} = 0$

Here we are told that V is a function of $(x^2 + y^2)$ but the precise nature of the function is not given. However, we can treat this as a 'function of a function' and write $f'(x^2 + y^2)$ to represent the derivative of the function w.r.t. its own combined variable $(x^2 + y^2)$.

$$\therefore \dfrac{\partial V}{\partial x} = f'(x^2 + y^2) \times \dfrac{\partial}{\partial x}(x^2 + y^2) = f'(x^2 + y^2) \cdot 2x$$

$$\dfrac{\partial V}{\partial y} = f'(x^2 + y^2) \cdot \dfrac{\partial}{\partial y}(x^2 + y^2) = f'(x^2 + y^2) \cdot 2y$$

$$\therefore x\dfrac{\partial V}{\partial y} - y\dfrac{\partial V}{\partial x} = x \cdot f'(x^2 + y^2) \cdot 2y - y \cdot f'(x^2 + y^2) \cdot 2x$$

$$= 2xy \cdot f'(x^2 + y^2) - 2xy \cdot f'(x^2 + y^2)$$

$$= 0$$

Let us have another one of that kind in the next frame

Partial differentiation

Example 2

If $z = f\left\{\dfrac{y}{x}\right\}$, show that $x\dfrac{\partial z}{\partial x} + y\dfrac{\partial z}{\partial y} = 0$

Much the same as before:

$$\dfrac{\partial z}{\partial x} = f'\left\{\dfrac{y}{x}\right\} \cdot \dfrac{\partial}{\partial x}\left\{\dfrac{y}{x}\right\} = f'\left\{\dfrac{y}{x}\right\}\left(-\dfrac{y}{x^2}\right) = -\dfrac{y}{x^2}f'\left\{\dfrac{y}{x}\right\}$$

$$\dfrac{\partial z}{\partial y} = f'\left\{\dfrac{y}{x}\right\} \cdot \dfrac{\partial}{\partial y}\left\{\dfrac{y}{x}\right\} = f'\left\{\dfrac{y}{x}\right\} \cdot \dfrac{1}{x} = \dfrac{1}{x}f'\left\{\dfrac{y}{x}\right\}$$

$$\therefore x\dfrac{\partial z}{\partial x} + y\dfrac{\partial z}{\partial y} = x\left(-\dfrac{y}{x^2}\right)f'\left\{\dfrac{y}{x}\right\} + y\dfrac{1}{x}f'\left\{\dfrac{y}{x}\right\}$$

$$= -\dfrac{y}{x}f'\left\{\dfrac{y}{x}\right\} + \dfrac{y}{x}f'\left\{\dfrac{y}{x}\right\}$$

$$= 0$$

And one for you, just to get your hand in:

If $V = f(ax + by)$, show that $b\dfrac{\partial V}{\partial x} - a\dfrac{\partial V}{\partial y} = 0$

When you have done it, check your working against that in Frame 24

Here is the working; this is how it goes.

$V = f(ax + by)$

$\therefore \dfrac{\partial V}{\partial x} = f'(ax + by) \cdot \dfrac{\partial}{\partial x}(ax + by)$

$\qquad = f'(ax + by) \cdot a = a \cdot f'(ax + by)$ (a)

$\dfrac{\partial z}{\partial y} = f'(ax + by) \cdot \dfrac{\partial}{\partial y}(ax + by)$

$\qquad = f'(ax + by) \cdot b = b \cdot f'(ax + by)$ (b)

$\therefore b\dfrac{\partial V}{\partial x} - a\dfrac{\partial V}{\partial y} = ab \cdot f'(ax + by) - ab \cdot f'(ax + by)$

$\qquad = 0$

Move on to Frame 25

25

So to sum up so far.

Partial differentiation is easy, no matter how complicated the expression to be differentiated may seem.

To differentiate partially w.r.t. x, all independent variables other than x are constant for the time being.

To differentiate partially w.r.t. y, all independent variables other than y are constant for the time being.

So that, if z is a function of x and y, i.e. if $z = f(x, y)$, we can find:

$$\frac{\partial z}{\partial x} \qquad \frac{\partial z}{\partial y}$$

$$\frac{\partial^2 z}{\partial x^2} \qquad \frac{\partial^2 z}{\partial y^2}$$

$$\frac{\partial^2 z}{\partial y \cdot \partial x} \qquad \frac{\partial^2 z}{\partial x \cdot \partial y} \qquad \text{And also:} \qquad \frac{\partial^2 z}{\partial y \cdot \partial x} = \frac{\partial^2 z}{\partial x \cdot \partial y}$$

Now for a review exercise

26

Review exercise

1 Find all first and second partial derivatives for each of the following functions:

(a) $z = 3x^2 + 2xy + 4y^2$

(b) $z = \sin xy$

(c) $z = \dfrac{x+y}{x-y}$

2 If $z = \ln(e^x + e^y)$, show that $\dfrac{\partial z}{\partial x} + \dfrac{\partial z}{\partial y} = 1$.

3 If $z = x \cdot f(xy)$, express $x\dfrac{\partial z}{\partial x} - y\dfrac{\partial z}{\partial y}$ in its simplest form.

When you have finished, check with the solutions in Frame 27

27

1 (a) $z = 3x^2 + 2xy + 4y^2$

$$\frac{\partial z}{\partial x} = 6x + 2y \qquad\qquad \frac{\partial z}{\partial y} = 2x + 8y$$

$$\frac{\partial^2 z}{\partial x^2} = 6 \qquad\qquad \frac{\partial^2 z}{\partial y^2} = 8$$

$$\frac{\partial^2 z}{\partial y \cdot \partial x} = 2 \qquad\qquad \frac{\partial^2 z}{\partial x \cdot \partial y} = 2$$

Partial differentiation

(b) $z = \sin xy$

$$\frac{\partial z}{\partial x} = y \cos xy \qquad \frac{\partial z}{\partial y} = x \cos xy$$

$$\frac{\partial^2 z}{\partial x^2} = -y^2 \sin xy \qquad \frac{\partial^2 z}{\partial y^2} = -x^2 \sin xy$$

$$\frac{\partial^2 z}{\partial y \cdot \partial x} = y(-x \sin xy) + \cos xy \qquad \frac{\partial^2 z}{\partial x \cdot \partial y} = x(-y \sin xy) + \cos xy$$

$$= \cos xy - xy \sin xy \qquad = \cos xy - xy \sin xy$$

(c) $z = \dfrac{x+y}{x-y}$

$$\frac{\partial z}{\partial x} = \frac{(x-y)1 - (x+y)1}{(x-y)^2} = \frac{-2y}{(x-y)^2}$$

$$\frac{\partial z}{\partial y} = \frac{(x-y)1 - (x+y)(-1)}{(x-y)^2} = \frac{2x}{(x-y)^2}$$

$$\frac{\partial^2 z}{\partial x^2} = (-2y)\frac{(-2)}{(x-y)^3} = \frac{4y}{(x-y)^3}$$

$$\frac{\partial^2 z}{\partial y^2} = 2x\frac{(-2)}{(x-y)^3}(-1) = \frac{4x}{(x-y)^3}$$

$$\frac{\partial^2 z}{\partial y \cdot \partial x} = \frac{(x-y)^2(-2) - (-2y)2(x-y)(-1)}{(x-y)^4}$$

$$= \frac{-2(x-y)^2 - 4y(x-y)}{(x-y)^4}$$

$$= \frac{-2}{(x-y)^2} - \frac{4y}{(x-y)^3}$$

$$= \frac{-2x + 2y - 4y}{(x-y)^3} = \frac{-2x - 2y}{(x-y)^3}$$

$$\frac{\partial^2 z}{\partial x \cdot \partial y} = \frac{(x-y)^2(2) - 2x \cdot 2(x-y)1}{(x-y)^4}$$

$$= \frac{2(x-y)^2 - 4x(x-y)}{(x-y)^4}$$

$$= \frac{2}{(x-y)^2} - \frac{4x}{(x-y)^3}$$

$$= \frac{2x - 2y - 4x}{(x-y)^3} = \frac{-2x - 2y}{(x-y)^3}$$

16 Vector Analysis

2 $z = \ln(e^x + e^y)$

$$\frac{\partial z}{\partial x} = \frac{1}{e^x + e^y} \cdot e^x \qquad \frac{\partial z}{\partial y} = \frac{1}{e^x + e^y} \cdot e^y$$

$$\frac{\partial z}{\partial x} + \frac{\partial z}{\partial y} = \frac{e^x}{e^x + e^y} + \frac{e^y}{e^x + e^y}$$

$$= \frac{e^x + e^y}{e^x + e^y} = 1$$

$$\frac{\partial z}{\partial x} + \frac{\partial z}{\partial y} = 1$$

3 $z = x \cdot f(xy)$

$$\frac{\partial z}{\partial x} = x \cdot f'(xy) \cdot y + f(xy)$$

$$\frac{\partial z}{\partial y} = x \cdot f'(xy) \cdot x$$

$$x\frac{\partial z}{\partial x} - y\frac{\partial z}{\partial y} = x^2 y f'(xy) + x f(xy) - x^2 y f'(xy)$$

$$x\frac{\partial z}{\partial x} - y\frac{\partial z}{\partial y} = x f(xy) = z$$

That was a pretty good review test. Do not be unduly worried if you made a slip or two in your working. Try to avoid doing so, of course, but you are doing fine. Now on to the next part of the Program.

So far we have been concerned with the technique of partial differentiation. Now let us look at one of its applications.

So move on to Frame 28

Small increments

28

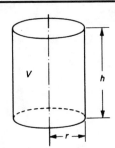

If we return to the volume of the cylinder with which we started this Program, we have once again that $V = \pi r^2 h$. We have seen that we can find $\dfrac{\partial V}{\partial r}$ with h constant, and $\dfrac{\partial V}{\partial h}$ with r constant.

$$\frac{\partial V}{\partial r} = 2\pi rh; \quad \frac{\partial V}{\partial h} = \pi r^2$$

Now let us see what we get if r and h both change simultaneously.

▶

Partial differentiation

If r becomes $r + \delta r$, and h becomes $h + \delta h$, let V become $V + \delta V$. Then the new volume is given by:

$$V + \delta V = \pi(r + \delta r)^2(h + \delta h)$$
$$= \pi(r^2 + 2r\delta r + [\delta r]^2)(h + \delta h)$$
$$= \pi(r^2 h + 2rh\delta r + h[\delta r]^2 + r^2 \delta h + 2r\delta r \delta h + [\delta r]^2 \delta h)$$

Subtract $V = \pi r^2 h$ from each side, giving:

$$\delta V = \pi(2rh\delta r + h[\delta r]^2 + r^2 \delta h + 2r\delta r \delta h + [\delta r]^2 \delta h)$$

$$\approx \pi(2rh\delta r + r^2 \delta h) \quad \text{since } \delta r \text{ and } \delta h \text{ are small and all the remaining terms are of a higher degree of smallness.}$$

Therefore

$$\delta V \approx 2\pi rh\delta r + \pi r^2 \delta h, \quad \text{that is:}$$

$$\delta V \approx \frac{\partial V}{\partial r}\delta r + \frac{\partial V}{\partial h}\delta h$$

Let us now do a numerical example to see how it all works out.

On to Frame 29

A cylinder has dimensions $r = 5$ cm, $h = 10$ cm. Find the approximate increase in volume when r increases by 0.2 cm and h decreases by 0.1 cm. Well now

$$V = \pi r^2 h \text{ so } \frac{\partial V}{\partial r} = 2\pi rh \quad \frac{\partial V}{\partial h} = \pi r^2$$

In this case, when $r = 5$ cm, $h = 10$ cm so

$$\frac{\partial V}{\partial r} = 2\pi 5 \cdot 10 = 100\pi \quad \frac{\partial V}{\partial h} = \pi r^2 = \pi 5^2 = 25\pi$$

$\delta r = 0.2$ and $\delta h = -0.1$ (minus because h is decreasing)

$$\therefore \delta V \approx \frac{\partial V}{\partial r} \cdot \delta r + \frac{\partial V}{\partial h} \cdot \delta h$$

$$\delta V = 100\pi(0.2) + 25\pi(-0.1)$$

$$= 20\pi - 2.5\pi = 17.5\pi$$

$$\therefore \delta V \approx 54.98 \text{ cm}^3$$

i.e. the volume increases by 54.98 cm^3

Just like that!

30

This kind of result applies not only to the volume of the cylinder, but to any function of two independent variables. Here is an example:

If z is a function of x and y, i.e. $z = f(x, y)$ and if x and y increase by small amounts δx and δy, the increase δz will also be relatively small. If we expand δz in powers of δx and δy, we get:

$$\delta z = A\delta x + B\delta y + \text{higher powers of } \delta x \text{ and } \delta y,$$

where A and B are functions of x and y.

If y remains constant, so that $\delta y = 0$, then:

$$\delta z = A\delta x + \text{higher powers of } \delta x$$

$$\therefore \frac{\delta z}{\delta x} = A. \text{ So that if } \delta x \to 0, \text{ this becomes } A = \frac{\partial z}{\partial x}$$

Similarly, if x remains constant, making $\delta y \to 0$ gives $B = \dfrac{\partial z}{\partial y}$

$$\therefore \delta z = \frac{\partial z}{\partial x}\delta x + \frac{\partial z}{\partial y}\delta y + \text{higher powers of very small quantities which can be ignored}$$

$$\delta z = \frac{\partial z}{\partial x}\delta x + \frac{\partial z}{\partial y}\delta y$$

31

So, if $z = f(x, y)$

$$\delta z = \frac{\partial z}{\partial x}\delta x + \frac{\partial z}{\partial y}\delta y$$

This is the key to all the forthcoming applications and will be quoted over and over again.

The result is quite general and a similar result applies for a function of three independent variables. For example:

If $z = f(x, y, w)$

then $\delta z = \dfrac{\partial z}{\partial x}\delta x + \dfrac{\partial z}{\partial y}\delta y + \dfrac{\partial z}{\partial w}\delta w$

If we remember the rule for a function of two independent variables, we can easily extend it when necessary.

Here it is once again:

$$\text{If } z = f(x, y) \text{ then } \delta z = \frac{\partial z}{\partial x}\delta x + \frac{\partial z}{\partial y}\delta y$$

Copy this result into your record book in a prominent position, such as it deserves!

Partial differentiation

Now for a couple of examples

32

Example 1

If $I = \dfrac{V}{R}$, and $V = 250$ volts and $R = 50$ ohms, find the change in I resulting from an increase of 1 volt in V and an increase of 0.5 ohm in R.

$$I = f(V, R) \qquad \therefore \ \delta I = \frac{\partial I}{\partial V} \delta V + \frac{\partial I}{\partial R} \delta R$$

$$\frac{\partial I}{\partial V} = \frac{1}{R} \quad \text{and} \quad \frac{\partial I}{\partial R} = -\frac{V}{R^2}$$

$$\therefore \ \delta I = \frac{1}{R} \delta V - \frac{V}{R^2} \delta R$$

So when $R = 50$, $V = 250$, $\delta V = 1$ and $\delta R = 0.5$:

$$\delta I = \frac{1}{50}(1) - \frac{250}{2500}(0.5)$$

$$= \frac{1}{50} - \frac{1}{20}$$

$$= 0.02 - 0.05 = -0.03$$

i.e. I decreases by 0.03 amperes

Here is another example.

33

Example 2

If $y = \dfrac{ws^3}{d^4}$, find the percentage increase in y when w increases by 2 per cent, s decreases by 3 per cent and d increases by 1 per cent.

Notice that, in this case, y is a function of three variables, w, s and d. The formula therefore becomes:

$$\delta y = \frac{\partial y}{\partial w} \delta w + \frac{\partial y}{\partial s} \delta s + \frac{\partial y}{\partial d} \delta d$$

We have

$$\frac{\partial y}{\partial w} = \frac{s^3}{d^4}; \quad \frac{\partial y}{\partial s} = \frac{3ws^2}{d^4}; \quad \frac{\partial y}{\partial d} = -\frac{4ws^3}{d^5}$$

$$\therefore \ \delta y = \frac{s^3}{d^4} \delta w + \frac{3ws^2}{d^4} \delta s + \frac{-4ws^3}{d^5} \delta d$$

Now then, what are the values of $\delta w, \delta s$ and δd?

Is it true to say that $\delta w = \dfrac{2}{100}$; $\delta s = \dfrac{-3}{100}$; $\delta d = \dfrac{1}{100}$?

If not, why not?

Next frame

34

> No. It is not correct

Because δw is not $\dfrac{2}{100}$ of a unit, but 2 per cent of w, i.e. $\delta w = \dfrac{2}{100}$ of $w = \dfrac{2w}{100}$

Similarly, $\delta s = \dfrac{-3}{100}$ of $s = \dfrac{-3s}{100}$ and $\delta d = \dfrac{d}{100}$. Now that we have cleared that point up, we can continue with the problem.

$$\delta y = \frac{s^3}{d^4}\left(\frac{2w}{100}\right) + \frac{3ws^2}{d^4}\left(\frac{-3s}{100}\right) - \frac{4ws^3}{d^5}\left(\frac{d}{100}\right)$$

$$= \frac{ws^3}{d^4}\left(\frac{2}{100}\right) - \frac{ws^3}{d^4}\left(\frac{9}{100}\right) - \frac{ws^3}{d^4}\left(\frac{4}{100}\right)$$

$$= \frac{ws^3}{d^4}\left\{\frac{2}{100} - \frac{9}{100} - \frac{4}{100}\right\}$$

$$= y\left\{-\frac{11}{100}\right\} = -11 \text{ per cent of } y$$

i.e. y decreases by 11 per cent

Remember that where the increment of w is given as 2 per cent, it is *not* $\dfrac{2}{100}$ of a unit, but $\dfrac{2}{100}$ of w, and the symbol w must be included.

Move on to Frame 35

35

Now here is an exercise for you to do.

$P = w^2hd$. If errors of up to 1 per cent (plus or minus) are possible in the measured values of w, h and d, find the maximum possible percentage error in the calculated values of P.

This is very much like the previous example, so you will be able to deal with it without any trouble. Work it right through and then go on to Frame 36 and check your result.

36

$P = w^2hd$. $\therefore \delta P = \dfrac{\partial P}{\partial w} \cdot \delta w + \dfrac{\partial P}{\partial h} \cdot \delta h + \dfrac{\partial P}{\partial d} \cdot \delta d$

$\dfrac{\partial P}{\partial w} = 2whd;\quad \dfrac{\partial P}{\partial h} = w^2d;\quad \dfrac{\partial P}{\partial d} = w^2h$

$\delta P = 2whd \cdot \delta w + w^2d \cdot \delta h + w^2h \cdot \delta d$

Now $\quad \delta w = \pm\dfrac{w}{100};\quad \delta h = \pm\dfrac{h}{100};\quad \delta d = \pm\dfrac{d}{100}$

$$\delta P = 2whd\left(\pm\frac{w}{100}\right) + w^2d\left(\pm\frac{h}{100}\right) + w^2h\left(\pm\frac{d}{100}\right)$$

$$= \pm\frac{2w^2hd}{100} \pm \frac{w^2dh}{100} \pm \frac{w^2hd}{100}$$

The greatest possible error in P will occur when the signs are chosen so that they are all of the same kind, i.e. all plus or minus. If they were mixed, they would tend to cancel each other out.

Partial differentiation

$$\therefore \delta P = \pm w^2 hd \left\{ \frac{2}{100} + \frac{1}{100} + \frac{1}{100} \right\} = \pm P \left(\frac{4}{100} \right)$$

∴ Maximum possible error in P is 4 per cent of P

Finally, here is one last example for you to do. Work right through it and then check your results with those in Frame 37.

The two sides forming the right-angle of a right-angled triangle are denoted by a and b. The hypotenuse is h. If there are possible errors of ± 0.5 per cent in measuring a and b, find the maximum possible error in calculating (a) the area of the triangle and (b) the length of h.

37

(a) $\delta A = 1$ per cent of A
(b) $\delta h = 0.5$ per cent of h

Here is the working in detail:

(a) $A = \dfrac{a \cdot b}{2}$ $\qquad \delta A = \dfrac{\partial A}{\partial a} \cdot \delta a + \dfrac{\partial A}{\partial b} \cdot \delta b$

$\dfrac{\partial A}{\partial a} = \dfrac{b}{2}; \quad \dfrac{\partial A}{\partial b} = \dfrac{a}{2}; \quad \delta a = \pm \dfrac{a}{200}; \quad \delta b = \pm \dfrac{b}{200}$

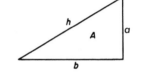

$\delta A = \dfrac{b}{2} \left(\pm \dfrac{a}{200} \right) + \dfrac{a}{2} \left(\pm \dfrac{b}{200} \right)$

$= \pm \dfrac{a \cdot b}{2} \left[\dfrac{1}{200} + \dfrac{1}{200} \right] = \pm A \cdot \dfrac{1}{100}$

∴ $\delta A = 1$ per cent of A

(b) $h = \sqrt{a^2 + b^2} = (a^2 + b^2)^{\frac{1}{2}}$

$\delta h = \dfrac{\partial h}{\partial a} \delta a + \dfrac{\partial h}{\partial b} \delta b$

$\dfrac{\partial h}{\partial a} = \dfrac{1}{2}(a^2 + b^2)^{-\frac{1}{2}}(2a) = \dfrac{a}{\sqrt{a^2 + b^2}}$

$\dfrac{\partial h}{\partial b} = \dfrac{1}{2}(a^2 + b^2)^{-\frac{1}{2}}(2b) = \dfrac{b}{\sqrt{a^2 + b^2}}$

Also $\qquad \delta a = \pm \dfrac{a}{200}; \quad \delta b = \pm \dfrac{b}{200}$

$\therefore \delta h = \dfrac{a}{\sqrt{a^2 + b^2}} \left(\pm \dfrac{a}{200} \right) + \dfrac{b}{\sqrt{a^2 + b^2}} \left(\pm \dfrac{b}{200} \right)$

$= \pm \dfrac{1}{200} \dfrac{a^2 + b^2}{\sqrt{a^2 + b^2}}$

$= \pm \dfrac{1}{200} \sqrt{a^2 + b^2} = \pm \dfrac{1}{200} (h)$

∴ $\delta h = 0.5$ per cent of h

▶

✅ Can You?

Checklist 1

Check this list before and after you try the end of Program test.

On a scale of 1 to 5 how confident are you that you can: Frames

- Find the first partial derivatives of a function of two real variables?   **1** to **14**
  Yes ☐ ☐ ☐ ☐ ☐ No

- Find second-order partial derivatives of a function of two real variables?   **15** to **25**
  Yes ☐ ☐ ☐ ☐ ☐ No

- Calculate errors using partial differentiation?   **28** to **37**
  Yes ☐ ☐ ☐ ☐ ☐ No

🚴 Test exercise 1

Take your time over the questions; do them carefully.

1. Find all first and second partial derivatives of the following:
   (a) $z = 4x^3 - 5xy^2 + 3y^3$
   (b) $z = \cos(2x + 3y)$
   (c) $z = e^{x^2 - y^2}$
   (d) $z = x^2 \sin(2x + 3y)$

2. (a) If $V = x^2 + y^2 + z^2$, express in its simplest form
   $$x\frac{\partial V}{\partial x} + y\frac{\partial V}{\partial y} + z\frac{\partial V}{\partial z}.$$

   (b) If $z = f(x + ay) + F(x - ay)$, find $\dfrac{\partial^2 z}{\partial x^2}$ and $\dfrac{\partial^2 z}{\partial y^2}$ and hence prove that
   $$\frac{\partial^2 z}{\partial y^2} = a^2 \cdot \frac{\partial^2 z}{\partial x^2}.$$

Partial differentiation

3 The power P dissipated in a resistor is given by $P = \dfrac{E^2}{R}$.

If $E = 200$ volts and $R = 8$ ohms, find the change in P resulting from a drop of 5 volts in E and an increase of 0.2 ohm in R.

4 If $\theta = kHLV^{-\frac{1}{2}}$, where k is a constant, and there are possible errors of ± 1 per cent in measuring H, L and V, find the maximum possible error in the calculated value of θ.

That's it

Further problems 1

1 If $z = \dfrac{1}{x^2 + y^2 - 1}$, show that $x\dfrac{\partial z}{\partial x} + y\dfrac{\partial z}{\partial y} = -2z(1+z)$.

2 Prove that, if $V = \ln(x^2 + y^2)$, then $\dfrac{\partial^2 V}{\partial x^2} + \dfrac{\partial^2 V}{\partial y^2} = 0$.

3 If $z = \sin(3x + 2y)$, verify that $3\dfrac{\partial^2 z}{\partial y^2} - 2\dfrac{\partial^2 z}{\partial x^2} = 6z$.

4 If $u = \dfrac{x+y+z}{(x^2+y^2+z^2)^{\frac{1}{2}}}$, show that $x\dfrac{\partial u}{\partial x} + y\dfrac{\partial u}{\partial y} + z\dfrac{\partial u}{\partial z} = 0$.

5 Show that the equation $\dfrac{\partial^2 z}{\partial x^2} + \dfrac{\partial^2 z}{\partial y^2} = 0$, is satisfied by

$$z = \ln\sqrt{x^2+y^2} + \dfrac{1}{2}\tan^{-1}\left(\dfrac{y}{x}\right)$$

6 If $z = e^x(x \cos y - y \sin y)$, show that $\dfrac{\partial^2 z}{\partial x^2} + \dfrac{\partial^2 z}{\partial y^2} = 0$.

7 If $u = (1+x)\sinh(5x - 2y)$, verify that $4\dfrac{\partial^2 u}{\partial x^2} + 20\dfrac{\partial^2 u}{\partial x \cdot \partial y} + 25\dfrac{\partial^2 u}{\partial y^2} = 0$.

8 If $z = f\left(\dfrac{y}{x}\right)$, show that $x^2\dfrac{\partial^2 z}{\partial x^2} + 2xy\dfrac{\partial^2 z}{\partial x.\partial y} + y^2\dfrac{\partial^2 z}{\partial y^2} = 0$.

9 If $z = (x+y) \cdot f\left(\dfrac{y}{x}\right)$, where f is an arbitrary function, show that

$$x\dfrac{\partial z}{\partial x} + y\dfrac{\partial z}{\partial y} = z.$$

10 In the formula $D = \dfrac{Eh^3}{12(1-v^2)}$, h is given as 0.1 ± 0.002 and v as 0.3 ± 0.02. Express the approximate maximum error in D in terms of E.

11 The formula $z = \dfrac{a^2}{x^2 + y^2 - a^2}$ is used to calculate z from observed values of x and y. If x and y have the same percentage error p, show that the percentage error in z is approximately $-2p(1+z)$.

40

12 In a balanced bridge circuit, $R_1 = R_2 R_3 / R_4$. If R_2, R_3, R_4 have known tolerances of $\pm x$ per cent, $\pm y$ per cent, $\pm z$ per cent respectively, determine the maximum percentage error in R_1, expressed in terms of x, y and z.

13 The deflection y at the centre of a circular plate suspended at the edge and uniformly loaded is given by $y = \dfrac{kwd^4}{t^3}$, where $w =$ total load, $d =$ diameter of plate, $t =$ thickness and k is a constant.

Calculate the approximate percentage change in y if w is increased by 3 per cent, d is decreased by $2\tfrac{1}{2}$ per cent and t is increased by 4 per cent.

14 The coefficient of rigidity (n) of a wire of length (L) and uniform diameter (d) is given by $n = \dfrac{AL}{d^4}$, where A is a constant. If errors of ± 0.25 per cent and ± 1 per cent are possible in measuring L and d respectively, determine the maximum percentage error in the calculated value of n.

15 If $k/k_0 = (T/T_0)^n \cdot p/760$, show that the change in k due to small changes of a per cent in T and b per cent in p is approximately $(na + b)$ per cent.

16 The deflection y at the centre of a rod is known to be given by $y = \dfrac{kwl^3}{d^4}$, where k is a constant. If w increases by 2 per cent, l by 3 per cent, and d decreases by 2 per cent, find the percentage increase in y.

17 The displacement y of a point on a vibrating stretched string, at a distance x from one end, at time t, is given by

$$\frac{\partial^2 y}{\partial t^2} = c^2 \cdot \frac{\partial^2 y}{\partial x^2}$$

Show that one solution of this equation is $y = A \sin\dfrac{px}{c} \cdot \sin(pt + a)$, where A, p, c and a are constants.

18 If $y = A \sin(px + a) \cos(qt + b)$, find the error in y due to small errors δx and δt in x and t respectively.

19 Show that $\phi = A e^{-kt/2} \sin pt \cos qx$, satisfies the equation

$$\frac{\partial^2 \phi}{\partial x^2} = \frac{1}{c^2}\left\{\frac{\partial^2 \phi}{\partial t^2} + k\frac{\partial \phi}{\partial t}\right\}, \text{ provided that } p^2 = c^2 q^2 - \frac{k^2}{4}.$$

20 Show that (a) the equation $\dfrac{\partial^2 V}{\partial x^2} + \dfrac{\partial^2 V}{\partial y^2} + \dfrac{\partial^2 V}{\partial z^2} = 0$ is satisfied by

$V = \dfrac{1}{\sqrt{x^2 + y^2 + z^2}}$, and that (b) the equation $\dfrac{\partial^2 V}{\partial x^2} + \dfrac{\partial^2 V}{\partial y^2} = 0$

is satisfied by $V = \tan^{-1}\left(\dfrac{y}{x}\right)$.

Program 2

Applications of partial differentiation

Frames
1 to 59

Learning outcomes

When you have completed this Program you will be able to:

- Derive the first- and second-order partial derivatives of a function of two real variables
- Apply partial differentiation to rate-of-change problems
- Apply partial differentiation to change-of-variable problems
- Use the Jacobian to obtain the derivatives of inverse functions of two or more variables

Introduction

1

In the first part of the Program on partial differentiation, we established a result which, we said, would be the foundation of most of the applications of partial differentiation to follow.

You surely remember it: it went like this:

If z is a function of two independent variables, x and y, i.e. if $z = f(x,y)$, then

$$\delta z = \frac{\partial z}{\partial x}\delta x + \frac{\partial z}{\partial y}\delta y$$

We were able to use it, just as it stands, to work out certain problems on small increments, errors and tolerances. It is also the key to much of the work of this Program, so copy it down into your record book, thus:

If $z = f(x,y)$ then $\delta z = \dfrac{\partial z}{\partial x}\delta x + \dfrac{\partial z}{\partial y}\delta y$

2

If $z = f(x,y)$, then $\delta z = \dfrac{\partial z}{\partial x}\delta x + \dfrac{\partial z}{\partial y}\delta y$

In this expression, $\dfrac{\partial z}{\partial x}$ and $\dfrac{\partial z}{\partial y}$ are the partial derivatives of z with respect to x and y respectively, and you will remember that to find:

(a) $\dfrac{\partial z}{\partial x}$, we differentiate the function z, with respect to x, keeping all independent variables other than x, for the time being, ...

(b) $\dfrac{\partial z}{\partial y}$, we differentiate the function z with respect to y, keeping all independent variables other than y, for the time being, ...

3

> constant (in both cases)

An example, just to remind you:

If $\quad z = x^3 + 4x^2y - 3y^3$

then $\quad \dfrac{\partial z}{\partial x} = 3x^2 + 8xy - 0 \qquad$ (y is constant)

and $\quad \dfrac{\partial z}{\partial y} = 0 + 4x^2 - 9y^2 \qquad$ (x is constant)

In practice, of course, we do not write down the zero terms.

Before we tackle any further applications, we must be expert at finding partial derivatives, so with the reminder above, have a go at this one.

(1) If $z = \tan(x^2 - y^2)$, find $\dfrac{\partial z}{\partial x}$ and $\dfrac{\partial z}{\partial y}$

When you have finished it, check with the next frame

Applications of partial differentiation

$$\boxed{\frac{\partial z}{\partial x} = 2x\sec^2(x^2 - y^2); \quad \frac{\partial z}{\partial y} = -2y\sec^2(x^2 - y^2)}$$

Because $z = \tan(x^2 - y^2)$

$$\therefore \frac{\partial z}{\partial x} = \sec^2(x^2 - y^2) \times \frac{\partial}{\partial x}(x^2 - y^2)$$
$$= \sec^2(x^2 - y^2)(2x) = 2x\sec^2(x^2 - y^2)$$

and $\dfrac{\partial z}{\partial y} = \sec^2(x^2 - y^2) \times \dfrac{\partial}{\partial y}(x^2 - y^2)$
$$= \sec^2(x^2 - y^2)(-2y) = -2y\sec^2(x^2 - y^2)$$

That was easy enough. Now do this one:

(2) If $z = e^{2x-3y}$, find $\dfrac{\partial^2 z}{\partial x^2}, \dfrac{\partial^2 z}{\partial y^2}, \dfrac{\partial^2 z}{\partial x \cdot \partial y}$

Finish them all. Then move on to Frame 5 and check your results

Here are the results in detail:

$z = e^{2x-3y}$ $\quad \therefore \dfrac{\partial z}{\partial x} = e^{2x-3y} \cdot 2 = 2 \cdot e^{2x-3y}$

$\dfrac{\partial z}{\partial y} = e^{2x-3y}(-3) = -3 \cdot e^{2x-3y}$

$\dfrac{\partial^2 z}{\partial x^2} = 2 \cdot e^{2x-3y} \cdot 2 = 4 \cdot e^{2x-3y}$

$\dfrac{\partial^2 z}{\partial y^2} = -3 \cdot e^{2x-3y}(-3) = 9 \cdot e^{2x-3y}$

$\dfrac{\partial^2 z}{\partial x \cdot \partial y} = -3 \cdot e^{2x-3y} \cdot 2 = -6 \cdot e^{2x-3y}$

All correct?

You remember, too, that in the 'mixed' second partial derivative, the order of differentiating does not matter. So in this case, since

$\dfrac{\partial^2 z}{\partial x \cdot \partial y} = -6 \cdot e^{2x-3y}$, then $\dfrac{\partial^2 z}{\partial y \cdot \partial x} = \ldots\ldots\ldots\ldots$

$$\boxed{\dfrac{\partial^2 z}{\partial x \cdot \partial y} = \dfrac{\partial^2 z}{\partial y \cdot \partial x} = -6 \cdot e^{2x-3y}}$$

Well now, before we move on to new work, see what you make of these. Find all the first and second partial derivatives of the following:

(a) $z = x\sin y$
(b) $z = (x + y)\ln(xy)$

When you have found all the derivatives,
check your work with the solutions in the next frame

7

Here they are. Check your results carefully.

(a) $z = x \sin y$

$$\therefore \frac{\partial z}{\partial x} = \sin y \qquad \frac{\partial z}{\partial y} = x \cos y$$

$$\frac{\partial^2 z}{\partial x^2} = 0 \qquad \frac{\partial^2 z}{\partial y^2} = -x \sin y$$

$$\frac{\partial^2 z}{\partial y \cdot \partial x} = \cos y \qquad \frac{\partial^2 z}{\partial x \cdot \partial y} = \cos y$$

(b) $z = (x+y) \ln(xy)$

$$\therefore \frac{\partial z}{\partial x} = (x+y)\frac{1}{xy} \cdot y + \ln(xy) = \frac{(x+y)}{x} + \ln(xy)$$

$$\frac{\partial z}{\partial y} = (x+y)\frac{1}{xy} \cdot x + \ln(xy) = \frac{(x+y)}{y} + \ln(xy)$$

$$\therefore \frac{\partial^2 z}{\partial x^2} = \frac{x - (x+y)}{x^2} + \frac{1}{xy} \cdot y = \frac{x - x - y}{x^2} + \frac{1}{x}$$

$$= \frac{x - y}{x^2}$$

$$\frac{\partial^2 z}{\partial y^2} = \frac{y - (x+y)}{y^2} + \frac{1}{xy} \cdot x = \frac{y - x - y}{y^2} + \frac{1}{y}$$

$$= \frac{y - x}{y^2}$$

$$\frac{\partial^2 z}{\partial y \cdot \partial x} = \frac{1}{x} + \frac{1}{xy} \cdot x = \frac{1}{x} + \frac{1}{y}$$

$$= \frac{y + x}{xy}$$

$$\frac{\partial^2 z}{\partial x \cdot \partial y} = \frac{1}{y} + \frac{1}{xy} \cdot y = \frac{1}{y} + \frac{1}{x}$$

$$= \frac{x + y}{xy}$$

Well now, that was just by way of warming up with work you have done before. Let us now move on to the next section of this Program.

Applications of partial differentiation

Rate-of-change problems

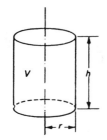

Let us consider a cylinder of radius r and height h as before. Then the volume is given by
$$V = \pi r^2 h$$
$$\therefore \frac{\partial V}{\partial r} = 2\pi r h \text{ and } \frac{\partial V}{\partial h} = \pi r^2$$

[8]

Since V is a function of r and h, we also know that
$$\delta V = \frac{\partial V}{\partial r} \delta r + \frac{\partial V}{\partial h} \delta h$$
(Here it is, popping up again!)

Now divide both sides by δt: $\quad \dfrac{\delta V}{\delta t} = \dfrac{\partial V}{\partial r} \cdot \dfrac{\delta r}{\delta t} + \dfrac{\partial V}{\partial h} \cdot \dfrac{\delta h}{\delta t}$

Then if $\delta t \to 0$, $\dfrac{\delta V}{\delta t} \to \dfrac{\mathrm{d}V}{\mathrm{d}t}$, $\dfrac{\delta r}{\delta t} \to \dfrac{\mathrm{d}r}{\mathrm{d}t}$, $\dfrac{\delta h}{\delta t} \to \dfrac{\mathrm{d}h}{\mathrm{d}t}$, but the partial derivatives, which do not contain δt, will remain unchanged.

So our result now becomes $\dfrac{\mathrm{d}V}{\mathrm{d}t} = \ldots\ldots\ldots$

$$\boxed{\dfrac{\mathrm{d}V}{\mathrm{d}t} = \dfrac{\partial V}{\partial r} \cdot \dfrac{\mathrm{d}r}{\mathrm{d}t} + \dfrac{\partial V}{\partial h} \cdot \dfrac{\mathrm{d}h}{\mathrm{d}t}}$$

[9]

This result is really the key to problems of the kind we are about to consider. If we know the rate at which r and h are changing, we can now find the corresponding rate of change of V. Like this:

Example 1

The radius of a cylinder increases at the rate of 0.2 cm/s while the height decreases at the rate of 0.5 cm/s. Find the rate at which the volume is changing at the instant when $r = 8$ cm and $h = 12$ cm.

Warning: The first inclination is to draw a diagram and to put in the given values for its dimensions, i.e. $r = 8$ cm, $h = 12$ cm. This we *must NOT do*, for the radius and height are changing and the given values are instantaneous values only. Therefore on the diagram we keep the symbols r and h to indicate that they are variables.

10

Here it is then:

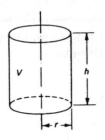

$$V = \pi r^2 h$$

$$\delta V = \frac{\partial V}{\partial r}\delta r + \frac{\partial V}{\partial h}\delta h$$

$$\therefore \frac{dV}{dt} = \frac{\partial V}{\partial r}\cdot\frac{dr}{dt} + \frac{\partial V}{\partial h}\cdot\frac{dh}{dt}$$

$$\frac{\partial V}{\partial r} = 2\pi r h; \quad \frac{\partial V}{\partial h} = \pi r^2$$

$$\therefore \frac{dV}{dt} = 2\pi rh\frac{dr}{dt} + \pi r^2\frac{dh}{dt}$$

Now at the instant we are considering:

$$r = 8,\; h = 12,\; \frac{dr}{dt} = 0.2,\; \frac{dh}{dt} = -0.5 \text{ (minus since } h \text{ is decreasing)}$$

So you can now substitute these values in the last statement and finish off the calculation, giving:

$$\frac{dV}{dt} = \ldots\ldots\ldots$$

11

$$\boxed{\frac{dV}{dt} = 20.1 \text{ cm}^3/\text{s}}$$

Because
$$\frac{dV}{dt} = 2\pi rh\cdot\frac{dr}{dt} + \pi r^2\frac{dh}{dt}$$
$$= 2\pi 8\cdot 12\cdot(0.2) + \pi 64(-0.5)$$
$$= 38.4\pi - 32\pi$$
$$= 6.4\pi = 20.1 \text{ cm}^3/\text{s}$$

Now another one.

Example 2

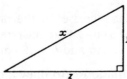

In the right-angled triangle shown, x is increasing at 2 cm/s while y is decreasing at 3 cm/s. Calculate the rate at which z is changing when $x = 5$ cm and $y = 3$ cm.

The first thing to do, of course, is to express z in terms of x and y. That is not difficult.

$$z = \ldots\ldots\ldots$$

Applications of partial differentiation

$$z = \sqrt{x^2 - y^2}$$

$z = \sqrt{x^2 - y^2} = (x^2 - y^2)^{\frac{1}{2}}$

$\therefore \delta z = \dfrac{\partial z}{\partial x} \delta x + \dfrac{\partial z}{\partial y} \delta y$ (The key to the whole business)

$\therefore \dfrac{dz}{dt} = \dfrac{\partial z}{\partial x} \cdot \dfrac{dx}{dt} + \dfrac{\partial z}{\partial y} \cdot \dfrac{dy}{dt}$

In this case $\dfrac{\partial z}{\partial x} = \dfrac{1}{2}(x^2 - y^2)^{-\frac{1}{2}}(2x) = \dfrac{x}{\sqrt{x^2 - y^2}}$

$\dfrac{\partial z}{\partial y} = \dfrac{1}{2}(x^2 - y^2)^{-\frac{1}{2}}(-2y) = \dfrac{-y}{\sqrt{x^2 - y^2}}$

$\dfrac{dz}{dt} = \dfrac{x}{\sqrt{x^2 - y^2}} \cdot \dfrac{dx}{dt} - \dfrac{y}{\sqrt{x^2 - y^2}} \cdot \dfrac{dy}{dt}$

So far so good. Now for the numerical values:

$x = 5,\ y = 3,\ \dfrac{dx}{dt} = 2,\ \dfrac{dy}{dt} = -3$

$\dfrac{dz}{dt} = \ldots\ldots\ldots\ldots$

Finish it off, then move to Frame 13

$$\dfrac{dz}{dt} = 4.75 \text{ cm/s}$$

Because we have $\dfrac{dz}{dt} = \dfrac{5}{\sqrt{5^2 - 3^2}}(2) - \dfrac{3}{\sqrt{5^2 - 3^2}}(-3)$

$= \dfrac{5(2)}{4} + \dfrac{3(3)}{4} = \dfrac{10}{4} + \dfrac{9}{4} = \dfrac{19}{4} = 4.75$ cm/s

$\therefore$ Side z increases at the rate of 4.75 cm/s

Now here is

Example 3

The total surface area S of a cone of base radius r and perpendicular height h is given by

$$S = \pi r^2 + \pi r \sqrt{r^2 + h^2}$$

If r and h are each increasing at the rate of 0.25 cm/s, find the rate at which S is increasing at the instant when $r = 3$ cm and $h = 4$ cm.

Do that one entirely on your own. Take your time: there is no need to hurry. Be quite sure that each step you write down is correct.

Then move to Frame 14 and check your result

14 Here is the solution in detail:

$$S = \pi r^2 + \pi r\sqrt{r^2 + h^2} = \pi r^2 + \pi r(r^2 + h^2)^{\frac{1}{2}}$$

$$\delta S = \frac{\partial S}{\partial r} \cdot \delta r + \frac{\partial S}{\partial h} \cdot \delta h \quad \therefore \quad \frac{dS}{dt} = \frac{\partial S}{\partial r} \cdot \frac{dr}{dt} + \frac{\partial S}{\partial h} \cdot \frac{dh}{dt}$$

(1) $\dfrac{\partial S}{\partial r} = 2\pi r + \pi r \cdot \dfrac{1}{2}(r^2 + h^2)^{-\frac{1}{2}}(2r) + \pi(r^2 + h^2)^{\frac{1}{2}}$

$$= 2\pi r + \frac{\pi r^2}{\sqrt{r^2 + h^2}} + \pi\sqrt{r^2 + h^2}$$

When $r = 3$ and $h = 4$:

$$\frac{\partial S}{\partial r} = 2\pi 3 + \frac{\pi 9}{5} + \pi 5 = 11\pi + \frac{9\pi}{5} = \frac{64\pi}{5}$$

(2) $\dfrac{\partial S}{\partial h} = \pi r \dfrac{1}{2}(r^2 + h^2)^{-\frac{1}{2}}(2h) = \dfrac{\pi r h}{\sqrt{r^2 + h^2}}$

$$= \frac{\pi 3 \cdot 4}{5} = \frac{12\pi}{5}$$

Also we are given that $\dfrac{dr}{dt} = 0.25$ and $\dfrac{dh}{dt} = 0.25$

$$\therefore \quad \frac{dS}{dt} = \frac{64\pi}{5} \cdot \frac{1}{4} + \frac{12\pi}{5} \cdot \frac{1}{4}$$

$$= \frac{16\pi}{5} + \frac{3\pi}{5} = \frac{19\pi}{5}$$

$$= 3.8\pi = 11.94 \text{ cm}^2/\text{s}$$

---

15 So there we are. Rate-of-change problems are all very much the same. What you must remember is simply this:

(a) The basic statement

$$\text{If } z = f(x, y) \text{ then } \delta z = \frac{\partial z}{\partial x} \cdot \delta x + \frac{\partial z}{\partial y} \cdot \delta y \qquad \text{(a)}$$

(b) Divide this result by δt and make $\delta t \to 0$. This converts the result into the form for rate-of-change problems:

$$\frac{dz}{dt} = \frac{\partial z}{\partial x} \cdot \frac{dx}{dt} + \frac{\partial z}{\partial y} \cdot \frac{dy}{dt} \qquad \text{(b)}$$

The second result follows directly from the first. Make a note of both of these in your record book for future reference.

Then for the next part of the work, move on to Frame 16

Applications of partial differentiation

Partial differentiation can also be used with advantage in finding *derivatives of implicit functions*.

For example, suppose we are required to find an expression for $\dfrac{dy}{dx}$ when we are given that $x^2 + 2xy + y^3 = 0$.

We can set about it in this way:

Let z stand for the function of x and y, i.e. $z = x^2 + 2xy + y^3$. Again we use the basic relationship $\delta z = \dfrac{\partial z}{\partial x} \delta x + \dfrac{\partial z}{\partial y} \delta y$.

If we divide both sides by δx, we get:

$$\frac{\delta z}{\delta x} = \frac{\partial z}{\partial x} + \frac{\partial z}{\partial y} \cdot \frac{\delta y}{\delta x}$$

Now, if $\delta x \to 0$, $\dfrac{dz}{dx} = \dfrac{\partial z}{\partial x} + \dfrac{\partial z}{\partial y} \cdot \dfrac{dy}{dx}$

If we now find expressions for $\dfrac{\partial z}{\partial x}$ and $\dfrac{\partial z}{\partial y}$, we shall be quite a way towards finding $\dfrac{dy}{dx}$ (which you see at the end of the expression).

In this particular example, where $z = x^2 + 2xy + y^3$, $\dfrac{\partial z}{\partial x} = \ldots\ldots\ldots$ and $\dfrac{\partial z}{\partial y} = \ldots\ldots\ldots$

$$\frac{\partial z}{\partial x} = 2x + 2y; \quad \frac{\partial z}{\partial y} = 2x + 3y^2$$

Substituting these in our previous result gives us:

$$\frac{dz}{dx} = (2x + 2y) + (2x + 3y^2)\frac{dy}{dx}$$

If only we knew $\dfrac{dz}{dx}$, we could rearrange this result and obtain an expression for $\dfrac{dy}{dx}$. So where can we find out something about $\dfrac{dz}{dx}$?

Refer back to the beginning of the problem. We have used z to stand for $x^2 + 2xy + y^3$ and we were told initially that $x^2 + 2xy + y^3 = 0$. Therefore $z = 0$, i.e. z is a constant (in this case zero) and hence $\dfrac{dz}{dx} = 0$.

$$\therefore 0 = (2x + 2y) + (2x + 3y^2)\frac{dy}{dx}$$

From this we can find $\dfrac{dy}{dx}$. So finish it off.

$$\frac{dy}{dx} = \ldots\ldots\ldots$$

On to Frame 18

Vector Analysis

18

$$\boxed{\frac{dy}{dx} = -\frac{2x + 2y}{2x + 3y^2}}$$

This is almost a routine that always works. In general, we have:

If $f(x,y) = 0$, find $\frac{dy}{dx}$

Let $z = f(x,y)$ then $\delta z = \frac{\partial z}{\partial x}\delta x + \frac{\partial z}{\partial y}\delta y$. Divide by δx and make $\delta x \to 0$, in which case:

$$\frac{dz}{dx} = \frac{\partial z}{\partial x} + \frac{\partial z}{\partial y} \cdot \frac{dy}{dx}$$

But $z = 0$ (constant) $\therefore \frac{dz}{dx} = 0 \quad \therefore 0 = \frac{\partial z}{\partial x} + \frac{\partial z}{\partial y} \cdot \frac{dy}{dx}$

giving $\frac{dy}{dx} = -\frac{\partial z}{\partial x} \bigg/ \frac{\partial z}{\partial y}$

The easiest form to remember is the one that comes direct from the basic result:

$$\delta z = \frac{\partial z}{\partial x}\delta x + \frac{\partial z}{\partial y}\delta y$$

Divide by δx, etc.

$$\frac{dz}{dx} = \frac{\partial z}{\partial x} + \frac{\partial z}{\partial y} \cdot \frac{dy}{dx} \quad \left\{\frac{dz}{dx} = 0\right\}$$

Make a note of this result.

19

Now for some examples.

Example 1

If $e^{xy} + x + y = 1$, evaluate $\frac{dy}{dx}$ at $(0, 0)$, The function can be written $e^{xy} + x + y - 1 = 0$.

Let $z = e^{xy} + x + y - 1 \quad \delta z = \frac{\partial z}{\partial x} \cdot \delta x + \frac{\partial z}{\partial y} \cdot \delta y \quad \therefore \frac{dz}{dx} = \frac{\partial z}{\partial x} + \frac{\partial z}{\partial y} \cdot \frac{dy}{dx}$

$\frac{\partial z}{\partial x} = e^{xy} \cdot y + 1; \quad \frac{\partial z}{\partial y} = e^{xy} \cdot x + 1 \quad \therefore \frac{dz}{dx} = (y \cdot e^{xy} + 1) + (x.e^{xy} + 1)\frac{dy}{dx}$

But $z = 0 \quad \therefore \frac{dz}{dx} = 0 \quad \therefore \frac{dy}{dx} = -\left\{\frac{y \cdot e^{xy} + 1}{x \cdot e^{xy} + 1}\right\}$

At $x = 0, y = 0, \frac{dy}{dx} = -\frac{1}{1} = -1 \quad \therefore \frac{dy}{dx} = -1$

All very easy so long as you can find partial derivatives correctly.

On to Frame 20

Applications of partial differentiation

Now here is:

Example 2

If $xy + \sin y = 2$, find $\dfrac{dy}{dx}$

Let $z = xy + \sin y - 2 = 0$

$$\delta z = \frac{\partial z}{\partial x}\delta x + \frac{\partial z}{\partial y}\delta y$$

$$\frac{dz}{dx} = \frac{\partial z}{\partial x} + \frac{\partial z}{\partial y}\cdot\frac{dy}{dx}$$

$$\frac{\partial z}{\partial x} = y; \quad \frac{\partial z}{\partial y} = x + \cos y$$

$$\therefore \frac{dz}{dx} = y + (x + \cos y)\frac{dy}{dx}$$

But $z = 0$ $\therefore \dfrac{dz}{dx} = 0$

$$\therefore \frac{dy}{dx} = \frac{-y}{x + \cos y}$$

Here is one for you to do:

Example 3

Find an expression for $\dfrac{dy}{dx}$ when $x\tan y = y\sin x$. Do it all on your own.

Then check your working with that in Frame 21

$$\boxed{\frac{dy}{dx} = -\frac{\tan y - y\cos x}{x\sec^2 y - \sin x}}$$

Did you get that? If so, go straight on to Frame 22. If not, here is the working below. Follow it through and see where you have gone astray!

$x\tan y = y\sin x$ $\therefore x\tan y - y\sin x = 0$

Let $z = x\tan y - y\sin x = 0$

$$\delta z = \frac{\partial z}{\partial x}\delta x + \frac{\partial z}{\partial y}\delta y$$

$$\frac{dz}{dx} = \frac{\partial z}{\partial x} + \frac{\partial z}{\partial y}\cdot\frac{dy}{dx}$$

$$\frac{\partial z}{\partial x} = \tan y - y\cos x; \quad \frac{\partial z}{\partial y} = x\sec^2 y - \sin x$$

$$\therefore \frac{dz}{dx} = (\tan y - y\cos x) + (x\sec^2 y - \sin x)\frac{dy}{dx}$$

But $z = 0$ $\therefore \dfrac{dz}{dx} = 0$

$$\frac{dy}{dx} = -\frac{\tan y - y\cos x}{x\sec^2 y - \sin x}$$

On now to Frame 22

22

Right. Now here is just one more for you to do. They are really very much the same.

Example 4

If $e^{x+y} = x^2y^2$, find an expression for $\dfrac{dy}{dx}$

$e^{x+y} - x^2y^2 = 0.$ Let $z = e^{x+y} - x^2y^2 = 0$

$$\delta z = \frac{\partial z}{\partial x}\delta x + \frac{\partial z}{\partial y}\delta y$$

$$\frac{dz}{dx} = \frac{\partial z}{\partial x} + \frac{\partial z}{\partial y}\cdot\frac{dy}{dx}$$

So continue with the good work and finish it off, finally getting that

$$\frac{dy}{dx} = \ldots\ldots\ldots\ldots$$

Then move to Frame 23

23

$$\boxed{\frac{dy}{dx} = \frac{2xy^2 - e^{x+y}}{e^{x+y} - 2x^2y}}$$

Because $z = e^{x+y} - x^2y^2 = 0$

$$\frac{\partial z}{\partial x} = e^{x+y} - 2xy^2; \quad \frac{\partial z}{\partial y} = e^{x+y} - 2x^2y$$

$$\therefore \frac{dz}{dx} = (e^{x+y} - 2xy^2) + (e^{x+y} - 2x^2y)\frac{dy}{dx}$$

But $z = 0 \quad \therefore \dfrac{dz}{dx} = 0$

$$\therefore \frac{dy}{dx} = -\frac{(e^{x+y} - 2xy^2)}{(e^{x+y} - 2x^2y)}$$

$$\therefore \frac{dy}{dx} = \frac{2xy^2 - e^{x+y}}{(e^{x+y} - 2x^2y)}$$

That is how they are all done. But now there is one more process that you must know how to tackle.

So on to Frame 24

Applications of partial differentiation

Change of variables

24

If z is a function of x and y, i.e. $z = f(x, y)$, and x and y are themselves functions of two other variables u and v, then z is also a function of u and v. We may therefore need to find $\dfrac{\partial z}{\partial u}$ and $\dfrac{\partial z}{\partial v}$. How do we go about it?

$$z = f(x, y) \qquad \therefore \ \delta z = \frac{\partial z}{\partial x}\delta x + \frac{\partial z}{\partial y}\delta y$$

Divide both sides by δu:

$$\frac{\delta z}{\delta u} = \frac{\partial z}{\partial x}\cdot\frac{\delta x}{\delta u} + \frac{\partial z}{\partial y}\cdot\frac{\delta y}{\delta u}$$

If v is kept constant for the time being, then $\dfrac{\delta x}{\delta u}$ when $\delta u \to 0$ becomes $\dfrac{\partial x}{\partial u}$ and $\dfrac{\delta y}{\delta u}$ becomes $\dfrac{\partial y}{\partial u}$.

$$\therefore \ \frac{\partial z}{\partial u} = \frac{\partial z}{\partial x}\cdot\frac{\partial x}{\partial u} + \frac{\partial z}{\partial y}\cdot\frac{\partial y}{\partial u}$$

and $\ \dfrac{\partial z}{\partial v} = \dfrac{\partial z}{\partial x}\cdot\dfrac{\partial x}{\partial v} + \dfrac{\partial z}{\partial y}\cdot\dfrac{\partial y}{\partial v}$

Note these

Next frame

Here is an example of this work.

25

Example 1

If $z = x^2 + y^2$, where $x = r\cos\theta$ and $y = r\sin 2\theta$, find $\dfrac{\partial z}{\partial r}$ and $\dfrac{\partial z}{\partial \theta}$

$$\frac{\partial z}{\partial r} = \frac{\partial z}{\partial x}\cdot\frac{\partial x}{\partial r} + \frac{\partial z}{\partial y}\cdot\frac{\partial y}{\partial r}$$

and $\ \dfrac{\partial z}{\partial \theta} = \dfrac{\partial z}{\partial x}\cdot\dfrac{\partial x}{\partial \theta} + \dfrac{\partial z}{\partial y}\cdot\dfrac{\partial y}{\partial \theta}$

Now, $\dfrac{\partial z}{\partial x} = 2x \qquad \left[\dfrac{\partial z}{\partial y} = 2y\right]$

$\dfrac{\partial x}{\partial r} = \cos\theta \qquad \left[\dfrac{\partial y}{\partial r} = \sin 2\theta\right]$

$\therefore \ \dfrac{\partial z}{\partial r} = 2x\cos\theta + 2y\sin 2\theta$

and $\ \dfrac{\partial x}{\partial \theta} = -r\sin\theta$ and $\dfrac{\partial y}{\partial \theta} = 2r\cos 2\theta$

$\therefore \ \dfrac{\partial z}{\partial \theta} = 2x(-r\sin\theta) + 2y(2r\cos 2\theta)$

$\dfrac{\partial z}{\partial \theta} = 4yr\cos 2\theta - 2xr\sin\theta$

And in these two results, the symbols x and y can be replaced by $r\cos\theta$ and $r\sin 2\theta$ respectively.

26

One more example:

If $z = e^{xy}$ where $x = \ln(u+v)$ and $y = \sin(u-v)$, find $\dfrac{\partial z}{\partial u}$ and $\dfrac{\partial z}{\partial v}$.

We have $\dfrac{\partial z}{\partial u} = \dfrac{\partial z}{\partial x} \cdot \dfrac{\partial x}{\partial u} + \dfrac{\partial z}{\partial y} \cdot \dfrac{\partial y}{\partial u} = y \cdot e^{xy} \cdot \dfrac{1}{u+v} + x \cdot e^{xy} \cdot \cos(u-v)$

$$= e^{xy} \left\{ \dfrac{y}{u+v} + x \cdot \cos(u-v) \right\}$$

and $\dfrac{\partial z}{\partial v} = \dfrac{\partial z}{\partial x} \cdot \dfrac{\partial x}{\partial v} + \dfrac{\partial z}{\partial y} \cdot \dfrac{\partial y}{\partial v} = y \cdot e^{xy} \cdot \dfrac{1}{u+v} + x \cdot e^{xy} \cdot \{-\cos(u-v)\}$

$$= e^{xy} \left\{ \dfrac{y}{u+v} - x \cdot \cos(u-v) \right\}$$

Now move on to Frame 27

27

Here is one for you to do on your own. All that it entails is to find the various partial derivatives and to substitute them in the established results:

$$\dfrac{\partial z}{\partial u} = \dfrac{\partial z}{\partial x} \cdot \dfrac{\partial x}{\partial u} + \dfrac{\partial z}{\partial y} \cdot \dfrac{\partial y}{\partial u} \quad \text{and} \quad \dfrac{\partial z}{\partial v} = \dfrac{\partial z}{\partial x} \cdot \dfrac{\partial x}{\partial v} + \dfrac{\partial z}{\partial y} \cdot \dfrac{\partial y}{\partial v}$$

So you do this one:

If $z = \sin(x+y)$, where $x = u^2 + v^2$ and $y = 2uv$, find $\dfrac{\partial z}{\partial u}$ and $\dfrac{\partial z}{\partial v}$

The method is the same as before.

When you have completed the work, check with the result in Frame 28

28

$z = \sin(x+y); \quad x = u^2 + v^2; \quad y = 2uv$

$\dfrac{\partial z}{\partial x} = \cos(x+y); \qquad \dfrac{\partial z}{\partial y} = \cos(x+y)$

$\dfrac{\partial x}{\partial u} = 2u \qquad\qquad \dfrac{\partial y}{\partial u} = 2v$

$\dfrac{\partial z}{\partial u} = \dfrac{\partial z}{\partial x} \cdot \dfrac{\partial x}{\partial u} + \dfrac{\partial z}{\partial y} \cdot \dfrac{\partial y}{\partial u}$

$= \cos(x+y) \cdot 2u + \cos(x+y) \cdot 2v$

$= 2(u+v) \cos(x+y)$

Also $\dfrac{\partial z}{\partial v} = \dfrac{\partial z}{\partial x} \cdot \dfrac{\partial x}{\partial v} + \dfrac{\partial z}{\partial y} \cdot \dfrac{\partial y}{\partial v}$

$\dfrac{\partial x}{\partial v} = 2v; \qquad \dfrac{\partial y}{\partial v} = 2u$

$\dfrac{\partial z}{\partial v} = \cos(x+y) \cdot 2v + \cos(x+y) \cdot 2u$

$= 2(u+v) \cos(x+y)$

Applications of partial differentiation

Example 2

29

If $z = x^2 - y^2$ and $x = r\cos\theta$ and $y = r\sin\theta$, then

$$\frac{\partial z}{\partial r} = \frac{\partial z}{\partial x}\cdot\frac{\partial x}{\partial r} + \frac{\partial z}{\partial y}\cdot\frac{\partial y}{\partial r}$$

and

$$\frac{\partial z}{\partial \theta} = \frac{\partial z}{\partial x}\cdot\frac{\partial x}{\partial \theta} + \frac{\partial z}{\partial y}\cdot\frac{\partial y}{\partial \theta}$$

We now need the various partial derivatives

$$\frac{\partial z}{\partial x} = \ldots\ldots ; \quad \frac{\partial x}{\partial r} = \ldots\ldots ; \quad \frac{\partial y}{\partial r} = \ldots\ldots$$

$$\frac{\partial z}{\partial y} = \ldots\ldots ; \quad \frac{\partial x}{\partial \theta} = \ldots\ldots ; \quad \frac{\partial y}{\partial \theta} = \ldots\ldots$$

30

$$\frac{\partial z}{\partial x} = 2x; \quad \frac{\partial x}{\partial r} = \cos\theta; \quad \frac{\partial y}{\partial r} = \sin\theta$$

$$\frac{\partial z}{\partial y} = -2y; \quad \frac{\partial x}{\partial \theta} = -r\sin\theta; \quad \frac{\partial y}{\partial \theta} = r\cos\theta$$

Substituting in the two equations and simplifying:

$$\frac{\partial z}{\partial r} = \ldots\ldots ; \quad \frac{\partial z}{\partial \theta} = \ldots\ldots$$

31

$$\frac{\partial z}{\partial r} = 2x\cos\theta - 2y\sin\theta; \quad \frac{\partial z}{\partial \theta} = -(2xr\sin\theta + 2yr\cos\theta)$$

Finally, we can express x and y in terms of r and θ as given, so that, after tidying up, we obtain

$$\frac{\partial z}{\partial r} = \ldots\ldots ; \quad \frac{\partial z}{\partial \theta} = \ldots\ldots$$

32

$$\frac{\partial z}{\partial r} = 2r(\cos^2\theta - \sin^2\theta); \quad \frac{\partial z}{\partial \theta} = -4r^2\sin\theta\cos\theta$$

Of course, we could express these as

$$\frac{\partial z}{\partial r} = 2r\cos 2\theta \quad \text{and} \quad \frac{\partial z}{\partial \theta} = -2r^2\sin 2\theta$$

From these results, we can, if necessary, find the second partial derivatives in the normal manner.

$$\frac{\partial^2 z}{\partial r^2} = \frac{\partial}{\partial r}\left(\frac{\partial z}{\partial r}\right) = \frac{\partial}{\partial r}(2r\cos 2\theta) = 2\cos 2\theta$$

Similarly $\dfrac{\partial^2 z}{\partial \theta^2} = \ldots\ldots\ldots$ and $\dfrac{\partial^2 z}{\partial r \partial \theta} = \ldots\ldots\ldots$

33

$$\frac{\partial^2 z}{\partial \theta^2} = -4r^2 \cos 2\theta; \quad \frac{\partial^2 z}{\partial r \partial \theta} = -4r \sin 2\theta$$

Because

$$\frac{\partial^2 z}{\partial \theta^2} = \frac{\partial}{\partial \theta}\left(\frac{\partial z}{\partial \theta}\right) = \frac{\partial}{\partial \theta}(-2r^2 \sin 2\theta) = -4r^2 \cos 2\theta$$

and

$$\frac{\partial^2 z}{\partial r \partial \theta} = \frac{\partial}{\partial r}\left(\frac{\partial z}{\partial \theta}\right) = \frac{\partial}{\partial r}(-2r^2 \sin 2\theta) = -4r \sin 2\theta$$

Example 3

If $z = f(x, y)$ and $x = \frac{1}{2}(u^2 - v^2)$ and $y = uv$, show that

$$u\frac{\partial z}{\partial v} - v\frac{\partial z}{\partial u} = 2\left(x\frac{\partial z}{\partial y} - y\frac{\partial z}{\partial x}\right)$$

Although this is much the same as the previous example, there is, at least, one difference. In this case, we are not told the precise nature of $f(x, y)$. We must remember that z is a function of x and y and, therefore, of u and v. With that in mind, we set off with the usual two equations.

$$\frac{\partial z}{\partial u} = \ldots\ldots\ldots$$

$$\frac{\partial z}{\partial v} = \ldots\ldots\ldots$$

34

$$\frac{\partial z}{\partial u} = \frac{\partial z}{\partial x} \cdot \frac{\partial x}{\partial u} + \frac{\partial z}{\partial y} \cdot \frac{\partial y}{\partial u}$$

$$\frac{\partial z}{\partial v} = \frac{\partial z}{\partial x} \cdot \frac{\partial x}{\partial v} + \frac{\partial z}{\partial y} \cdot \frac{\partial y}{\partial v}$$

From the given information:

$$\frac{\partial x}{\partial u} = \ldots\ldots\ldots; \quad \frac{\partial y}{\partial u} = \ldots\ldots\ldots$$

$$\frac{\partial x}{\partial v} = \ldots\ldots\ldots; \quad \frac{\partial y}{\partial v} = \ldots\ldots\ldots$$

Applications of partial differentiation

$$\frac{\partial x}{\partial u} = u; \quad \frac{\partial y}{\partial u} = v$$
$$\frac{\partial x}{\partial v} = -v; \quad \frac{\partial y}{\partial v} = u$$

[35]

Whereupon $\dfrac{\partial z}{\partial u} = \ldots\ldots\ldots\ldots$

$\dfrac{\partial z}{\partial v} = \ldots\ldots\ldots\ldots$

$$\frac{\partial z}{\partial u} = u\frac{\partial z}{\partial x} + v\frac{\partial z}{\partial y}$$
$$\frac{\partial z}{\partial v} = -v\frac{\partial z}{\partial x} + u\frac{\partial z}{\partial y}$$

[36]

If we now multiply the first of these by $(-v)$ and the second by u and add the two equations, we get the desired result.

$$-v\frac{\partial z}{\partial u} = -uv\frac{\partial z}{\partial x} - v^2\frac{\partial z}{\partial y}$$

$$u\frac{\partial z}{\partial v} = -uv\frac{\partial z}{\partial x} + u^2\frac{\partial z}{\partial y}$$

Adding $\quad u\dfrac{\partial z}{\partial v} - v\dfrac{\partial z}{\partial u} = -2uv\dfrac{\partial z}{\partial x} + (u^2 - v^2)\dfrac{\partial z}{\partial y}$

$$= -2y\frac{\partial z}{\partial x} + 2x\frac{\partial z}{\partial y}$$

$$\therefore u\frac{\partial z}{\partial v} - v\frac{\partial z}{\partial u} = 2\left(x\frac{\partial z}{\partial y} - y\frac{\partial z}{\partial x}\right)$$

With the same given data, i.e.

$z = f(x, y)$ with $x = \dfrac{1}{2}(u^2 - v^2)$ and $y = uv$

we can now show that $\dfrac{\partial^2 z}{\partial u^2} + \dfrac{\partial^2 z}{\partial v^2} = (u^2 + v^2)\left(\dfrac{\partial^2 z}{\partial x^2} + \dfrac{\partial^2 z}{\partial y^2}\right).$

In determining the second partial derivatives, keep in mind that z is a function of u and v and that both of these variables also occur in $\dfrac{\partial z}{\partial x}$ and $\dfrac{\partial z}{\partial y}$.

$$\frac{\partial^2 z}{\partial u^2} = \ldots\ldots\ldots\ldots$$

37

$$\frac{\partial^2 z}{\partial u^2} = \frac{\partial z}{\partial x} + u^2 \frac{\partial^2 z}{\partial x^2} + 2uv \frac{\partial^2 z}{\partial x \partial y} + v^2 \frac{\partial^2 z}{\partial y^2}$$

Because

$$\frac{\partial}{\partial u} = \left(u \frac{\partial}{\partial x} + v \frac{\partial}{\partial y}\right) \quad \text{and} \quad \frac{\partial}{\partial v} = \left(-v \frac{\partial}{\partial x} + u \frac{\partial}{\partial y}\right)$$

$$\therefore \frac{\partial^2 z}{\partial u^2} = \frac{\partial}{\partial u}\left(u \frac{\partial z}{\partial x} + v \frac{\partial z}{\partial y}\right) = \frac{\partial z}{\partial x} + u \frac{\partial}{\partial u}\left(\frac{\partial z}{\partial x}\right) + v \frac{\partial}{\partial u}\left(\frac{\partial z}{\partial y}\right)$$

$$= \frac{\partial z}{\partial x} + u\left(u \frac{\partial}{\partial x} + v \frac{\partial}{\partial y}\right)\frac{\partial z}{\partial x} + v\left(u \frac{\partial}{\partial x} + v \frac{\partial}{\partial y}\right)\frac{\partial z}{\partial y}$$

$$= \frac{\partial z}{\partial x} + u^2 \frac{\partial^2 z}{\partial x^2} + uv \frac{\partial^2 z}{\partial x \partial y} + uv \frac{\partial^2 z}{\partial x \partial y} + v^2 \frac{\partial^2 z}{\partial y^2}$$

$$\therefore \frac{\partial^2 z}{\partial u^2} = \frac{\partial z}{\partial x} + u^2 \frac{\partial^2 z}{\partial x^2} + 2uv \frac{\partial^2 z}{\partial x \partial y} + v^2 \frac{\partial^2 z}{\partial y^2} \tag{1}$$

Likewise, $\dfrac{\partial^2 z}{\partial v^2} = \dfrac{\partial}{\partial v}\left(\dfrac{\partial z}{\partial v}\right) = \dfrac{\partial}{\partial v}\left(-v \dfrac{\partial z}{\partial x} + u \dfrac{\partial z}{\partial y}\right)$

$$= \ldots\ldots\ldots$$

38

$$\frac{\partial^2 z}{\partial v^2} = \frac{\partial z}{\partial x} + v^2 \frac{\partial^2 z}{\partial x^2} - 2uv \frac{\partial^2 z}{\partial x \partial y} + u^2 \frac{\partial^2 z}{\partial y^2}$$

Because

$$\frac{\partial^2 z}{\partial v^2} = \frac{\partial}{\partial v}\left(-v \frac{\partial z}{\partial x} + u \frac{\partial z}{\partial y}\right) = -\frac{\partial z}{\partial x} + \frac{\partial}{\partial v}\left(\frac{\partial z}{\partial x}\right) + u \frac{\partial}{\partial v}\left(\frac{\partial z}{\partial y}\right)$$

$$= \frac{\partial z}{\partial x} - v\left(-v \frac{\partial}{\partial x} + u \frac{\partial}{\partial y}\right)\frac{\partial z}{\partial x} + u\left(-v \frac{\partial}{\partial x} + u \frac{\partial}{\partial y}\right)\frac{\partial z}{\partial y}$$

$$= \frac{\partial z}{\partial x} + v^2 \frac{\partial^2 z}{\partial x^2} - uv \frac{\partial^2 z}{\partial x \partial y} - uv \frac{\partial^2 z}{\partial x \partial y} + u^2 \frac{\partial^2 z}{\partial y^2}$$

$$\therefore \frac{\partial^2 z}{\partial v^2} = \frac{\partial z}{\partial x} + v^2 \frac{\partial^2 z}{\partial x^2} - 2uv \frac{\partial^2 z}{\partial x \partial y} + u^2 \frac{\partial^2 z}{\partial y^2} \tag{2}$$

Adding together results (1) and (2), we get

$$\ldots\ldots\ldots$$

39

$$\frac{\partial^2 z}{\partial u^2} + \frac{\partial^2 z}{\partial v^2} = (u^2 + v^2)\left(\frac{\partial^2 z}{\partial x^2} + \frac{\partial^2 z}{\partial y^2}\right)$$

and that is it.

Now, for something slightly different, move on to the next frame

Applications of partial differentiation

Inverse functions

40

If $z = f(x, y)$ and x and y are functions of two independent variables u and v defined by $u = g(x, y)$ and $v = h(x, y)$, we can theoretically solve these two equations to obtain x and y in terms of u and v. Hence we can determine $\dfrac{\partial x}{\partial u}, \dfrac{\partial x}{\partial v}, \dfrac{\partial y}{\partial u}, \dfrac{\partial y}{\partial v}$ and then $\dfrac{\partial z}{\partial x}$ and $\dfrac{\partial z}{\partial y}$ as required.

In practice, however, the solution of $u = g(x, y)$ and $v = h(x, y)$ may well be difficult or even impossible by normal means. The following example shows how we can get over this difficulty.

Example 1

If $z = f(x, y)$ and $u = e^x \cos y$ and $v = e^{-x} \sin y$, we have to find $\dfrac{\partial x}{\partial u}, \dfrac{\partial x}{\partial v}, \dfrac{\partial y}{\partial u}, \dfrac{\partial y}{\partial v}$.

We start off once again with our standard relationships

$$\delta u = \frac{\partial u}{\partial x} \delta x + \frac{\partial u}{\partial y} \delta y \qquad (1)$$

$$\delta v = \frac{\partial v}{\partial x} \delta x + \frac{\partial v}{\partial y} \delta y \qquad (2)$$

Now $u = e^x \cos y$ and $v = e^{-x} \sin y$

So $\dfrac{\partial u}{\partial x} = \ldots\ldots\ldots$; $\dfrac{\partial u}{\partial y} = \ldots\ldots\ldots$

$\dfrac{\partial v}{\partial x} = \ldots\ldots\ldots$; $\dfrac{\partial v}{\partial y} = \ldots\ldots\ldots$

41

$$\frac{\partial u}{\partial x} = e^x \cos y; \qquad \frac{\partial u}{\partial y} = -e^x \sin y$$

$$\frac{\partial v}{\partial x} = -e^{-x} \sin y; \qquad \frac{\partial v}{\partial y} = e^{-x} \cos y$$

Substituting in equations (1) and (2) above, we have

$$\delta u = e^x \cos y \, \delta x - e^x \sin y \, \delta y \qquad (3)$$

$$\delta v = -e^{-x} \sin y \, \delta x + e^{-x} \cos y \, \delta y \qquad (4)$$

Eliminating δy from (3) and (4), we get

$$\delta x = \ldots\ldots\ldots$$

42

$$\delta x = \frac{e^{-x}\cos y}{\cos 2y}\delta u + \frac{e^x \sin y}{\cos 2y}\delta v$$

Because

(3) × $e^{-x}\cos y$: $e^{-x}\cos y\,\delta u = \cos^2 y\,\delta x - \sin y\cos y\,\delta y$

(4) × $e^x \sin y$: $e^x \sin y\,\delta v = -\sin^2 y\,\delta x + \sin y\cos y\,\delta y$

Adding: $e^{-x}\cos y\,\delta u + e^x \sin y\,\delta v = (\cos^2 y - \sin^2 y)\,\delta x$

$$\therefore\ \delta x = \frac{e^{-x}\cos y}{\cos 2y}\delta u + \frac{e^x \sin y}{\cos 2y}\delta v$$

But $\delta x = \dfrac{\partial x}{\partial u}\delta u + \dfrac{\partial x}{\partial v}\delta v$

$$\therefore\ \frac{\partial x}{\partial u} = \frac{e^{-x}\cos y}{\cos 2y} \quad \text{and} \quad \frac{\partial x}{\partial v} = \frac{e^x \sin y}{\cos 2y}$$

which are, of course, two of the expressions we have to find.
Starting again with equations (3) and (4), we can obtain

$$\delta y = \dots\dots\dots$$

43

$$\delta y = \frac{e^{-x}\sin y}{\cos 2y}\delta u + \frac{e^x \cos y}{\cos 2y}\delta v$$

Because

(3) × $e^{-x}\sin y$: $e^{-x}\sin y\,\delta u = \sin y\cos y\,\delta x - \sin^2 y\,\delta y$

(4) × $e^x \cos y$: $e^x \cos y\,\delta v = -\sin y\cos y\,\delta x + \cos^2 y\,\delta y$

Adding: $e^{-x}\sin y\,\delta u + e^x \cos y\,\delta v = (\cos^2 y - \sin^2 y)\,\delta y$

$$\therefore\ \delta y = \frac{e^{-x}\sin y}{\cos 2y}\delta u + \frac{e^x \cos y}{\cos 2y}\delta v$$

But, $\delta y = \dots\dots\dots$ Finish it off.

Applications of partial differentiation

> $$\delta y = \frac{\partial y}{\partial u}\delta u + \frac{\partial y}{\partial v}\delta v$$
> $$\therefore \frac{\partial y}{\partial u} = \frac{e^{-x}\sin y}{\cos 2y} \quad \text{and} \quad \frac{\partial y}{\partial v} = \frac{e^{x}\cos y}{\cos 2y}$$

So, collecting our four results together:

$$\frac{\partial x}{\partial u} = \frac{e^{-x}\cos y}{\cos 2y}; \quad \frac{\partial x}{\partial v} = \frac{e^{x}\sin y}{\cos 2y}$$
$$\frac{\partial y}{\partial u} = \frac{e^{-x}\sin y}{\cos 2y}; \quad \frac{\partial y}{\partial v} = \frac{e^{x}\cos y}{\cos 2y}$$

We can tackle most similar problems in the same way, but it is more efficient to investigate a general case and to streamline the results. Let us do that.

General case

If $z = f(x, y)$ with $u = g(x, y)$ and $v = h(x, y)$, then we have

$$\delta u = \frac{\partial u}{\partial x}\delta x + \frac{\partial u}{\partial y}\delta y \tag{1}$$

$$\delta v = \frac{\partial v}{\partial x}\delta x + \frac{\partial v}{\partial y}\delta y \tag{2}$$

We now solve these for δx and δy. Eliminating δy, we have

(1) $\times \dfrac{\partial v}{\partial y}$: $\qquad \dfrac{\partial v}{\partial y}\delta u = \dfrac{\partial v}{\partial y}\cdot\dfrac{\partial u}{\partial x}\delta x + \dfrac{\partial v}{\partial y}\cdot\dfrac{\partial u}{\partial y}\delta y$

(2) $\times \dfrac{\partial u}{\partial y}$: $\qquad \dfrac{\partial u}{\partial y}\delta v = \dfrac{\partial u}{\partial y}\cdot\dfrac{\partial v}{\partial x}\delta x + \dfrac{\partial u}{\partial y}\cdot\dfrac{\partial v}{\partial y}\delta y$

Subtracting: $\qquad \dfrac{\partial v}{\partial y}\delta u - \dfrac{\partial u}{\partial y}\delta v = \left(\dfrac{\partial u}{\partial x}\cdot\dfrac{\partial v}{\partial y} - \dfrac{\partial v}{\partial x}\cdot\dfrac{\partial u}{\partial y}\right)\delta x$

$$\therefore \delta x = \frac{\dfrac{\partial v}{\partial y}\delta u - \dfrac{\partial u}{\partial y}\delta v}{\dfrac{\partial u}{\partial x}\cdot\dfrac{\partial v}{\partial y} - \dfrac{\partial v}{\partial x}\cdot\dfrac{\partial u}{\partial y}}$$

Starting afresh from (1) and (2) and eliminating δx, we have

$$\delta y = \ldots\ldots\ldots\ldots$$

$$\delta y = \frac{\dfrac{\partial u}{\partial x}\delta v - \dfrac{\partial v}{\partial x}\delta u}{\dfrac{\partial u}{\partial x}\cdot\dfrac{\partial v}{\partial y} - \dfrac{\partial v}{\partial x}\cdot\dfrac{\partial u}{\partial y}}$$

The two results so far are therefore

$$\delta x = \frac{\dfrac{\partial v}{\partial y}\delta u - \dfrac{\partial u}{\partial y}\delta v}{\dfrac{\partial u}{\partial x}\cdot\dfrac{\partial v}{\partial y} - \dfrac{\partial v}{\partial x}\cdot\dfrac{\partial u}{\partial y}} \quad \text{and} \quad \delta y = \frac{\dfrac{\partial u}{\partial x}\delta v - \dfrac{\partial v}{\partial x}\delta u}{\dfrac{\partial u}{\partial x}\cdot\dfrac{\partial v}{\partial y} - \dfrac{\partial v}{\partial x}\cdot\dfrac{\partial u}{\partial y}}$$

You will notice that the denominator is the same in each case and that it can be expressed in determinant form

$$\frac{\partial u}{\partial x}\cdot\frac{\partial v}{\partial y} - \frac{\partial v}{\partial x}\cdot\frac{\partial u}{\partial y} = \begin{vmatrix} \dfrac{\partial u}{\partial x} & \dfrac{\partial v}{\partial x} \\ \dfrac{\partial u}{\partial y} & \dfrac{\partial v}{\partial y} \end{vmatrix}$$

This determinant is called the *Jacobian* of u, v with respect to x, y and is denoted by the symbol J:

i.e. $\quad J = \begin{vmatrix} \dfrac{\partial u}{\partial x} & \dfrac{\partial v}{\partial x} \\ \dfrac{\partial u}{\partial y} & \dfrac{\partial v}{\partial y} \end{vmatrix} \quad$ and is often written as $\quad \dfrac{\partial(u,v)}{\partial(x,y)}$

So $\quad J = \dfrac{\partial(u,v)}{\partial(x,y)} = \begin{vmatrix} \dfrac{\partial u}{\partial x} & \dfrac{\partial v}{\partial x} \\ \dfrac{\partial u}{\partial y} & \dfrac{\partial v}{\partial y} \end{vmatrix}$

Our last two results can therefore be written

$$\delta x = \ldots\ldots\ldots\ldots; \quad \delta y = \ldots\ldots\ldots\ldots$$

Applications of partial differentiation

$$\delta x = \cfrac{\cfrac{\partial v}{\partial y}\delta u - \cfrac{\partial u}{\partial y}\delta v}{J} = \cfrac{\begin{vmatrix} \delta u & \delta v \\ \cfrac{\partial u}{\partial y} & \cfrac{\partial v}{\partial y} \end{vmatrix}}{\begin{vmatrix} \cfrac{\partial u}{\partial x} & \cfrac{\partial v}{\partial x} \\ \cfrac{\partial u}{\partial y} & \cfrac{\partial v}{\partial y} \end{vmatrix}}, \quad \delta y = \cfrac{\cfrac{\partial u}{\partial x}\delta v - \cfrac{\partial v}{\partial x}\delta u}{J} = \cfrac{\begin{vmatrix} \cfrac{\partial u}{\partial x} & \cfrac{\partial v}{\partial x} \\ \delta u & \delta v \end{vmatrix}}{\begin{vmatrix} \cfrac{\partial u}{\partial x} & \cfrac{\partial v}{\partial x} \\ \cfrac{\partial u}{\partial y} & \cfrac{\partial v}{\partial y} \end{vmatrix}}$$

We can now get a number of useful relationships.

(a) If v is kept constant, $\delta v = 0$ $\therefore \delta x = \dfrac{\partial v}{\partial y}\delta u \Big/ J$

Dividing by δu and letting $\delta u \to 0$ $\dfrac{\partial x}{\partial u} = \dfrac{\partial v}{\partial y}\Big/J$

Similarly $\dfrac{\partial y}{\partial u} = -\dfrac{\partial v}{\partial x}\Big/J$

(b) If u is kept constant, $\delta u = 0$ $\therefore \delta x = -\dfrac{\partial u}{\partial y}\delta v \Big/ J$

Dividing by δv and letting $\delta v \to 0$ $\dfrac{\partial x}{\partial v} = -\dfrac{\partial u}{\partial y}\Big/J$

Similarly $\dfrac{\partial y}{\partial v} = \dfrac{\partial u}{\partial x}\Big/J$

So, at this stage, we had better summarize the results.

Summary

If $z = f(x, y)$ and $u = g(x, y)$ and $v = h(x, y)$ then

$$\dfrac{\partial x}{\partial u} = \dfrac{\partial v}{\partial y}\Big/J \qquad \dfrac{\partial x}{\partial v} = -\dfrac{\partial u}{\partial y}\Big/J$$

$$\dfrac{\partial y}{\partial u} = -\dfrac{\partial v}{\partial x}\Big/J \qquad \dfrac{\partial y}{\partial v} = \dfrac{\partial u}{\partial x}\Big/J$$

where, in each case

$$J = \dfrac{\partial(u,v)}{\partial(x,y)} = \begin{vmatrix} \dfrac{\partial u}{\partial x} & \dfrac{\partial v}{\partial x} \\ \dfrac{\partial u}{\partial y} & \dfrac{\partial v}{\partial y} \end{vmatrix}$$

Let us put this into practice by doing again the same example that we started with (Example 1 on page 43), but by the new method. First of all, however, make a note of the important summary listed above for future reference.

Example 1A

If $z = f(x, y)$ and $u = e^x \cos y$ and $v = e^{-x} \sin y$, find the derivatives $\dfrac{\partial x}{\partial u}, \dfrac{\partial x}{\partial v}, \dfrac{\partial y}{\partial u}, \dfrac{\partial y}{\partial v}$.

$u = e^x \cos y$ $\qquad v = e^{-x} \sin y$

$\dfrac{\partial u}{\partial x} = e^x \cos y \qquad \dfrac{\partial v}{\partial x} = -e^{-x} \sin y$

$\dfrac{\partial u}{\partial y} = -e^x \sin y \qquad \dfrac{\partial v}{\partial y} = e^{-x} \cos y$

$$J = \dfrac{\partial(u, v)}{\partial(x, y)} = \begin{vmatrix} \dfrac{\partial u}{\partial x} & \dfrac{\partial v}{\partial x} \\ \dfrac{\partial u}{\partial y} & \dfrac{\partial v}{\partial y} \end{vmatrix} = \begin{vmatrix} e^x \cos y & -e^{-x} \sin y \\ -e^x \sin y & e^{-x} \cos y \end{vmatrix}$$

$$= (e^x \cos y)(e^{-x} \cos y) - (-e^x \sin y)(-e^{-x} \sin y)$$

$$= \cos^2 y \qquad - \qquad \sin^2 y \qquad = \cos 2y$$

Then $\dfrac{\partial x}{\partial u} = \dfrac{\partial v}{\partial y} \Big/ J = \dfrac{e^{-x} \cos y}{\cos 2y}; \qquad \dfrac{\partial x}{\partial v} = -\dfrac{\partial u}{\partial y} \Big/ J = \dfrac{e^x \sin y}{\cos 2y}$

$\dfrac{\partial y}{\partial u} = -\dfrac{\partial v}{\partial x} \Big/ J = \dfrac{e^{-x} \sin y}{\cos 2y}; \qquad \dfrac{\partial y}{\partial v} = \dfrac{\partial u}{\partial x} \Big/ J = \dfrac{e^x \cos y}{\cos 2y}$

which is a lot shorter than our first approach.

Move on for a further example

Example 2

If $z = f(x, y)$ with $u = x^2 - y^2$ and $v = xy$, find expressions for $\dfrac{\partial x}{\partial u}, \dfrac{\partial x}{\partial v}, \dfrac{\partial y}{\partial u}, \dfrac{\partial y}{\partial v}$.

First we need

$\dfrac{\partial u}{\partial x} = \ldots\ldots\ldots; \quad \dfrac{\partial u}{\partial y} = \ldots\ldots\ldots; \quad \dfrac{\partial v}{\partial x} = \ldots\ldots\ldots; \quad \dfrac{\partial v}{\partial y} = \ldots\ldots\ldots$

$$\dfrac{\partial u}{\partial x} = 2x; \quad \dfrac{\partial u}{\partial y} = -2y; \quad \dfrac{\partial v}{\partial x} = y; \quad \dfrac{\partial v}{\partial y} = x$$

Then we calculate J which, in this case, is $\ldots\ldots\ldots$

Applications of partial differentiation

$$J = 2(x^2 + y^2)$$

Because

$$J = \frac{\partial(u, v)}{\partial(x, y)} = \begin{vmatrix} \dfrac{\partial u}{\partial x} & \dfrac{\partial v}{\partial x} \\ \dfrac{\partial u}{\partial y} & \dfrac{\partial v}{\partial y} \end{vmatrix} = \begin{vmatrix} 2x & y \\ -2y & x \end{vmatrix} = 2x^2 + 2y^2$$

Finally, we have the four relationships

$$\frac{\partial x}{\partial u} = \frac{\partial v}{\partial y} \Big/ J = \ldots\ldots\ldots ; \qquad \frac{\partial x}{\partial v} = -\frac{\partial u}{\partial y} \Big/ J = \ldots\ldots\ldots$$

$$\frac{\partial y}{\partial u} = -\frac{\partial v}{\partial x} \Big/ J = \ldots\ldots\ldots ; \qquad \frac{\partial y}{\partial v} = \frac{\partial u}{\partial x} \Big/ J = \ldots\ldots\ldots$$

$$\frac{\partial x}{\partial u} = \frac{x}{2(x^2 + y^2)}; \qquad \frac{\partial x}{\partial v} = \frac{y}{x^2 + y^2}$$

$$\frac{\partial y}{\partial u} = \frac{-y}{2(x^2 + y^2)}; \qquad \frac{\partial y}{\partial v} = \frac{x}{x^2 + y^2}$$

And that is all there is to it.

If we know the details of the function $z = f(x, y)$ then we can go one stage further and use the results $\dfrac{\partial x}{\partial u}, \dfrac{\partial x}{\partial v}, \dfrac{\partial y}{\partial u}, \dfrac{\partial y}{\partial v}$ to find $\dfrac{\partial z}{\partial u}$ and $\dfrac{\partial z}{\partial v}$.

Let us see this in a further example.

Example 3

If $z = 2x^2 + 3xy + 4y^2$ and $u = x^2 + y^2$ and $v = x + 2y$, determine

(a) $\dfrac{\partial x}{\partial u}, \dfrac{\partial x}{\partial v}, \dfrac{\partial y}{\partial u}, \dfrac{\partial y}{\partial v}$ (b) $\dfrac{\partial z}{\partial u}$ and $\dfrac{\partial z}{\partial v}$.

Section (a) is just like the previous example. Complete that on your own.

53

$$\frac{\partial x}{\partial u} = \frac{1}{2x-y}; \quad \frac{\partial x}{\partial v} = \frac{-y}{2x-y}; \quad \frac{\partial y}{\partial u} = \frac{-1}{2(2x-y)}; \quad \frac{\partial y}{\partial v} = \frac{x}{2x-y}$$

Because if $u = x^2 + y^2$ and $v = x + 2y$

$$\frac{\partial u}{\partial x} = 2x; \quad \frac{\partial u}{\partial y} = 2y; \quad \frac{\partial v}{\partial x} = 1; \quad \frac{\partial v}{\partial y} = 2$$

$$J = \frac{\partial(u,v)}{\partial(x,y)} = \begin{vmatrix} \frac{\partial u}{\partial x} & \frac{\partial v}{\partial x} \\ \frac{\partial u}{\partial y} & \frac{\partial v}{\partial y} \end{vmatrix} = \begin{vmatrix} 2x & 1 \\ 2y & 2 \end{vmatrix} = 4x - 2y = 2(2x-y)$$

Then $\quad \dfrac{\partial x}{\partial u} = \dfrac{\partial v}{\partial y} \Big/ J = 2 \Big/ 2(2x-y) = \dfrac{1}{2x-y}$

$\dfrac{\partial x}{\partial v} = -\dfrac{\partial u}{\partial y} \Big/ J = -2y \Big/ 2(2x-y) = \dfrac{-y}{2x-y}$

$\dfrac{\partial y}{\partial u} = -\dfrac{\partial v}{\partial x} \Big/ J = -1 \Big/ 2(2x-y) = \dfrac{-1}{2(2x-y)}$

$\dfrac{\partial y}{\partial v} = \dfrac{\partial u}{\partial x} \Big/ J = 2x \Big/ 2(2x-y) = \dfrac{x}{2x-y}$

$$\therefore \frac{\partial x}{\partial u} = \frac{1}{2x-y}; \quad \frac{\partial x}{\partial v} = \frac{-y}{2x-y}; \quad \frac{\partial y}{\partial u} = \frac{-1}{2(2x-y)}; \quad \frac{\partial y}{\partial v} = \frac{x}{2x-y}$$

Now for part (b).

Since z is also a function of u and v, the expressions for $\dfrac{\partial z}{\partial u}$ and $\dfrac{\partial z}{\partial v}$ are

$$\frac{\partial z}{\partial u} = \ldots\ldots\ldots$$

$$\frac{\partial z}{\partial v} = \ldots\ldots\ldots$$

54

$$\frac{\partial z}{\partial u} = \frac{\partial z}{\partial x} \cdot \frac{\partial x}{\partial u} + \frac{\partial z}{\partial y} \cdot \frac{\partial y}{\partial u}$$

$$\frac{\partial z}{\partial v} = \frac{\partial z}{\partial x} \cdot \frac{\partial x}{\partial v} + \frac{\partial z}{\partial y} \cdot \frac{\partial y}{\partial v}$$

The only remaining items of information we need are the expressions for $\dfrac{\partial z}{\partial x}$ and $\dfrac{\partial z}{\partial y}$ which we obtain from $z = 2x^2 + 3xy + 4y^2$

$$\frac{\partial z}{\partial x} = 4x + 3y \quad \text{and} \quad \frac{\partial z}{\partial y} = 3x + 8y$$

Using these and the previous set of derivatives, we now get

$$\frac{\partial z}{\partial u} = \ldots\ldots\ldots; \quad \frac{\partial z}{\partial v} = \ldots\ldots\ldots$$

Applications of partial differentiation

$$\frac{\partial z}{\partial u} = \frac{5x - 2y}{2(2x - y)}; \quad \frac{\partial z}{\partial v} = \frac{3x^2 + 4xy - 3y^2}{2x - y}$$

Because

$$\frac{\partial z}{\partial u} = \frac{\partial z}{\partial x} \cdot \frac{\partial x}{\partial u} + \frac{\partial z}{\partial y} \cdot \frac{\partial y}{\partial u}$$

$$\therefore \frac{\partial z}{\partial u} = (4x + 3y)\left\{\frac{1}{2x - y}\right\} + (3x + 8y)\left\{\frac{-1}{2(2x - y)}\right\}$$

$$= \frac{5x - 2y}{2(2x - y)} \qquad \therefore \frac{\partial z}{\partial u} = \frac{5x - 2y}{2(2x - y)}$$

and $\dfrac{\partial z}{\partial v} = \dfrac{\partial z}{\partial x} \cdot \dfrac{\partial x}{\partial v} + \dfrac{\partial z}{\partial y} \cdot \dfrac{\partial y}{\partial v}$

$$\therefore \frac{\partial z}{\partial v} = (4x + 3y)\left\{\frac{-y}{2x - y}\right\} + (3x + 8y)\left\{\frac{x}{2x - y}\right\}$$

$$= \frac{3x^2 + 4xy - 3y^2}{2x - y} \qquad \therefore \frac{\partial z}{\partial v} = \frac{3x^2 + 4xy - 3y^2}{2x - y}$$

They are all done in the same general way.

Review summary

You have now reached the end of this Program and know quite a bit about partial differentiation. We have established some important results during the work, so let us list them once more.

1 Small increments

$$z = f(x, y) \quad \delta z = \frac{\partial z}{\partial x} \delta x + \frac{\partial z}{\partial y} \delta y \qquad \text{(a)}$$

2 Rates of change

$$\frac{dz}{dt} = \frac{\partial z}{\partial x} \cdot \frac{dx}{dt} + \frac{\partial z}{\partial y} \cdot \frac{dy}{dt} \qquad \text{(b)}$$

3 Implicit functions

$$\frac{dz}{dx} = \frac{\partial z}{\partial x} + \frac{\partial z}{\partial y} \cdot \frac{dy}{dx} \qquad \text{(c)}$$

4 Change of variables

$$\frac{\partial z}{\partial u} = \frac{\partial z}{\partial x} \cdot \frac{\partial x}{\partial u} + \frac{\partial z}{\partial y} \cdot \frac{\partial y}{\partial u}$$

$$\frac{\partial z}{\partial v} = \frac{\partial z}{\partial x} \cdot \frac{\partial x}{\partial v} + \frac{\partial z}{\partial y} \cdot \frac{\partial y}{\partial v} \qquad \text{(d)}$$

5 Inverse functions

$$z = f(x,y) \quad u = g(x,y) \quad v = h(x,y)$$

$$\frac{\partial x}{\partial u} = \frac{\partial v}{\partial y}\bigg/J; \quad \frac{\partial x}{\partial v} = -\frac{\partial u}{\partial y}\bigg/J$$

$$\frac{\partial y}{\partial u} = -\frac{\partial v}{\partial x}\bigg/J; \quad \frac{\partial y}{\partial v} = \frac{\partial u}{\partial x}\bigg/J$$

where $J = \dfrac{\partial(u,v)}{\partial(x,y)} = \begin{vmatrix} \dfrac{\partial u}{\partial x} & \dfrac{\partial v}{\partial x} \\ \dfrac{\partial u}{\partial y} & \dfrac{\partial v}{\partial y} \end{vmatrix}$

All that now remains is the **Can You?** checklist and the **Test exercise**, so move to Frames 57 and 58 and work through them carefully at your own speed.

✓ Can You?

57 Checklist 2

Check this list before and after you try the end of Program test.

On a scale of 1 to 5 how confident are you that you can: Frames

- Derive the first- and second-order partial derivatives of a function of two real variables?   **1** to **7**
  Yes ☐ ☐ ☐ ☐ ☐ No

- Apply partial differentiation to rate-of-change problems?   **8** to **23**
  Yes ☐ ☐ ☐ ☐ ☐ No

- Apply partial differentiation to change-of-variable problems?   **24** to **39**
  Yes ☐ ☐ ☐ ☐ ☐ No

- Use the Jacobian to obtain the derivatives of inverse functions of two real variables?   **40** to **55**
  Yes ☐ ☐ ☐ ☐ ☐ No

Applications of partial differentiation

 Test exercise 2

Take your time and work carefully. The questions are just like those you have been doing quite successfully.

1. Use partial differentiation to determine expressions for $\dfrac{dy}{dx}$ in the following cases:
   (a) $x^3 + y^3 - 2x^2 y = 0$
   (b) $e^x \cos y = e^y \sin x$
   (c) $\sin^2 x - 5 \sin x \cos y + \tan y = 0$

2. The base radius of a cone, r, is decreasing at the rate of 0.1 cm/s while the perpendicular height, h, is increasing at the rate of 0.2 cm/s. Find the rate at which the volume, V, is changing when $r = 2$ cm and $h = 3$ cm.

3. If $z = 2xy - 3x^2 y$ and x is increasing at 2 cm/s, determine at what rate y must be changing in order that z shall be neither increasing nor decreasing at the instant when $x = 3$ cm and $y = 1$ cm.

4. If $z = x^4 + 2x^2 y + y^3$ and $x = r \cos \theta$ and $y = r \sin \theta$, find $\dfrac{\partial z}{\partial r}$ and $\dfrac{\partial z}{\partial \theta}$ in their simplest forms.

5. If $z = \dfrac{xy}{x - y}$, show that
   (a) $x \dfrac{\partial z}{\partial x} + y \dfrac{\partial z}{\partial y} = z$
   (b) $x^2 \dfrac{\partial^2 z}{\partial x^2} - y^2 \dfrac{\partial^2 z}{\partial y^2} = 0$
   (c) $z \dfrac{\partial^2 z}{\partial x \partial y} = 2 \dfrac{\partial z}{\partial x} \cdot \dfrac{\partial z}{\partial y}$.

6. Two sides of a triangular plate are measured as 125 mm and 160 mm, each to the nearest millimeter. The included angle is quoted as $60° \pm 1°$. Calculate the length of the remaining side and the maximum possible error in the result.

7. If $z = (x^2 - y^2)^{1/2}$ and x is increasing at 3.5 m/s, determine at what rate y must change in order that z shall be neither increasing nor decreasing at the instant when $x = 5$ m and $y = 3$ m.

8. If $2x^2 + 4xy + 3y^2 = 1$, obtain expressions for $\dfrac{dy}{dx}$ and $\dfrac{d^2 y}{dx^2}$.

9. If $u = x^2 + y^2$ and $v = 4xy$, determine
   $\dfrac{\partial x}{\partial u},\ \dfrac{\partial x}{\partial v},\ \dfrac{\partial y}{\partial u},\ \dfrac{\partial y}{\partial v}$.

Further problems 2

1. If $F = f(x,y)$ where $x = e^u \cos v$ and $y = e^u \sin v$, show that
$$\frac{\partial F}{\partial u} = x\frac{\partial F}{\partial x} + y\frac{\partial F}{\partial y} \quad \text{and} \quad \frac{\partial F}{\partial v} = -y\frac{\partial F}{\partial x} + x\frac{\partial F}{\partial y}.$$

2. Given that $z = x^3 + y^3$ and $x^2 + y^2 = 1$, determine an expression for $\dfrac{dz}{dx}$ in terms of x and y.

3. If $z = f(x,y) = 0$, show that $\dfrac{dy}{dx} = -\dfrac{\partial z}{\partial x}\bigg/\dfrac{\partial z}{\partial y}$. The curves $2y^2 + 3x - 8 = 0$ and $x^3 + 2xy^3 + 3y - 1 = 0$ intersect at the point $(2, -1)$. Find the tangent of the angle between the tangents to the curves at this point.

4. If $u = (x^2 - y^2)f(t)$ where $t = xy$ and f denotes an arbitrary function, prove that $\dfrac{\partial^2 u}{\partial x \cdot \partial y} = (x^2 - y^2)\{t \cdot f''(t) + 3f'(t)\}$. [Note: $f''(t)$ is the second derivative of $f(t)$ w.r.t. t.]

5. If $V = xy/(x^2 + y^2)^2$ and $x = r\cos\theta$, $y = r\sin\theta$, show that
$$\frac{\partial^2 V}{\partial r^2} + \frac{1}{r}\frac{\partial V}{\partial r} + \frac{1}{r^2}\frac{\partial^2 V}{\partial \theta^2} = 0.$$

6. If $u = f(x,y)$ where $x = r^2 - s^2$ and $y = 2rs$, prove that
$$r\frac{\partial u}{\partial r} - s\frac{\partial u}{\partial s} = 2(r^2 + s^2)\frac{\partial u}{\partial x}.$$

7. If $f = F(x,y)$ and $x = re^\theta$ and $y = re^{-\theta}$, prove that
$$2x\frac{\partial f}{\partial x} = r\frac{\partial f}{\partial r} + \frac{\partial f}{\partial \theta} \quad \text{and} \quad 2y\frac{\partial f}{\partial y} = r\frac{\partial f}{\partial r} - \frac{\partial f}{\partial \theta}.$$

8. If $z = x\ln(x^2 + y^2) - 2y\tan^{-1}\left(\frac{y}{x}\right)$ verify that $x\dfrac{\partial z}{\partial x} + y\dfrac{\partial z}{\partial y} = z + 2x$.

9. By means of partial differentiation, determine $\dfrac{dy}{dx}$ in each of the following cases:
   (a) $xy + 2y - x = 4$
   (b) $x^3y^2 - 2x^2y + 3xy^2 - 8xy = 5$
   (c) $\dfrac{4y}{x} + \dfrac{2x}{y} = 3$

10. If $z = 3xy - y^3 + (y^2 - 2x)^{3/2}$, verify that:
    (a) $\dfrac{\partial^2 z}{\partial x \cdot \partial y} = \dfrac{\partial^2 z}{\partial y \cdot \partial x}$ and (b) $\dfrac{\partial^2 z}{\partial x^2} \cdot \dfrac{\partial^2 z}{\partial y^2} = \left(\dfrac{\partial^2 z}{\partial x \cdot \partial y}\right)^2$

11. If $f = \dfrac{1}{\sqrt{1 - 2xy + y^2}}$, show that $y\dfrac{\partial f}{\partial y} = (x - y)\dfrac{\partial f}{\partial x}$.

12. If $z = x \cdot f\left(\dfrac{y}{x}\right) + F\left(\dfrac{y}{x}\right)$, prove that:
    (a) $x\dfrac{\partial z}{\partial x} + y\dfrac{\partial z}{\partial y} = z - F\left(\dfrac{y}{x}\right)$
    (b) $x^2\dfrac{\partial^2 z}{\partial x^2} + 2xy\dfrac{\partial^2 z}{\partial x \cdot \partial y} + y^2\dfrac{\partial^2 z}{\partial y^2} = 0$

Applications of partial differentiation

13 If $z = e^{k(r-x)}$, where k is a constant, and $r^2 = x^2 + y^2$, prove:

(a) $\left(\dfrac{\partial z}{\partial x}\right)^2 + \left(\dfrac{\partial z}{\partial y}\right)^2 + 2zk\dfrac{\partial z}{\partial x} = 0$

(b) $\dfrac{\partial^2 z}{\partial x^2} + \dfrac{\partial^2 z}{\partial y^2} + 2k\dfrac{\partial z}{\partial x} = \dfrac{kz}{r}$

14 If $z = f(x - 2y) + F(3x + y)$, where f and F are arbitrary functions, and if

$$\dfrac{\partial^2 z}{\partial x^2} + a\dfrac{\partial^2 z}{\partial x \cdot \partial y} + b\dfrac{\partial^2 z}{\partial y^2} = 0,$$ find the values of a and b.

15 If $z = xy/(x^2 + y^2)^2$, verify that $\dfrac{\partial^2 z}{\partial x^2} + \dfrac{\partial^2 z}{\partial y^2} = 0$.

16 If $\sin^2 x - 5 \sin x \cos y + \tan y = 0$, find $\dfrac{dy}{dx}$ by using partial differentiation.

17 Find $\dfrac{dy}{dx}$ by partial differentiation, when $x \tan y = y \sin x$.

18 If $V = \tan^{-1}\left\{\dfrac{2xy}{x^2 - y^2}\right\}$, prove that:

(a) $x\dfrac{\partial V}{\partial x} + y\dfrac{\partial V}{\partial y} = 0$

(b) $\dfrac{\partial^2 V}{\partial x^2} + \dfrac{\partial^2 V}{\partial y^2} = 0$

19 Prove that, if $z = 2xy + x \cdot f\left(\dfrac{y}{x}\right)$ then $x\dfrac{\partial z}{\partial x} + y\dfrac{\partial z}{\partial y} = z + 2xy$.

20 (a) Find $\dfrac{dy}{dx}$ given that $x^2 y + \sin xy = 0$.

(b) Find $\dfrac{dy}{dx}$ given that $x \sin xy = 1$.

21 If $z = 2x^2 - 3y$ with $u = x^2 \sin y$ and $v = 2y \cos x$, determine expressions for $\dfrac{\partial z}{\partial u}$ and $\dfrac{\partial z}{\partial v}$.

22 If $u = x^2 + e^{-3y}$ and $v = 2x + e^{3y}$, determine $\dfrac{\partial x}{\partial u}, \dfrac{\partial x}{\partial v}, \dfrac{\partial y}{\partial u}, \dfrac{\partial y}{\partial v}$.

23 If $z = f(x, y)$ where $x = uv$ and $y = u^2 - v^2$, show that

(a) $2x\dfrac{\partial z}{\partial x} + 2y\dfrac{\partial z}{\partial y} = u\dfrac{\partial z}{\partial u} + v\dfrac{\partial z}{\partial v}$

(b) $2\dfrac{\partial z}{\partial y} = \dfrac{1}{u^2 + v^2}\left\{u\dfrac{\partial z}{\partial u} - v\dfrac{\partial z}{\partial v}\right\}$.

24 If $z = \cosh 2x \sin 3y$ and $u = e^x(1 + y^2)$ and $v = 2ye^{-x}$, determine expressions for $\dfrac{\partial x}{\partial u}, \dfrac{\partial x}{\partial v}, \dfrac{\partial y}{\partial u}, \dfrac{\partial y}{\partial v}$, and hence find $\dfrac{\partial z}{\partial u}$ and $\dfrac{\partial z}{\partial v}$.

25 If $z = f(u, v)$ where $u = \tfrac{1}{2}(x^2 - y^2)$ and $v = xy$, prove that

$$\dfrac{\partial^2 z}{\partial x^2} - \dfrac{\partial^2 z}{\partial y^2} = 2u\left(\dfrac{\partial^2 z}{\partial u^2} - \dfrac{\partial^2 z}{\partial v^2}\right) + 4v\dfrac{\partial^2 z}{\partial u \partial v} + 2\dfrac{\partial z}{\partial u}.$$

Program 3

Polar coordinates

Frames
1 to 50

Learning outcomes

When you have completed this Program you will be able to:
- Convert expressions from Cartesian coordinates to polar coordinates and vice versa
- Plot the graphs of polar curves
- Recognize equations of standard polar curves
- Evaluate the areas enclosed by polar curves
- Evaluate the volumes of revolution generated by polar curves
- Evaluate the lengths of polar curves
- Evaluate the surface of revolution generated by polar curves

Polar coordinates

Introduction

We already know that there are two main ways in which the position of a point in a plane can be represented:

(a) by Cartesian coordinates, i.e. (x, y)
(b) by polar coordinates, i.e. (r, θ).

The relationship between the two systems can be seen from a diagram:

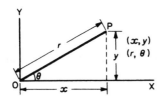

For instance, x and y can be expressed in terms of r and θ.

$x = \ldots\ldots\ldots$; $y = \ldots\ldots\ldots$

$$x = r\cos\theta; \quad y = r\sin\theta$$

Or, working in the reverse direction, the coordinates r and θ can be found if we know the values of x and y.

$r = \ldots\ldots\ldots$; $\theta = \ldots\ldots\ldots$

$$r = \sqrt{x^2 + y^2}; \quad \theta = \tan^{-1}\left(\frac{y}{x}\right)$$

This is just by way of review. You should already be familiar with polar coordinates from past studies. In this Program, we are going to focus attention more to the *polar coordinates system* and its applications.

First of all, some easy examples to warm up.

Example 1

Express in polar coordinates the position $(-5, 2)$.

Important hint: *always* draw a diagram; it will enable you to see which quadrant you are dealing with and prevent your making an initial slip.

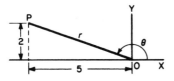

Remember that θ is measured from the positive OX direction.

In this case, the polar coordinates of P are $\ldots\ldots\ldots$

4

$$(5.385, 158°12')$$

Because

(a) $r^2 = 2^2 + 5^2 = 4 + 25 = 29$

$\therefore r = \sqrt{29} = 5.385.$

(b) $\tan E = \dfrac{2}{5} = 0.4$

$\therefore E = 21°48'$

$\therefore \theta = 158°12'$

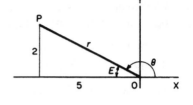

Position of P is $(5.385, 158°12')$

A sketch diagram will help you to check that θ is in the correct quadrant.

Example 2

Express $(4, -3)$ in polar coordinates. Draw a sketch and you cannot go wrong!

When you are ready, move to Frame 5

5

$$5, 323°8'$$

Here it is:

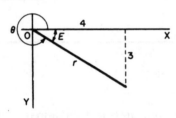

(a) $r^2 = 3^2 + 4^2 = 25 \quad \therefore r = 5$

(b) $\tan E = \dfrac{3}{4} = 0.75 \quad \therefore E = 36°52'$

$\therefore \theta = 323°8'$

$(4, -3) = (5, 323°8')$

Example 3

Express in polar coordinates $(-2, -3)$.

Finish it off and then move to Frame 6

6

$$3.606, \ 236°19'$$

Check your result:

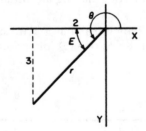

(a) $r^2 = 2^2 + 3^2 = 4 + 9 = 13$

$r = \sqrt{13} = 3.606$

(b) $\tan E = \dfrac{3}{2} = 1.5 \quad \therefore E = 56°19'$

$\therefore \theta = 236°19'$

$(-2, -3) = (3.606, 236°19')$

Polar coordinates

Of course conversion in the opposite direction is just a matter of evaluating $x = r\cos\theta$ and $y = r\sin\theta$. Here is an example.

Example 4

Express $(5, 124°)$ in Cartesian coordinates.

Do that, and then move on to Frame 7

$$(-2.796, \ 4.145)$$

Working:

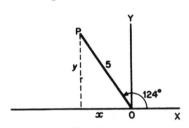

(a) $x = 5\cos 124° = -5\cos 56°$
$ = -5(0.5592) = -2.7960$
(b) $y = 5\sin 124° = 5\sin 56°$
$ = 5(0.8290) = 4.1450$

$\therefore \ (5, 124°) = (-2.796, 4.145)$

That was all very easy.

Now, on to the next frame

Polar curves

In Cartesian coordinates, the equation of a curve is given as the general relationship between x and y, i.e. $y = f(x)$.

Similarly, in the polar coordinate system, the equation of a curve is given in the form $r = f(\theta)$. We can then take spot values for θ, calculate the corresponding values of r, plot r against θ, and join the points up with a smooth curve to obtain the graph of $r = f(\theta)$.

Example 1

To plot the polar graph of $r = 2\sin\theta$ between $\theta = 0$ and $\theta = 2\pi$.

We take values of θ at convenient intervals and build up a table of values giving the corresponding values of r:

θ	0	30	60	90	120	150	180
$\sin\theta$	0	0.5	0.866	1	0.866	0.5	0
$r = 2\sin\theta$	0	1.0	1.732	2	1.732	1.0	0

θ	210	240	270	300	330	360
$\sin\theta$						
$r = 2\sin\theta$						

Complete the table, being careful of signs.

When you have finished, move on to Frame 9

60 Vector Analysis

9 Here is the complete table:

θ	0	30	60	90	120	150	180
$\sin\theta$	0	0.5	0.866	1	0.866	0.5	0
$r = 2\sin\theta$	0	1.0	1.732	2	1.732	1.0	0

θ	210	240	270	300	330	360
$\sin\theta$	−0.5	−0.866	−1	−0.866	−0.5	0
$r = 2\sin\theta$	−1.0	−1.732	−2	−1.732	−1.0	0

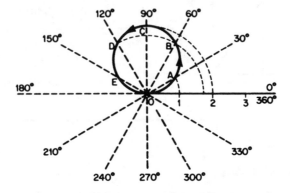

(a) We choose a linear scale for r and indicate it along the initial line.

(b) The value of r is then laid off along each direction in turn, points plotted, and finally joined up with a smooth curve. The resulting graph is as shown above.

Note: When we are dealing with the 210° direction, the value of r is negative (−1) and this distance is therefore laid off in the reverse direction which once again brings us to the point A. So for values of θ between $\theta = 180°$ and $\theta = 360°$, r is negative and the first circle is retraced exactly. The graph, therefore, looks like one circle, but consists, in fact, of two circles, one on top of the other.

Now here is an example for you to do.

Example 2

In the same way, you can plot the graph of $r = 2\sin^2\theta$.

Compile a table of values at 30° intervals between $\theta = 0°$ and $\theta = 360°$ and proceed as we did above.

Take a little time over it.

When you have finished, move on to Frame 10

Polar coordinates

Here is the result in detail:

θ	0	30	60	90	120	150	180
$\sin \theta$	0	0.5	0.866	1	0.866	0.5	0
$\sin^2 \theta$	0	0.25	0.75	1	0.75	0.25	0
$r = 2\sin^2 \theta$	0	0.5	1.5	2	1.5	0.5	0

θ	210	240	270	300	330	360
$\sin \theta$	-0.5	-0.866	-1	-0.866	-0.5	0
$\sin^2 \theta$	0.25	0.75	1	0.75	0.25	0
$r = 2\sin^2 \theta$	0.5	1.5	2	1.5	0.5	0

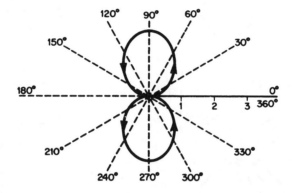

This time, r is always positive and so there are, in fact, two distinct loops.

Now on to the next frame

Standard polar curves

Polar curves can always be plotted from sample points as we have done above. However, it is often useful to know something of the shape of the curve without the rather tedious task of plotting points in detail.

In the next few frames, we will look at some of the more common polar curves.

So on to Frame 12

Typical polar curves

1 $r = a\sin\theta$

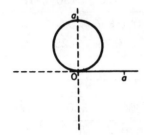

2 $r = a\sin^2\theta$

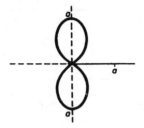

3 $r = a\cos\theta$

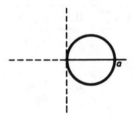

4 $r = a\cos^2\theta$

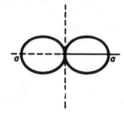

5 $r = a\sin 2\theta$

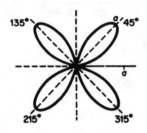

6 $r = a\sin 3\theta$

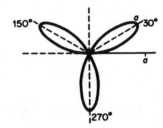

7 $r = a\cos 2\theta$

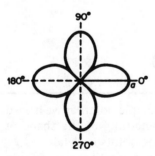

8 $r = a\cos 3\theta$

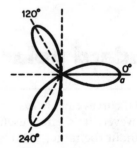

There are some more interesting polar curves worth seeing, so move on to Frame 13

9 $r = a(1 + \cos\theta)$

10 $r = a(1 + 2\cos\theta)$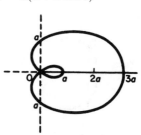

11 $r^2 = a^2 \cos 2\theta$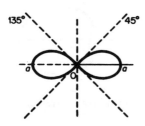

12 $r = a\theta$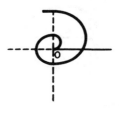

Sketch these 12 standard curves in your record book. They are quite common in use and worth remembering.

Then on to the next frame

The graphs of $r = a + b\cos\theta$ give three interesting results, according to the relative values of a and b.

(a) If $a = b$, we get 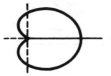 (cardioid)

(b) If $a < b$, we get (re-entrant loop)

(c) If $a > b$, we get 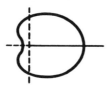 (no cusp or re-entrant loop)

So sketch the graphs of the following. Do *not* compile tables of values.
(a) $r = 2 + 2\cos\theta$ (c) $r = 1 + 2\cos\theta$
(b) $r = 5 + 3\cos\theta$ (d) $r = 2 + \cos\theta$

15

Here they are. See how closely you agree.

(a) $r = 2 + 2\cos\theta$ $(a = b)$ (b) $r = 5 + 3\cos\theta$ $(a > b)$

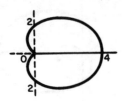 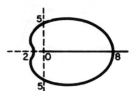

(c) $r = 1 + 2\cos\theta$ $(a < b)$ (d) $r = 2 + \cos\theta$ $(a > b)$

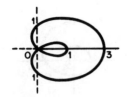 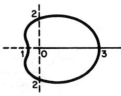

If you have slipped up with any of them, it would be worthwhile to plot a few points to confirm how the curve goes.

On to Frame 16

16

To find the area of the plane figure bounded by the polar curve $r = f(\theta)$ and the radius vectors at $\theta = \theta_1$ and $\theta = \theta_2$.

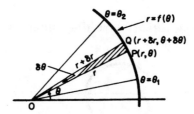

Area of sector $OPQ = \delta A \approx \dfrac{1}{2} r(r + \delta r) \sin \delta\theta$

$$\therefore \quad \frac{\delta A}{\delta \theta} \approx \frac{1}{2} r(r + \delta r) \frac{\sin \delta\theta}{\delta \theta}$$

If $\delta\theta \to 0$, $\dfrac{\delta A}{\delta \theta} \to \dfrac{dA}{d\theta}$, $\delta r \to 0$, $\dfrac{\sin \delta\theta}{\delta \theta} \to \ldots\ldots\ldots$

Next frame

Polar coordinates

$$\boxed{\dfrac{\sin \delta\theta}{\delta\theta} \to 1}$$

$$\therefore \dfrac{dA}{d\theta} = \dfrac{1}{2}r(r+0)1 = \dfrac{1}{2}r^2$$

$$\therefore A = \int_{\theta_1}^{\theta_2} \dfrac{1}{2} r^2 \, d\theta$$

Example 1

To find the area enclosed by the curve $r = 5 \sin \theta$ and the radius vectors at $\theta = 0$ and $\theta = \pi/3$.

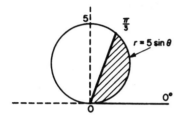

$$A = \int_0^{\pi/3} \dfrac{1}{2} r^2 \, d\theta$$

$$A = \int_0^{\pi/3} \dfrac{25}{2} \sin^2 \theta \, d\theta$$

$$\therefore A = \dfrac{25}{2} \int_0^{\pi/3} \dfrac{1}{2}(1 - \cos 2\theta) \, d\theta$$

$$= \ldots\ldots\ldots \quad \text{Finish it off.}$$

$$\boxed{A = \dfrac{25}{4}\left[\dfrac{\pi}{3} - \dfrac{\sqrt{3}}{4}\right] = 3.84}$$

Because

$$A = \dfrac{25}{4}\int_0^{\pi/3}(1 - \cos 2\theta)d\theta = \dfrac{25}{4}\left[\theta - \dfrac{\sin 2\theta}{2}\right]_0^{\pi/3}$$

$$= \dfrac{25}{4}\left(\dfrac{\pi}{3} - \dfrac{\sin 2\pi/3}{2}\right)$$

$$= \dfrac{25}{4}\left(\dfrac{\pi}{3} - \dfrac{\sqrt{3}}{4}\right) = 3.8386$$

$A = 3.84$ to 2 decimal places

Now this one:

Example 2

Find the area enclosed by the curve $r = 1 + \cos \theta$ and the radius vectors at $\theta = 0$ and $\theta = \pi/2$.

First of all, what does the curve look like?

19

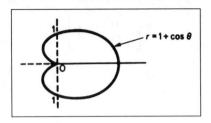

Right. So now calculate the value of A between $\theta = 0$ and $\theta = \pi/2$.

When you have finished, move on to Frame 20

20

$$A = \frac{3\pi}{8} + 1 = 2.178$$

Because

$$A = \frac{1}{2}\int_0^{\pi/2} r^2 d\theta = \frac{1}{2}\int_0^{\pi/2}(1 + 2\cos\theta + \cos^2\theta)d\theta$$

$$= \frac{1}{2}\left[\theta + 2\sin\theta + \frac{\theta}{2} + \frac{\sin 2\theta}{4}\right]_0^{\pi/2}$$

$$= \frac{1}{2}\left\{\left(\frac{3\pi}{4} + 2 + 0\right) - (0)\right\}$$

$$\therefore A = \frac{3\pi}{8} + 1 = 2.178$$

So the area of a polar sector is easy enough to obtain. It is simply

$$A = \int_{\theta_1}^{\theta_2} \frac{1}{2} r^2 d\theta$$

Make a note of this general result in your record book, if you have not already done so.

Next frame

21

Example 3

Find the total area enclosed by the curve $r = 2\cos 3\theta$. Notice that no limits are given, so we had better sketch the curve to see what is implied.

This was in fact one of the standard polar curves that we listed earlier in this Program. Do you remember how it goes? If not, refer to your notes: it should be there.

Then on to Frame 22

22

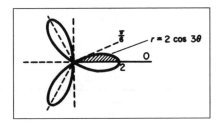

Since we are dealing with $r = 2\cos 3\theta$, r will become zero when $\cos 3\theta = 0$, i.e. when $3\theta = \pi/2$, i.e. when $\theta = \pi/6$.

We see that the figure consists of 3 equal loops, so that the total area, A, is given by:

$A = 3$ (area of one loop)

$\quad = 6$ (area between $\theta = 0$ and $\theta = \pi/6$)

$$A = 6\int_0^{\pi/6} \frac{1}{2}r^2 d\theta = 3\int_0^{\pi/6} 4\cos^2 3\theta\, d\theta = \ldots\ldots\ldots\ldots$$

23

$$\boxed{\pi \text{ units}^2}$$

Because

$$A = 12\int_0^{\pi/6} \frac{1}{2}(1 + \cos 6\theta)\, d\theta$$

$$= 6\left[\theta + \frac{\sin 6\theta}{6}\right]_0^{\pi/6} = \pi \text{ units}^2$$

Now here is one loop for you to do on your own.

Example 4

Find the area enclosed by one loop of the curve $r = a\sin 2\theta$.

First sketch the graph.

24

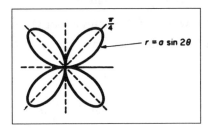

Arguing as before, $r = 0$ when $a\sin 2\theta = 0$, i.e. $\sin 2\theta = 0$, i.e. $2\theta = 0$, so that $2\theta = 0$, π, 2π, etc.

$\therefore \theta = 0$, $\pi/2$, π, etc.

So the integral denoting the area of the loop in the first quadrant will be $A = \ldots\ldots\ldots\ldots$

25

$$A = \frac{1}{2}\int_0^{\pi/2} r^2 \, d\theta$$

Correct. Now go ahead and calculate the area.

26

$$A = \pi a^2/8 \text{ units}^2$$

Here is the working: check yours.

$$A = \frac{1}{2}\int_0^{\pi/2} r^2 \, d\theta = \frac{a^2}{2}\int_0^{\pi/2} \sin^2 2\theta \, d\theta$$

$$= \frac{a^2}{4}\int_0^{\pi/2} (1 - \cos 4\theta) \, d\theta$$

$$= \frac{a^2}{4}\left[\theta - \frac{\sin 4\theta}{4}\right]_0^{\pi/2} = \frac{\pi a^2}{8} \text{ units}^2$$

Now on to Frame 27

27

To find the volume generated when the plane figure bounded by $r = f(\theta)$ and the radius vectors at $\theta = \theta_1$ and $\theta = \theta_2$, rotates about the initial line.

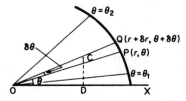

If we regard the elementary sector OPQ as approximately equal to the $\triangle$OPQ, then the centroid C is distance $\dfrac{2r}{3}$ from O.

We have: Area OPQ $\approx \dfrac{1}{2}r(r + \delta r) \sin \delta\theta$

Volume generated when OPQ rotates about OX $= \delta V$

$\therefore \delta V =$ area OPQ $\times$ distance travelled by its centroid (Pappus)

$$= \frac{1}{2}r(r + \delta r) \sin \delta\theta \cdot 2\pi \cdot CD$$

$$= \frac{1}{2}r(r + \delta r) \sin \delta\theta \cdot 2\pi \cdot \frac{2}{3}r \sin \theta = \frac{2}{3}\pi r^2(r + \delta r)\sin \delta\theta \cdot \sin \theta$$

$$\therefore \frac{\delta V}{\delta\theta} = \frac{2}{3}\pi r^2(r + \delta r)\frac{\sin \delta\theta}{\delta\theta} \cdot \sin \theta$$

Then when $\delta\theta \to 0$, $\dfrac{dV}{d\theta} = \ldots\ldots\ldots$

Polar coordinates

$$\frac{dV}{d\theta} = \frac{2}{3}\pi r^3 \sin\theta$$

and therefore $V = \ldots\ldots\ldots\ldots$

$$V = \int_{\theta_1}^{\theta_2} \frac{2}{3}\pi r^3 \sin\theta\, d\theta$$

Correct. This is another standard result, so add it to your notes.

Then move to the next frame for an example

Example 1

Find the volume of the solid formed when the plane figure bounded by $r = 2\sin\theta$ and the radius vectors at $\theta = 0$ and $\theta = \pi/2$, rotates about the initial line.

Well now, $\displaystyle V = \int_0^{\pi/2} \frac{2}{3}\pi r^3 \sin\theta\, d\theta$

$$= \int_0^{\pi/2} \frac{2}{3} \cdot \pi \cdot (2\sin\theta)^3 \cdot \sin\theta\, d\theta = \int_0^{\pi/2} \frac{16}{3}\pi \sin^4\theta\, d\theta$$

Since the limits are between 0 and $\pi/2$, we can use Wallis's formula for this. (Remember?)

So $V = \ldots\ldots\ldots\ldots$

$$V = \pi^2 \text{ units}^3$$

Because

$$V = \frac{16\pi}{3}\int_0^{\pi/2} \sin^4\theta\, d\theta$$

$$= \frac{16\pi}{3} \cdot \frac{3 \cdot 1}{4 \cdot 2} \cdot \frac{\pi}{2} = \pi^2 \text{ units}^3$$

Example 2

Find the volume of the solid formed when the plane figure bounded by $r = 2a\cos\theta$ and the radius vectors at $\theta = 0$ and $\theta = \pi/2$, rotates about the initial line.

Do that one entirely on your own.

When you have finished it, move on to the next frame

32

$$V = \frac{4\pi a^3}{3} \text{ units}^3$$

Because

$$V = \int_0^{\pi/2} \frac{2}{3} \cdot \pi \cdot r^3 \sin\theta \, d\theta \text{ and } r = 2a\cos\theta$$

$$= \int_0^{\pi/2} \frac{2}{3} \cdot \pi \cdot 8a^3 \cos^3\theta \cdot \sin\theta \, d\theta$$

$$= -\frac{16\pi a^3}{3} \int_0^{\pi/2} \cos^3\theta(-\sin\theta) \, d\theta$$

$$= -\frac{16\pi a^3}{3} \left[\frac{\cos^4\theta}{4}\right]_0^{\pi/2} = -\frac{16\pi a^3}{3}\left[-\frac{1}{4}\right]$$

$$V = \frac{4\pi a^3}{3} \text{ units}^3$$

So far then, we have had:

(a) $A = \int_{\theta_1}^{\theta_2} \frac{1}{2} r^2 \, d\theta$

(b) $V = \int_{\theta_1}^{\theta_2} \frac{2}{3} \pi r^3 \sin\theta \, d\theta$

Check that you have noted these results in your record book.

33

To find the length of arc of the polar curve $r = f(\theta)$ between $\theta = \theta_1$ and $\theta = \theta_2$.

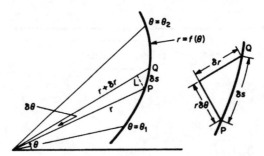

With the usual figure $\delta s^2 \approx r^2 \cdot \delta\theta^2 + \delta r^2$ $\quad \therefore \quad \dfrac{\delta s^2}{\delta\theta^2} \approx r^2 + \dfrac{\delta r^2}{\delta\theta^2}$

If $\delta\theta \to 0$, $\left(\dfrac{ds}{d\theta}\right)^2 = r^2 + \left(\dfrac{dr}{d\theta}\right)^2$ $\quad \therefore \quad \dfrac{ds}{d\theta} = \sqrt{r^2 + \left(\dfrac{dr}{d\theta}\right)^2}$

$\therefore s = \ldots\ldots\ldots\ldots$

Polar coordinates

34

$$S = \int_{\theta_1}^{\theta_2} \sqrt{r^2 + \left(\frac{dr}{d\theta}\right)^2} \, d\theta$$

Example 1

Find the length of arc of the spiral $r = ae^{3\theta}$ from $\theta = 0$ to $\theta = 2\pi$.

Now, $r = ae^{3\theta}$ $\therefore \dfrac{dr}{d\theta} = 3ae^{3\theta}$

$\therefore r^2 + \left(\dfrac{dr}{d\theta}\right)^2 = a^2 e^{6\theta} + 9a^2 e^{6\theta} = 10a^2 e^{6\theta}$

$\therefore S = \int_0^{2\pi} \sqrt{r^2 + \left(\dfrac{dr}{d\theta}\right)^2} \, d\theta = \int_0^{2\pi} \sqrt{10} \cdot ae^{3\theta} \, d\theta$

$= \ldots \ldots \ldots \ldots$

35

$$S = \frac{a\sqrt{10}}{3}\{e^{6\pi} - 1\}$$

Because

$$\int_0^{2\pi} \sqrt{10} \cdot a \cdot e^{3\theta} \, d\theta = \frac{\sqrt{10}a}{3} \left[e^{3\theta}\right]_0^{2\pi} = \frac{a\sqrt{10}}{3}\{e^{6\pi} - 1\}$$

As you can see, the method is very much the same every time. It is merely a question of substituting in the standard result, and, as usual, a knowledge of the shape of the polar curves is a very great help.

Here is our last result again:

$$S = \int_{\theta_1}^{\theta_2} \sqrt{r^2 + \left(\frac{dr}{d\theta}\right)^2} \, d\theta$$

Make a note of it: add it to the list.

Now here is an example for you to do.

36

Example 2

Find the length of the cardioid $r = a(1 + \cos\theta)$ between $\theta = 0$ and $\theta = \pi$.

Finish it completely, and then check with the next frame

37

$$s = 4a \text{ units}$$

Here is the working:

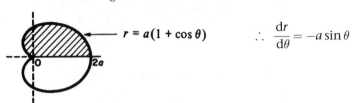

$r = a(1 + \cos\theta)$ $\quad\therefore \dfrac{dr}{d\theta} = -a\sin\theta$

$$\therefore r^2 + \left(\dfrac{dr}{d\theta}\right)^2 = a^2\{1 + 2\cos\theta + \cos^2\theta + \sin^2\theta\}$$
$$= a^2\{2 + 2\cos\theta\} = 2a^2(1 + \cos\theta)$$

Now $\cos\theta$ can be rewritten as $\left(2\cos^2\dfrac{\theta}{2} - 1\right)$

$$\therefore r^2 + \left(\dfrac{dr}{d\theta}\right)^2 = 2a^2 \cdot 2\cos^2\dfrac{\theta}{2}$$

$$\therefore \sqrt{r^2 + \left(\dfrac{dr}{d\theta}\right)^2} = 2a\cos\dfrac{\theta}{2}$$

$$\therefore s = \int_0^\pi 2a\cos\dfrac{\theta}{2}\,d\theta = 2a\left[2\sin\dfrac{\theta}{2}\right]_0^\pi$$
$$= 4a(1 - 0) = 4a \text{ units}$$

Next frame

38

Let us pause for a moment and think back. So far we have established three useful results relating to polar curves. Without looking back in this Program, or at your notes, complete the following.

If $r = f(\theta)$
(a) $A = \ldots\ldots\ldots$
(b) $V = \ldots\ldots\ldots$
(c) $s = \ldots\ldots\ldots$

To see how well you have got on, move on to Frame 39

39

(a) $A = \displaystyle\int_{\theta_1}^{\theta_2} \dfrac{1}{2} r^2 \, d\theta$

(b) $V = \displaystyle\int_{\theta_1}^{\theta_2} \dfrac{2}{3} \cdot \pi \cdot r^3 \sin\theta \, d\theta$

(c) $s = \displaystyle\int_{\theta_1}^{\theta_2} \sqrt{r^2 + \left(\dfrac{dr}{d\theta}\right)^2} \, d\theta$

If you were uncertain of any of them, be sure to review that particular result now. When you are ready, move on to the next section of the Program.

Polar coordinates

40

Finally, we come to this topic.

To find the area of the surface generated when the arc of the curve $r = f(\theta)$ between $\theta = \theta_1$ and $\theta = \theta_2$, rotates about the initial line.

Once again, we refer to our usual figure:

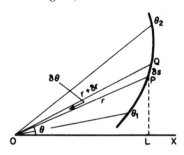

If the elementary arc PQ rotates about OX, then, by the theorem of Pappus, the surface generated, δS, is given by (length of arc) × (distance travelled by its centroid).

$$\therefore \; \delta S \approx \delta s \cdot 2\pi \cdot \text{PL} \approx \delta s \cdot 2\pi r \sin\theta$$

$$\therefore \; \frac{\delta S}{\delta \theta} \approx 2\pi r \sin\theta \frac{\delta s}{\delta \theta}$$

From our previous work, we know that $\dfrac{\delta s}{\delta \theta} \approx \sqrt{r^2 + \left(\dfrac{\delta r}{\delta \theta}\right)^2}$

so that $\dfrac{\delta S}{\delta \theta} \approx 2\pi r \sin\theta \sqrt{r^2 + \left(\dfrac{\delta r}{\delta \theta}\right)^2}$

And now, if $\delta\theta \to 0$, $\dfrac{dS}{d\theta} = 2\pi r \sin\theta \sqrt{r^2 + \left(\dfrac{dr}{d\theta}\right)^2}$

$$\therefore \; S = \int_{\theta_1}^{\theta_2} 2\pi r \sin\theta \sqrt{r^2 + \left(\frac{dr}{d\theta}\right)^2} \, d\theta$$

This is also an important result, so add it to your list.

41

$$\boxed{S = \int_{\theta_1}^{\theta_2} 2\pi r \sin\theta \sqrt{r^2 + \left(\frac{dr}{d\theta}\right)^2} \, d\theta}$$

This looks a little more involved, but the method of attack is much the same. An example will show.

Example 1

Find the surface area generated when the arc of the curve $r = 5(1 + \cos\theta)$ between $\theta = 0$ and $\theta = \pi$, rotates completely about the initial line.

Now, $r = 5(1 + \cos\theta) \quad \therefore \; \dfrac{dr}{d\theta} = -5\sin\theta$

$$\therefore \; r^2 + \left(\frac{dr}{d\theta}\right)^2 = \ldots\ldots\ldots\ldots$$

42

$$50(1+\cos\theta)$$

Because

$$r^2 + \left(\frac{dr}{d\theta}\right)^2 = 25(1+2\cos\theta+\cos^2\theta+\sin^2\theta)$$
$$= 25(2+2\cos\theta)$$
$$= 50(1+\cos\theta)$$

We would like to express this as a square, since we have to take its root, so we now write $\cos\theta$ in terms of its half angle.

$$\therefore r^2 + \left(\frac{dr}{d\theta}\right)^2 = 50\left(1+2\cos^2\frac{\theta}{2}-1\right)$$
$$= 100\cos^2\frac{\theta}{2}$$

$$\therefore \sqrt{r^2 + \left(\frac{dr}{d\theta}\right)^2} = 10\cos\frac{\theta}{2}$$

So the formula in this case now becomes

$$S = \ldots\ldots\ldots\ldots$$

43

$$S = \int_0^\pi 2\pi \cdot 5(1+\cos\theta)\sin\theta \cdot 10\cos\frac{\theta}{2} \cdot d\theta$$

$$\therefore S = 100\pi \int_0^\pi (1+\cos\theta)\sin\theta\cos\frac{\theta}{2} d\theta$$

We can make this more convenient if we express $(1+\cos\theta)$ and $\sin\theta$ also in terms of $\frac{\theta}{2}$.

What do we get?

44

$$S = 400\pi \int_0^\pi \cos^4\frac{\theta}{2}\sin\frac{\theta}{2} d\theta$$

Because

$$S = 100\pi \int_0^\pi (1+\cos\theta)\sin\theta\cos\frac{\theta}{2} d\theta$$
$$= 100\pi \int_0^\pi 2\cos^2\frac{\theta}{2} \cdot 2\sin\frac{\theta}{2}\cos\frac{\theta}{2} \cdot \cos\frac{\theta}{2} d\theta$$
$$= 400\pi \int_0^\pi \cos^4\frac{\theta}{2}\sin\frac{\theta}{2} d\theta$$

▶

Polar coordinates

Now the derivative of $\cos\dfrac{\theta}{2}$ is $\left\{-\dfrac{\sin\dfrac{\theta}{2}}{2}\right\}$

$$\therefore S = -800\pi \int_0^\pi \cos^4\dfrac{\theta}{2}\left\{-\dfrac{\sin\dfrac{\theta}{2}}{2}\right\}d\theta$$

$= \ldots\ldots\ldots\ldots$ Finish it off.

$$\boxed{S = 160\pi \text{ units}^2}$$

Because

$$S = -800\pi \int_0^\pi \cos^4\dfrac{\theta}{2}\left\{-\dfrac{\sin\dfrac{\theta}{2}}{2}\right\}d\theta$$

$$= -800\pi \left[\dfrac{\cos^5\dfrac{\theta}{2}}{5}\right]_0^\pi = \dfrac{-800\pi}{5}(0-1)$$

$S = 160\pi \text{ units}^2$

And finally, here is one for you to do.

Example 2

Find the area of the surface generated when the arc of the curve $r = ae^\theta$ between $\theta = 0$ and $\theta = \pi/2$, rotates about the initial line.

Finish it completely and then check with the next frame

$$\boxed{S = \dfrac{2\sqrt{2}}{5} \cdot \pi a^2(2e^\pi + 1)}$$

Because we have:

$$S = \int_0^{\pi/2} 2\pi r \sin\theta \sqrt{r^2 + \left(\dfrac{dr}{d\theta}\right)^2}\, d\theta$$

And, in this case:

$r = ae^\theta \quad \therefore \quad \dfrac{dr}{d\theta} = ae^\theta$

$\therefore r^2 + \left(\dfrac{dr}{d\theta}\right)^2 = a^2 e^{2\theta} + a^2 e^{2\theta} = 2a^2 e^{2\theta}$

$\therefore \sqrt{r^2 + \left(\dfrac{dr}{d\theta}\right)^2} = \sqrt{2}\cdot a \cdot e^\theta$

▶

$$\therefore S = \int_0^{\pi/2} 2\pi a e^{\theta} \sin\theta \cdot \sqrt{2} a e^{\theta} \, d\theta$$

$$= 2\sqrt{2}\pi a^2 \int_0^{\pi/2} e^{2\theta} \sin\theta \, d\theta$$

Let
$$I = \int e^{2\theta} \sin\theta \, d\theta = e^{2\theta}(-\cos\theta) + 2\int \cos\theta \, e^{2\theta} \, d\theta$$

$$= -e^{2\theta}\cos\theta + 2\left\{ e^{2\theta}\sin\theta - 2\int \sin\theta \, e^{2\theta} \, d\theta \right\}$$

$$I = -e^{2\theta}\cos\theta + 2e^{2\theta}\sin\theta - 4I$$

$$\therefore 5I = e^{2\theta}\left\{ 2\sin\theta - \cos\theta \right\}$$

$$I = \frac{e^{2\theta}}{5}\left\{ 2\sin\theta - \cos\theta \right\}$$

$$\therefore S = 2\sqrt{2} \cdot \pi \cdot a^2 \left[\frac{e^{2\theta}}{5}\left\{ 2\sin\theta - \cos\theta \right\} \right]_0^{\pi/2}$$

$$= \frac{2\sqrt{2} \cdot \pi \cdot a^2}{5}\left\{ e^{\pi}(2-0) - 1(0-1) \right\}$$

$$S = \frac{2\sqrt{2} \cdot \pi \cdot a^2}{5}(2e^{\pi} + 1) \text{ units}^2$$

We are almost at the end, but before we finish the Program, let us collect our results together.

So move to Frame 47

---

47 Review summary
Polar curves – applications.

1. Area $\qquad A = \int_{\theta_1}^{\theta_2} \frac{1}{2} r^2 \, d\theta$

2. Volume $\qquad V = \int_{\theta_1}^{\theta_2} \frac{2}{3}\pi r^3 \sin\theta \, d\theta$

3. Length of arc $\qquad s = \int_{\theta_1}^{\theta_2} \sqrt{r^2 + \left(\frac{dr}{d\theta}\right)^2} \, d\theta$

4. Surface of revolution $\qquad S = \int_{\theta_1}^{\theta_2} 2\pi r \sin\theta \sqrt{r^2 + \left(\frac{dr}{d\theta}\right)^2} \, d\theta$

It is important to know these. The detailed working will depend on the particular form of the function $r = f(\theta)$, but as you have seen, the method of approach is mainly consistent.

The **Can You?** checklist and **Test exercise** now remain to be worked. Brush up any points on which you are not perfectly clear.

Polar coordinates

✓ Can You?

Checklist 3

48

Check this list before and after you try the end of Program test.

On a scale of 1 to 5 how confident are you that you can: Frames

- Convert expressions from Cartesian coordinates to polar coordinates and vice versa? **1** to **7**
  Yes ☐ ☐ ☐ ☐ ☐ No
- Plot the graphs of polar curves? **8** to **10**
  Yes ☐ ☐ ☐ ☐ ☐ No
- Recognize equations of standard polar curves? **11** to **15**
  Yes ☐ ☐ ☐ ☐ ☐ No
- Evaluate the areas enclosed by polar curves? **16** to **26**
  Yes ☐ ☐ ☐ ☐ ☐ No
- Evaluate the volumes of revolution generated by polar curves? **27** to **32**
  Yes ☐ ☐ ☐ ☐ ☐ No
- Evaluate the lengths of polar curves? **33** to **39**
  Yes ☐ ☐ ☐ ☐ ☐ No
- Evaluate the surface of revolution generated by polar curves? **40** to **46**
  Yes ☐ ☐ ☐ ☐ ☐ No

🚴 Test exercise 3

All the questions are quite straightforward: there are no tricks. But take your time and work carefully. **49**

1. Calculate the area enclosed by the curve $r\theta^2 = 4$ and the radius vectors at $\theta = \pi/2$ and $\theta = \pi$.

2. Sketch the polar curves:
   - (a) $r = 2\sin\theta$
   - (b) $r = 5\cos^2\theta$
   - (c) $r = \sin 2\theta$
   - (d) $r = 1 + \cos\theta$
   - (e) $r = 1 + 3\cos\theta$
   - (f) $r = 3 + \cos\theta$

▶

3 The plane figure bounded by the curve $r = 2 + \cos\theta$ and the radius vectors at $\theta = 0$ and $\theta = \pi$, rotates about the initial line through a complete revolution. Determine the volume of the solid generated.

4 Find the length of the polar curve $r = 4\sin^2\dfrac{\theta}{2}$ between $\theta = 0$ and $\theta = \pi$.

5 Find the area of the surface generated when the arc of the curve $r = a(1 - \cos\theta)$ between $\theta = 0$ and $\theta = \pi$, rotates about the initial line.

That completes the work on polar curves. You are now ready for the next Program

Further problems 3

50

1 Sketch the curve $r = \cos^2\theta$. Find (a) the area of one loop and (b) the volume of the solid formed by rotating the curve about the initial line.

2 Show that $\sin^4\theta = \dfrac{3}{8} - \dfrac{1}{2}\cos 2\theta + \dfrac{1}{8}\cos 4\theta$. Hence find the area bounded by the curve $r = 4\sin^2\theta$ and the radius vectors at $\theta = 0$ and $\theta = \pi$.

3 Find the area of the plane figure enclosed by the curve $r = a\sec^2\left(\dfrac{\theta}{2}\right)$ and the radius vectors at $\theta = 0$ and $\theta = \pi/2$.

4 Determine the area bounded by the curve $r = 2\sin\theta + 3\cos\theta$ and the radius vectors at $\theta = 0$ and $\theta = \pi/2$.

5 Find the area enclosed by the curve $r = \dfrac{2}{1 + \cos 2\theta}$ and the radius vectors at $\theta = 0$ and $\theta = \pi/4$.

6 Plot the graph of $r = 1 + 2\cos\theta$ at intervals of 30 and show that it consists of a small loop within a larger loop. The area between the two loops is rotated about the initial line through two right-angles. Find the volume generated.

7 Find the volume generated when the plane figure enclosed by the curve $r = 2a\sin^2\left(\dfrac{\theta}{2}\right)$ between $\theta = 0$ and $\theta = \pi$, rotates around the initial line.

8 The plane figure bounded by the cardioid $r = 2a(1 + \cos\theta)$ and the parabola $r(1 + \cos\theta) = 2a$, rotates around the initial line. Show that the volume generated is $18\pi a^3$.

9 Find the length of the arc of the curve $r = a\cos^3\left(\dfrac{\theta}{3}\right)$ between $\theta = 0$ and $\theta = 3\pi$.

10 Find the length of the arc of the curve $r = 3\sin\theta + 4\cos\theta$ between $\theta = 0$ and $\theta = \pi/2$.

Polar coordinates

11 Find the length of the spiral $r = a\theta$ between $\theta = 0$ and $\theta = 2\pi$.

12 Sketch the curve $r = a\sin^3\left(\dfrac{\theta}{3}\right)$ and calculate its total length.

13 Show that the length of arc of the curve $r = a\cos^2\theta$ between $\theta = 0$ and $\theta = \pi/2$ is $a[2\sqrt{3} + \ln(2+\sqrt{3})]/(2\sqrt{3})$.

14 Find the length of the spiral $r = ae^{b\theta}$ between $\theta = 0$ and $\theta = \theta_1$, and the area swept out by the radius vector between these two limits.

15 Find the area of the surface generated when the arc of the curve $r^2 = a^2\cos 2\theta$ between $\theta = 0$ and $\theta = \pi/4$, rotates about the initial line.

Program 4

Multiple integrals

Frames
1 to 49

Learning outcomes

When you have completed this Program you will be able to:
- Determine the area of a rectangle using a double integral
- Evaluate double integrals over general areas
- Evaluate triple integrals over general volumes
- Apply double integrals to find areas and second moments of area
- Apply triple integrals to find volumes

Multiple integrals

Summation in two directions

Let us consider the rectangle bounded by the straight lines, $x = r$, $x = s$, $y = k$ and $y = m$, as shown:

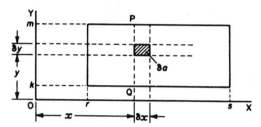

Then the area of the shaded element, $\delta a = \ldots\ldots\ldots\ldots$

$$\delta a = \delta y \cdot \delta x$$

If we add together all the elements of area, like δa, to form the vertical strip PQ, then δA, the area of the strip, can be expressed as $\delta A = \ldots\ldots\ldots\ldots$

$$\delta A = \sum_{y=k}^{y=m} \delta y \cdot \delta x$$

Did you remember to include the limits?
 Note that during this summation in the y-direction, δx is constant.

If we now sum all the strips across the figure from $x = r$ to $x = s$, we shall obtain the total area of the rectangle, A.

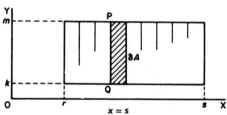

$$\therefore A = \sum_{x=r}^{x=s} (\text{all vertical strips like PQ})$$

$$= \sum_{x=r}^{x=s} \left\{ \sum_{y=k}^{y=m} \delta y \cdot \delta x \right\}$$

Removing the brackets, this becomes:

$$A = \sum_{x=r}^{x=s} \sum_{y=k}^{y=m} \delta y \cdot \delta x$$

If now $\delta y \to 0$ and $\delta x \to 0$, the finite summations become integrals, so the expression becomes $A = \ldots\ldots\ldots\ldots$

4

$$A = \int_{x=r}^{x=s} \int_{y=k}^{y=m} dy \cdot dx$$

To evaluate this expression, we start from the inside and work outwards.

$$A = \int_{x=r}^{x=s} \left[\int_{y=k}^{y=m} dy \right] dx = \int_{x=r}^{x=s} \left[y \right]_{y=k}^{y=m} dx$$

$$= \int_{x=r}^{x=s} (m-k) dx$$

and since m and k are constants, this gives $A = \ldots\ldots\ldots$

5

$$A = (m-k) \cdot (s-r)$$

Because

$$A = \left[(m-k)x \right]_{x=r}^{x=s} = (m-k) \left[x \right]_{x=r}^{x=s}$$

$$A = (m-k) \cdot (s-r)$$

which we know is correct, for it is merely $A = $ length $\times$ breadth.

That may seem a tedious way to find the area of a rectangle, but we have done it to introduce the method we are going to use.

First we define an element of area $\delta y \cdot \delta x$.
Then we sum in the y-direction to obtain the area of a $\ldots\ldots\ldots$
Finally, we sum the result in the x-direction to obtain the area of the $\ldots\ldots\ldots$

6

vertical strip;   whole figure

We could have worked slightly differently:

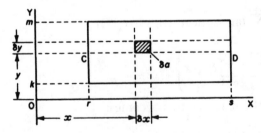

As before $\delta a = \delta x \cdot \delta y$. If we sum the elements in the x-direction this time, we get the area δA_1 of the horizontal strip, CD.

$$\therefore \delta A_1 = \ldots\ldots\ldots$$

Multiple integrals

7

$$\delta A_1 = \sum_{x=r}^{x=s} \delta x \cdot \delta y$$

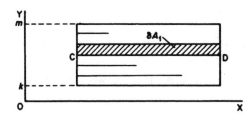

Now sum the strips vertically and we obtain once again the area of the whole rectangle.

$$A_1 = \sum_{y=k}^{y=m} \text{(all horizontal strips like CD)} = \sum_{y=k}^{y=m} \left\{ \sum_{x=r}^{x=s} \delta x \cdot \delta y \right\}$$

As before, if we now remove the brackets and consider what this becomes when $\delta x \to 0$ and $\delta y \to 0$, we get:

$A_1 = \ldots\ldots\ldots\ldots$

8

$$A_1 = \int_{y=k}^{y=m} \int_{x=r}^{x=s} dx \, dy$$

To evaluate this we start from the center:

$$A_1 = \int_{y=k}^{y=m} \left[\int_{x=r}^{x=s} dx \right] dy$$

$= \ldots\ldots\ldots\ldots$

Complete the working to find A_1 and then move on to Frame 9

9

$$A_1 = (s-r) \cdot (m-k)$$

Because

$$A_1 = \int_{y=k}^{y=m} \left[x \right]_r^s dy = \int_k^m (s-r) dy = (s-r) \left[y \right]_k^m$$

$\therefore A_1 = (s-r) \cdot (m-k)$ which is the same result as before.

So the order in which we carry out our two summations appears not to matter.

Remember:

(a) We work from the inside integral.

(b) We integrate with respect to x when the limits are values of x.

(c) We integrate with respect to y when the limits are values of y.

Move to the next frame

Vector Analysis

Double integrals

10

The expression $\int_{y_1}^{y_2} \int_{x_1}^{x_2} f(x,y)\,dx\,dy$ is called a *double integral* (for obvious reasons!) and indicates that:

(a) $f(x,y)$ is first integrated with respect to x (regarding y as being constant) between the limits $x = x_1$ and $x = x_2$

(b) the result is then integrated with respect to y between the limits $y = y_1$ and $y = y_2$.

Example 1

Evaluate $\quad I = \int_1^2 \int_2^4 (x + 2y)\,dx\,dy$

So $(x + 2y)$ is first integrated with respect to x between $x = 2$ and $x = 4$, with y regarded as constant for the time being.

$$I = \int_1^2 \left[\int_2^4 (x+2y)\,dx \right] dy$$

$$= \int_1^2 \left[\frac{x^2}{2} + 2xy \right]_2^4 dy$$

$$= \int_1^2 \left\{ (8 + 8y) - (2 + 4y) \right\} dy$$

$$= \int_1^2 (6 + 4y)\,dy = \ldots\ldots\ldots\ldots$$

Finish it off

11

$$I = 12$$

Because

$$I = \int_1^2 (6 + 4y)\,dy = \left[6y + 2y^2 \right]_1^2$$

$$= (12 + 8) - (6 + 2) = 20 - 8 = 12$$

Here is another.

Example 2

Evaluate $\quad I = \int_1^2 \int_0^3 x^2 y\,dx\,dy$

Do this one on your own. Remember to start with $\int_0^3 x^2 y\,dx$ with y constant.

Finish the double integral completely and then move on to Frame 12

Multiple integrals

$$I = 13.5$$

Check your working:

$$I = \int_1^2 \int_0^3 x^2 y \, dx \, dy = \int_1^2 \left[\int_0^3 x^2 y \, dx \right] dy$$

$$= \int_1^2 \left[\frac{x^3}{3} \cdot y \right]_{x=0}^{x=3} dy$$

$$= \int_1^2 (9y) \, dy = \left[\frac{9y^2}{2} \right]_1^2$$

$$= 18 - 4.5 = 13.5$$

Now do this one in just the same way.

Example 3

Evaluate $\quad I = \int_1^2 \int_0^\pi (3 + \sin \theta) \, d\theta \, dr$

When you have finished, check with the next frame

$$I = 3\pi + 2$$

Here it is:

$$I = \int_1^2 \int_0^\pi (3 + \sin \theta) \, d\theta \, dr$$

$$= \int_1^2 \left[3\theta - \cos \theta \right]_0^\pi dr$$

$$= \int_1^2 \left\{ (3\pi + 1) - (-1) \right\} dr$$

$$= \int_1^2 (3\pi + 2) \, dr$$

$$= \left[(3\pi + 2)r \right]_1^2$$

$$= (3\pi + 2)(2 - 1) = 3\pi + 2$$

On to the next frame

Vector Analysis

Triple integrals

14

Sometimes we have to deal with expressions such as

$$I = \int_a^b \int_c^d \int_e^f f(x,y,z)\,dx\,dy\,dz$$

but the rules are as before. Start with the innermost integral and work outwards.

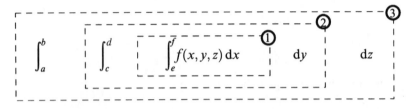

All symbols are regarded as constant for the time being, except the one variable with respect to which stage of integration is taking place. So try this one on your own straight away.

Example 1

Evaluate $\quad I = \int_1^3 \int_{-1}^1 \int_0^2 (x + 2y - z)\,dx\,dy\,dz$

15

$$\boxed{I = -8}$$

Did you manage it first time? Here is the working in detail.

$$I = \int_1^3 \int_{-1}^1 \int_0^2 (x + 2y - z)\,dx\,dy\,dz$$

$$= \int_1^3 \int_{-1}^1 \left[\frac{x^2}{2} + 2xy - xz\right]_0^2 dy\,dz$$

$$= \int_1^3 \int_{-1}^1 (2 + 4y - 2z)\,dy\,dz = \int_1^3 \left[2y + 2y^2 - 2yz\right]_{-1}^1 dz$$

$$= \int_1^3 \left\{(2 + 2 - 2z) - (-2 + 2 + 2z)\right\} dz = \int_1^3 (4 - 4z)\,dz$$

$$= \left[4z - 2z^2\right]_1^3 = (12 - 18) - (4 - 2) = -8$$

And another.

Example 2

Evaluate $\quad \int_1^2 \int_0^3 \int_0^1 (p^2 + q^2 - r^2)\,dp\,dq\,dr$

When you have finished it, move on to Frame 16

Multiple integrals

$$I = 3$$
16

Because

$$I = \int_1^2 \int_0^3 \int_0^1 (p^2 + q^2 - r^2) \, dp \, dq \, dr$$

$$= \int_1^2 \int_0^3 \left[\frac{p^3}{3} + pq^2 - pr^2 \right]_0^1 dq \, dr$$

$$= \int_1^2 \int_0^3 \left\{ \frac{1}{3} + q^2 - r^2 \right\} dq \, dr$$

$$= \int_1^2 \left[\frac{q}{3} + \frac{q^3}{3} - qr^2 \right]_0^3 dr$$

$$= \int_1^2 (1 + 9 - 3r^2) dr$$

$$= \left[10r - r^3 \right]_1^2 = (20 - 8) - (10 - 1)$$

$$= 12 - 9 = 3$$

It is all very easy if you take it steadily, step by step.

Now two quickies for review.
Evaluate:

(a) $\displaystyle\int_1^2 \int_3^5 dy \, dx$ (b) $\displaystyle\int_0^4 \int_1^{3x} 2y \, dy \, dx$

                            Finish them both and then move on to the next frame

$$\text{(a) } I = 2 \quad \text{(b) } I = 188$$
17

Here they are:

(a) $\displaystyle I = \int_1^2 \int_3^5 dy \, dx = \int_1^2 [y]_3^5 dx = \int_1^2 (5-3) dx = \int_1^2 2 \, dx = [2x]_1^2$

$= 4 - 2 = 2$

(b) $\displaystyle I = \int_0^4 \int_1^{3x} 2y \, dy \, dx = \int_0^4 [y^2]_1^{3x} dx = \int_0^4 (9x^2 - 1) dx$

$= \left[3x^3 - x \right]_0^4 = 192 - 4 = 188$

And finally, do this one.

$$I = \int_0^5 \int_1^2 (3x^2 - 4) \, dx \, dy = \ldots\ldots\ldots$$

18

$I = 15$

Check this working.

$$I = \int_0^5 \int_1^2 (3x^2 - 4)\,dx\,dy$$

$$= \int_0^5 \left[x^3 - 4x \right]_1^2 dy$$

$$= \int_0^5 \left\{ (8-8) - (1-4) \right\} dy$$

$$= \int_0^5 3\,dy = \left[3y \right]_0^5 = 15$$

Now let us see a few applications of multiple integrals.

Move on then to the next frame

Applications

19

Example 1

Find the area bounded by $y = \dfrac{4x}{5}$, the x-axis and the ordinate at $x = 5$.

Area of element $= \delta y \cdot \delta x$

$\therefore$ Area of strip $\displaystyle\sum_{y=0}^{y=y_1} \delta y \cdot \delta x$

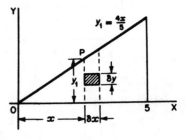

The sum of all such strips across the figure gives us:

$$A \approx \sum_{x=0}^{x=5} \left\{ \sum_{y=0}^{y=y_1} \delta y \cdot \delta x \right\}$$

$$\approx \sum_{x=0}^{x=5} \sum_{y=0}^{y=y_1} \delta y \cdot \delta x$$

Now, if $\delta y \to 0$ and $\delta x \to 0$, then:

$$A = \int_0^5 \int_0^{y_1} dy\,dx$$

$$= \int_0^5 \left[y \right]_0^{y_1} dx = \int_0^5 y_1\,dx$$

But $y_1 = \dfrac{4x}{5}$

So $A = \ldots\ldots\ldots\ldots$

Finish it off

$$A = 10 \text{ unit}^2$$

Because

$$A = \int_0^5 \frac{4x}{5} dx = \left[\frac{2x^2}{5}\right]_0^5 = 10$$

Right. Now what about this one?

Example 2

Find the area under the curve $y = 4\sin\frac{x}{2}$ between $x = \frac{\pi}{3}$ and $x = \pi$, by the double integral method.

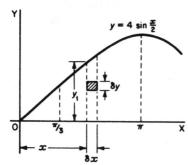

Steps as before:
Area of element $= \delta y \cdot \delta x$
Area of vertical strip

$$\sum_{y=0}^{y=y_1} \delta y \cdot \delta x$$

Total area of figure:

$$A \approx \sum_{x=\pi/3}^{x=\pi} \left\{ \sum_{y=0}^{y=y_1} \delta y \cdot \delta x \right\}$$

If $\delta y \to 0$ and $\delta x \to 0$, then:

$$A = \int_{\pi/3}^{\pi} \int_0^{y_1} dy\, dx = \ldots\ldots\ldots$$

Complete it, remembering that $y_1 = 4\sin\frac{x}{2}$.

$$A = 4\sqrt{3} \text{ unit}^2$$

Because you get:

$$A = \int_{\pi/3}^{\pi} \int_0^{y_1} dy\, dx = \int_{\pi/3}^{\pi} \left[y\right]_0^{y_1} dx = \int_{\pi/3}^{\pi} y_1\, dx$$

$$= \int_{\pi/3}^{\pi} 4\sin\frac{x}{2} dx = \left[-8\cos\frac{x}{2}\right]_{\pi/3}^{\pi}$$

$$= (-8\cos\pi/2) - (-8\cos\pi/6)$$

$$= 0 + 8 \cdot \frac{\sqrt{3}}{2} = 4\sqrt{3} \text{ unit}^2$$

Now for a rather more worthwhile example – on to Frame 22

Vector Analysis

22

Example 3

Find the area enclosed by the curves

$$y_1^2 = 9x \text{ and } y_2 = \frac{x^2}{9}$$

First we must find the points of intersection. For that, $y_1 = y_2$.

$$\therefore 9x = \frac{x^4}{81} \quad \therefore x = 0 \quad \text{or} \quad x^3 = 729, \quad \text{i.e. } x = 9.$$

So we have a diagram like this:

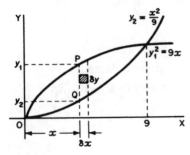

As usual:

Area of element $= \delta y \cdot \delta x$

$\therefore$ Area of strip PQ

$$\sum_{y=y_2}^{y=y_1} \delta y \cdot \delta x$$

Summing all strips between $x = 0$ and $x = 9$:

$$A \approx \sum_{x=0}^{x=9} \left\{ \sum_{y=y_2}^{y=y_1} \delta y \cdot \delta x \right\} = \sum_{x=0}^{x=9} \sum_{y=y_2}^{y=y_1} \delta y \cdot \delta x$$

If $\delta y \to 0$ and $\delta x \to 0$, $A = \int_0^9 \int_{y_2}^{y_1} dy \, dx$

Now finish it off, remembering that $y_1^2 = 9x$ and $y_2 = \frac{x^2}{9}$.

23

$$\boxed{A = 27 \text{ unit}^2}$$

Here it is.

$$A = \int_0^9 \int_{y_2}^{y_1} dy \, dx = \int_0^9 \left[y \right]_{y_2}^{y_1} dx$$

$$= \int_0^9 (y_1 - y_2) dx$$

$$= \int_0^9 \left\{ 3x^{\frac{1}{2}} - \frac{x^2}{9} \right\} dx$$

$$= \left[2x^{3/2} - \frac{x^3}{27} \right]_0^9$$

$$= 54 - 27 = 27 \text{ unit}^2$$

Now for a different one. So move on to the next frame

Double integrals can conveniently be used for finding other values besides areas.

Example 4

Find the second moment of area of a rectangle 6 cm × 4 cm about an axis through one corner perpendicular to the plane of the figure.

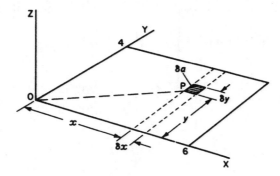

Second moment of element P about $OZ \approx \delta a(OP)^2$

$\approx \delta y \cdot \delta x \cdot (x^2 + y^2)$

Total second moment about OZ

$$I \approx \sum_{x=0}^{x=6} \sum_{y=0}^{y=4} (x^2 + y^2) \, dy \, dx$$

If $\delta x \to 0$ and $\delta y \to 0$, this becomes:

$$I = \int_0^6 \int_0^4 (x^2 + y^2) \, dy \, dx$$

Now complete the working, $I = \ldots\ldots\ldots$

$$\boxed{I = 416 \text{ cm}^4}$$

Because

$$I = \int_0^6 \int_0^4 (x^2 + y^2) \, dy \, dx = \int_0^6 \left[x^2 y + \frac{y^3}{3} \right]_0^4 dx = \int_0^6 \left\{ 4x^2 + \frac{64}{3} \right\} dx$$

$$= \left[\frac{4x^3}{3} + \frac{64x}{3} \right]_0^6 = 288 + 128 = 416 \text{ cm}^4$$

Now here is one for you to do on your own.

Example 5

Find the second moment of area of a rectangle 5 cm × 3 cm about one 5 cm side as axis.

Complete it and then on to Frame 26

26

$$I = 45 \text{ cm}^4$$

Here it is: check through the working.

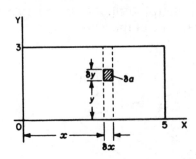

Area of element $= \delta a = \delta y \cdot \delta x$
Second moment of area of δa about OX
$= \delta a \cdot y^2$
$= y^2 \cdot \delta y \cdot \delta x$

Second moment of strip $\approx \sum\limits_{y=0}^{y=3} y^2 \cdot \delta y \cdot \delta x$

Second moment of whole figure $\approx \sum\limits_{x=0}^{x=5} \sum\limits_{y=0}^{y=3} y^2 \cdot \delta y \cdot \delta x$

If $\delta y \to 0$ and $\delta x \to 0$:

$$I = \int_0^5 \int_0^3 y^2 \, dy \, dx$$

$$\therefore I = \int_0^5 \left[\frac{y^3}{3}\right]_0^3 dx = \int_0^5 9 \, dx = \left[9x\right]_0^5$$

$$I = 45 \text{ cm}^4$$

On to Frame 27

27 **Review exercise**

Now a short review exercise. Finish both integrals, before moving on to the next frame. Here they are.

Evaluate the following:

(a) $\int_0^2 \int_1^3 (y^2 - xy) \, dy \, dx$

(b) $\int_0^3 \int_1^2 (x^2 + y^2) \, dy \, dx$.

When you have finished both, move on

Multiple integrals

> (a) $I = 9\frac{1}{3}$ (b) $I = 16$

Here they are in detail:

(a) $I = \int_0^2 \int_1^3 (y^2 - xy)\, dy\, dx = \int_0^2 \left[\frac{y^3}{3} - \frac{xy^2}{2}\right]_1^3 dx$

$= \int_0^2 \left\{\left(9 - \frac{9x}{2}\right) - \left(\frac{1}{3} - \frac{x}{2}\right)\right\} dx$

$= \int_0^2 \left(\frac{26}{3} - 4x\right) dx = \left[\frac{26x}{3} - 2x^2\right]_0^2$

$= 17\frac{1}{3} - 8 = 9\frac{1}{3}$

(b) $I = \int_0^3 \int_1^2 (x^2 + y^2)\, dy\, dx = \int_0^3 \left[x^2 y + \frac{y^3}{3}\right]_1^2 dx$

$= \int_0^3 \left\{\left(2x^2 + \frac{8}{3}\right) - \left(x^2 + \frac{1}{3}\right)\right\} dx$

$= \int_0^3 \left(x^2 + \frac{7}{3}\right) dx = \left[\frac{x^3}{3} + \frac{7x}{3}\right]_0^3$

$= 9 + 7 = 16$

Now on to Frame 29

Alternative notation

Sometimes double integrals are written in a slightly different way. For example, the last double integral $I = \int_0^3 \int_1^2 (x^2 + y^2)\, dy\, dx$ could have been written:

$$\int_0^3 dx \int_1^2 (x^2 + y^2)\, dy$$

The key now is that we start working from the *right-hand* side integral and gradually work back towards the front. Of course, we get the same result and the working is identical.

Let us have some examples, to get used to this notation.

Move on then to Frame 30

30

Example 1

$$I = \int_0^2 dx \int_0^{\pi/2} 5\cos\theta\, d\theta$$

$$= \int_0^2 dx \left[5\sin\theta\right]_0^{\pi/2}$$

$$= \int_0^2 dx \left[5\right] = \int_0^2 5\, dx = \left[5x\right]_0^2$$

$$= 10$$

It is all very easy, once you have seen the method.
You try this one.

Example 2

Evaluate $I = \int_3^6 dy \int_0^{\pi/2} 4\sin 3x\, dx$.

31

$$\boxed{I = 4}$$

Here it is:

$$I = \int_3^6 dy \int_0^{\pi/2} 4\sin 3x\, dx$$

$$= \int_3^6 dy \left[\frac{-4\cos 3x}{3}\right]_0^{\pi/2}$$

$$= \int_3^6 dy \left\{(0) - \left(-\frac{4}{3}\right)\right\} = \int_3^6 dy\, \frac{4}{3}$$

$$= \left[\frac{4y}{3}\right]_3^6 = (8) - (4) = 4$$

Now do these two.

Example 3

$$\int_0^3 dx \int_0^1 (x - x^2)\, dy$$

Example 4

$$\int_1^2 dy \int_y^{2y} (x - y)\, dx$$

(Take care with the second one!)

When you have finished them both, move on to the next frame

Multiple integrals

Example 3: $I = -4.5$ Example 4: $I = \dfrac{7}{6}$

Here is the working.

Example 3
$$I = \int_0^3 dx \int_0^1 (x - x^2)\, dy$$
$$= \int_0^3 dx \left[xy - x^2 y \right]_0^1$$
$$= \int_0^3 dx (x - x^2)$$
$$= \int_0^3 (x - x^2)\, dx$$
$$= \left[\dfrac{x^2}{2} - \dfrac{x^3}{3} \right]_0^3$$
$$= \dfrac{9}{2} - 9 = -4.5$$

Example 4
$$I = \int_1^2 dy \int_y^{2y} (x - y)\, dx$$
$$= \int_1^2 dy \left[\dfrac{x^2}{2} - xy \right]_{x=y}^{x=2y}$$
$$= \int_1^2 dy \left\{ (2y^2 - 2y^2) - \left(\dfrac{y^2}{2} - y^2 \right) \right\}$$
$$= \int_1^2 dy \dfrac{y^2}{2} = \int_1^2 \dfrac{y^2}{2}\, dy$$
$$= \left[\dfrac{y^3}{6} \right]_1^2 = \dfrac{8}{6} - \dfrac{1}{6} = \dfrac{7}{6}$$

Next frame

Now, by way of revision, evaluate these:

(a) $\displaystyle \int_0^4 \int_y^{2y} (2x + 3y)\, dx\, dy$

(b) $\displaystyle \int_1^4 dx \int_0^{\sqrt{x}} (2y - 5x)\, dy$

When you have completed both of them, move on to Frame 34

34

$$\boxed{\text{(a) } 128 \quad \text{(b) } -54.5}$$

Working:

(a) $I = \int_0^4 \int_y^{2y} (2x + 3y) \, dx \, dy$

$= \int_0^4 \left[x^2 + 3xy \right]_{x=y}^{x=2y} dy$

$= \int_0^4 \left\{ (4y^2 + 6y^2) - (y^2 + 3y^2) \right\} dy$

$= \int_0^4 \left\{ 10y^2 - 4y^2 \right\} dy$

$= \int_0^4 6y^2 \, dy$

$= \left[\frac{6y^3}{3} \right]_0^4$

$= \left[2y^3 \right]_0^4 = 128$

(b) $I = \int_1^4 dx \int_0^{\sqrt{x}} (2y - 5x) \, dy$

$= \int_1^4 dx \left[y^2 - 5xy \right]_{y=0}^{y=\sqrt{x}}$

$= \int_1^4 dx \left\{ x - 5x^{3/2} \right\}$

$= \int_1^4 (x - 5x^{3/2}) \, dx$

$= \left[\frac{x^2}{2} - 2x^{5/2} \right]_1^4$

$= (8 - 64) - \left(\frac{1}{2} - 2 \right)$

$= -56 + 1.5 = -54.5$

So it is just a question of being able to recognize and to interpret the two notations.

Now let us look at one or two further examples of the use of multiple integrals.

Move on then to Frame 35

Multiple integrals

Further example of use of multiple integrals　　　　　35

To find the area of the plane figure bounded by the polar curve $r = f(\theta)$, and the radius vectors at $\theta = \theta_1$ and $\theta = \theta_2$.

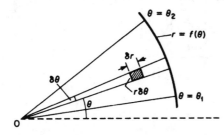

Small arc of a circle of radius r, subtending an angle $\delta\theta$ at the centre.

$$\therefore \text{arc} = r \cdot \delta\theta$$

We proceed very much as before:

Area of element $\approx \delta r \cdot r \delta\theta$

Area of thin sector $\approx \sum_{r=0}^{r=r_1} \delta r \cdot r \delta\theta$

Total area $\approx \sum_{\theta=\theta_1}^{\theta=\theta_2}$ (all such thin sectors)

$$\approx \sum_{\theta=\theta_1}^{\theta=\theta_2} \left\{ \sum_{r=0}^{r=r_1} r \cdot \delta r \cdot \delta\theta \right\}$$

$$\approx \sum_{\theta=\theta_1}^{\theta=\theta_2} \sum_{r=0}^{r=r_1} r \cdot \delta r \cdot \delta\theta$$

Then if $\delta\theta \to 0$ and $\delta r \to 0$:

$$A = \int_{\theta_1}^{\theta_2} \int_0^{r_1} r \, dr \, d\theta$$

$= \ldots\ldots\ldots\ldots$ Finish it off.

The working continues:　　　　　36

$$A = \int_{\theta_1}^{\theta_2} \left[\frac{r^2}{2} \right]_0^{r_1} d\theta$$

$$= \int_{\theta_1}^{\theta_2} \left(\frac{r_1^2}{2} \right) d\theta$$

i.e. in general, $A = \int_{\theta_1}^{\theta_2} \frac{1}{2} r^2 \, d\theta = \int_{\theta_1}^{\theta_2} \frac{1}{2} f^2(\theta) \, d\theta$

which is the result we have met before.

Let us work an actual example of this, so move on to Frame 37

37

By the use of double integrals, find the area enclosed by the polar curve $r = 4(1 + \cos\theta)$ and the radius vectors at $\theta = 0$ and $\theta = \pi$.

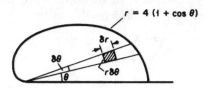

$$A \approx \sum_{\theta=0}^{\theta=\pi} \sum_{r=0}^{r=r_1} r\delta r \cdot \delta\theta$$

$$A = \int_0^\pi \int_0^{r_1} r\, dr\, d\theta$$

$$= \int_0^\pi \left[\frac{r^2}{2}\right]_0^{r_1} d\theta$$

$$= \int_0^\pi \left[\frac{r_1^2}{2}\right] d\theta \qquad \text{But } r_1 = f(\theta)$$
$$\qquad\qquad\qquad\qquad\qquad = 4(1+\cos\theta)$$

$$\therefore A = \int_0^\pi 8(1+\cos\theta)^2 \, d\theta$$

$$= \int_0^\pi 8(1 + 2\cos\theta + \cos^2\theta)\, d\theta$$

$$= \ldots\ldots\ldots\ldots$$

38

$$\boxed{A = 12\pi \text{ unit}^2}$$

Because

$$A = 8\int_0^\pi (1 + 2\cos\theta + \cos^2\theta)\, d\theta$$

$$= 8\left[\theta + 2\sin\theta + \frac{\theta}{2} + \frac{\sin 2\theta}{4}\right]_0^\pi$$

$$= 8\left(\pi + \frac{\pi}{2}\right) - (0)$$

$$= 8\pi + 4\pi = 12\pi \text{ unit}^2$$

Now let us deal with volumes by the same method, so move on to the next frame

Determination of volumes by multiple integrals

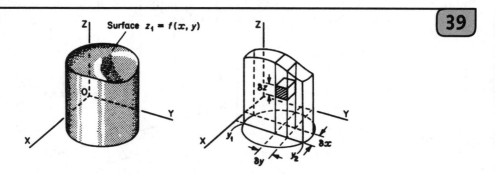

Element of volume $\delta v = \delta x \cdot \delta y \cdot \delta z$.
Summing the elements up the column, we have

$$\delta V_c = \sum_{z=0}^{z=z_1} \delta x \cdot \delta y \cdot \delta z$$

If we now sum the columns between $y = y_1$ and $y_1 = y_2$, we obtain the volume of the slice:

$$\delta V_s = \sum_{y=y_1}^{y=y_2} \sum_{z=0}^{z=z_1} \delta x \cdot \delta y \cdot \delta z$$

Then, summing all slices between $x = x_1$ and $x = x_2$, we have the total volume:

$$V = \sum_{x=x_1}^{x=x_2} \sum_{y=y_1}^{y=y_2} \sum_{z=z_0}^{z=z_1} \delta x \cdot \delta y \cdot \delta z$$

Then, as usual, if $\delta x \to 0$, $\delta y \to 0$ and $\delta z \to 0$:

$$V = \int_{x_1}^{x_2} \int_{y_1}^{y_2} \int_0^{z_1} dx\, dy\, dz$$

The result this time is a triple integral, but the development is very much the same as in our previous examples.

Let us see this in operation in the following examples.

Next frame

Example 1

A solid is enclosed by the plane $z = 0$, the planes $x = 1$, $x = 4$, $y = 2$, $y = 5$ and the surface $z = x + y$. Find the volume of the solid.

First of all, what does the figure look like? The plane $z = 0$ is the x–y plane and the plane $x = 1$ is positioned thus:

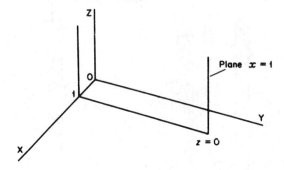

Working on the same lines, draw a sketch of the vertical sides.

The figure so far now looks like this:

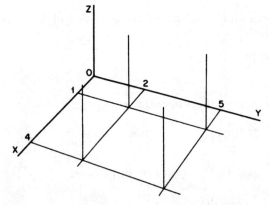

If we now mark in the calculated heights at each point of intersection $(z = x + y)$, we get:

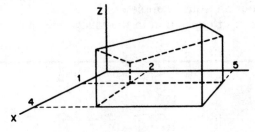

This is just preparing the problem, so that we can see how to develop the integral.

For the calculation stage, move on to the next frame

Multiple integrals

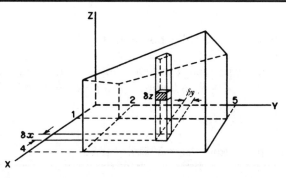

Volume element ≈ $\delta x \cdot \delta y \cdot \delta z$

Volume of column ≈ $\delta x \cdot \delta y \sum_{z=0}^{z=(x+y)} \delta z$

Volume of slice ≈ $\delta x \sum_{y=2}^{y=5} \delta y \sum_{z=0}^{z=x+y} \delta z$

Volume of total solid ≈ $\sum_{x=1}^{x=4} \delta x \sum_{y=2}^{y=5} \delta y \sum_{z=0}^{z=x+y} \delta z$

Then, as usual, if $\delta x \to 0$, $\delta y \to 0$, $\delta z \to 0$, this becomes:

$$V = \int_1^4 dx \int_2^5 dy \int_0^{x+y} dz$$

And this you can now finish off without any trouble. (With this form of notation, start at the right-hand end. Remember?)

So $V = \ldots\ldots\ldots$

$$\boxed{V = 54 \text{ unit}^3}$$

$$V = \int_1^4 dx \int_2^5 dy \int_0^{x+y} dz = \int_1^4 dx \int_2^5 dy(x+y)$$

$$= \int_1^4 dx \int_2^5 (x+y)\,dy = \int_1^4 dx \left[xy + \frac{y^2}{2}\right]_2^5$$

$$= \int_1^4 dx \left[5x + \frac{25}{2} - 2x - 2\right] = \int_1^4 \left(3x + \frac{21}{2}\right) dx$$

$$= \left[\frac{3x^2}{2} + \frac{21x}{2}\right]_1^4 = \frac{1}{2}\left[3x^2 + 21x\right]_1^4$$

$$= \frac{1}{2}\{(48+84) - (3+21)\} = \frac{1}{2}\{132 - 24\} = 54 \text{ unit}^3$$

44

Example 2

Find the volume of the solid bounded by the planes $z = 0$, $x = 1$, $x = 2$, $y = -1$, $y = 1$ and the surface $z = x^2 + y^2$.

In the light of the previous example, can you conjure up a mental picture of what this solid looks like? As before it will give rise to a triple integral.

$$V = \int_1^2 dx \int_{-1}^1 dy \int_0^{x^2+y^2} dz$$

Evaluate this and so find V. $V = \ldots\ldots\ldots\ldots$

45

$$\boxed{V = \frac{16}{3} \text{ unit}^3}$$

Because we have:

$$V = \int_1^2 dx \int_{-1}^1 dy \int_0^{x^2+y^2} dz = \int_1^2 dx \int_{-1}^1 dy(x^2 + y^2)$$

$$= \int_1^2 dx \left[x^2 y + \frac{y^3}{3} \right]_{-1}^1 = \int_1^2 \left\{ \left(x^2 + \frac{1}{3} \right) - \left(-x^2 - \frac{1}{3} \right) \right\} dx$$

$$= \int_1^2 \left\{ 2x^2 + \frac{2}{3} \right\} dx = \frac{2}{3} \left[x^3 + x \right]_1^2$$

$$= \frac{2}{3} \left\{ (8+2) - (1+1) \right\} = \frac{16}{3} \text{ unit}^3$$

Next frame

46 Review summary

Introductory work on double and triple integrals was covered in detail by this text's authors in Programme 23 of *Engineering Mathematics (Fifth Edition)*. Another look at the main points is well worth while.

You will no doubt recognize the following.

1 *Double integrals*

$$\int_{y_1}^{y_2} \int_{x_1}^{x_2} f(x, y) \, dx \, dy$$

is a double integral and is evaluated from the inside outwards, i.e.

$$\int_{y_1}^{y_2} \int_{x_1}^{x_2} f(x, y) \, dx \overset{(1)}{} dy \overset{(2)}{}$$

Multiple integrals

A double integral is sometimes expressed in the form

$$\int_{y_1}^{y_2} dy \int_{x_1}^{x_2} f(x, y)\, dx$$

in which case, we evaluate from the right-hand end, i.e.

$$\int_{y_1}^{y_2} dy \left[\int_{x_1}^{x_2} f(x, y)\, dx \right] \;①$$

then

$$\left[\int_{y_1}^{y_2} \left[\int_{x_1}^{x_2} f(x, y)\, dx \right] dy \right] \;②$$

2 Triple integrals

Triple integrals follow the same procedure.

$$\int_{z_1}^{z_2} \int_{y_1}^{y_2} \int_{x_1}^{x_2} f(x, y, z)\, dx\, dy\, dz \text{ is evaluated in the order}$$

$$\left[\int_{z_1}^{z_2} \left[\int_{y_1}^{y_2} \left[\int_{x_1}^{x_2} f(x,y,z)\, dx \;① \right] dy \;② \right] dz \;③ \right]$$

3 Applications

(a) *Areas of plane figures*

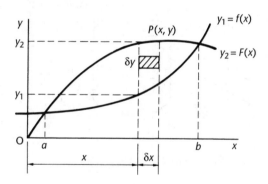

Area of element $\delta A = \delta x \delta y$

Area of strip $\approx \sum_{y=y_1}^{y=y_2} \delta x \delta y$

Area of all such strips $\approx \sum_{x=a}^{x=b} \left\{ \sum_{y=y_1}^{y=y_2} \delta x \delta y \right\}$

If $\delta x \to 0$ and $\delta y \to 0$, $A = \int_a^b \int_{y_1}^{y_2} dy\, dx$

▶

(b) *Areas of plane figures bounded by a polar curve $r = f(\theta)$ and radius vectors at $\theta = \theta_1$ and $\theta = \theta_2$*

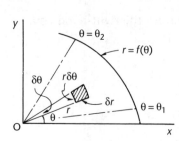

Small arc of circle of radius r, subtending angle $\delta\theta$ at center.

$\therefore$ Arc $= r\delta\theta$

Area of element $\delta A \approx r\delta\theta\,\delta r$

Area of thin sector $\approx \sum_{r=0}^{r=f(\theta)} r\,\delta\theta\,\delta r$

$\therefore$ Total area of all such sectors $\approx \sum_{\theta=\theta_1}^{\theta=\theta_2}\left\{\sum_{r=0}^{r=f(\theta)} r\,\delta r\,\delta\theta\right\}$

$\therefore$ If $\delta r \to 0$ and $\delta\theta \to 0$, $A = \int_{\theta_1}^{\theta_2}\int_0^{r=f(\theta)} r\,dr\,d\theta$

(c) *Volume of solids*

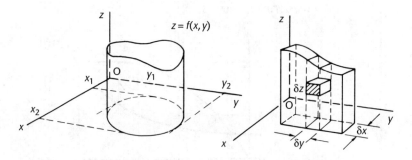

Volume of element $\delta V = \delta x\,\delta y\,\delta z$

Volume of column $\approx \sum_{z=0}^{z=f(x,\,y)} \delta x\,\delta y\,\delta z$

Volume of slice $\approx \sum_{y=y_1}^{y=y_2}\left\{\sum_{z=0}^{z=f(x,\,y)} \delta x\,\delta y\,\delta z\right\}$

$\therefore$ Total volume $V \approx$ sum of all such slices

i.e. $V \approx \sum_{x=x_1}^{x=x_2}\sum_{y=y_1}^{y=y_2}\sum_{z=0}^{z=f(x,\,y)} \delta x\,\delta y\,\delta z$

▶

Multiple integrals

Then, if $\delta x \to 0$, $\delta y \to 0$, $\delta z \to 0$,

$$V = \int_{x_1}^{x_2} \int_{y_1}^{y_2} \int_0^{z=f(x,y)} dz\,dy\,dx$$

If $z = f(x, y)$, this becomes

$$V = \int_{x_1}^{x_2} \int_{y_1}^{y_2} f(x, y)\,dy\,dx$$

✓ Can You?

Checklist 4   [47]

Check this list before and after you try the end of Program test.

On a scale of 1 to 5 how confident are you that you can: Frames

- Determine the area of a rectangle using a double integral?   [1] to [9]
  Yes ☐ ☐ ☐ ☐ ☐   No

- Evaluate double integrals over general areas?   [10] to [13]
  Yes ☐ ☐ ☐ ☐ ☐   No

- Evaluate triple integrals over general volumes?   [14] to [18]
  Yes ☐ ☐ ☐ ☐ ☐   No

- Apply double integrals to find areas and second moments of area?   [19] to [38]
  Yes ☐ ☐ ☐ ☐ ☐   No

- Apply triple integrals to find volumes?   [39] to [45]
  Yes ☐ ☐ ☐ ☐ ☐   No

🚴 Test exercise 4

The questions are just like those you have been doing quite successfully. They are all quite straightforward and should cause you no trouble.   [48]

1 Evaluate:

(a) $\displaystyle\int_1^3 \int_0^2 (y^3 - xy)\,dy\,dx$

(b) $\displaystyle\int_0^a dx \int_0^{y_1} (x - y)\,dy$, where $y_1 = \sqrt{a^2 - x^2}$

▶

2 Determine:

(a) $\int_0^{\sqrt{3}+2} \int_0^{\pi/3} (2\cos\theta - 3\sin 3\theta)\, d\theta\, dr$

(b) $\int_2^4 \int_1^2 \int_0^4 xy(z+2)\, dx\, dy\, dz$

(c) $\int_0^1 dz \int_1^2 dx \int_0^x (x+y+z)\, dy$

3 The line $y = 2x$ and the parabola $y^2 = 16x$ intersect at $x = 4$. Find by a double integral, the area enclosed by $y = 2x$, $y^2 = 16x$ and the ordinate at $x = 1$, and the point of intersection at $x = 4$.

4 A triangle is bounded by the x-axis, the line $y = 2x$ and the ordinate at $x = 4$. Build up a double integral representing the second moment of area of this triangle about the x-axis and evaluate the integral.

5 Form a double integral to represent the area of the plane figure bounded by the polar curve $r = 3 + 2\cos\theta$ and the radius vectors at $\theta = 0$ and $\theta = \pi/2$, and evaluate it.

6 A solid is enclosed by the planes $z = 0$, $y = 1$, $y = 3$, $x = 0$, $x = 3$ and the surface $z = x^2 + xy$. Calculate the volume of the solid.

That's it!

Further problems 4

49

1 Evaluate $\int_0^\pi \int_0^{\cos\theta} r\sin\theta\, dr\, d\theta$

2 Evaluate $\int_0^{2\pi} \int_0^3 r^3(9 - r^2)\, dr\, d\theta$

3 Evaluate $\int_{-2}^1 \int_{x^2+4x}^{3x+2} dy\, dx$

4 Evaluate $\int_0^a \int_0^b \int_0^c (x^2 + y^2)\, dx\, dy\, dz$

5 Evaluate $\int_0^\pi \int_0^{\pi/2} \int_0^r x^2 \sin\theta\, dx\, d\theta\, d\phi$

6 Find the area bounded by the curve $y = x^2$ and the line $y = x + 2$.

7 Find the area of the polar figure enclosed by the circle $r = 2$ and the cardioid $r = 2(1 + \cos\theta)$.

8 Evaluate $\int_0^2 dx \int_1^3 dy \int_1^2 xy^2 z\, dz$

Multiple integrals

9 Evaluate $\displaystyle\int_0^2 dx \int_1^2 (x^2 + y^2)\, dy$

10 Evaluate $\displaystyle\int_0^1 dr \int_0^{\pi/4} r\cos^2\theta\, d\theta$

11 Determine the area bounded by the curves $x = y^2$ and $x = 2y - y^2$.

12 Express as a double integral, the area contained by one loop of the curve $r = 2\cos 3\theta$, and evaluate the integral.

13 Evaluate $\displaystyle\int_0^{\pi/2} \int_{\pi/4}^{\tan^{-1}(2)} \int_0^4 x\sin y\, dx\, dy\, dz$

14 Evaluate $\displaystyle\int_0^\pi \int_0^{4\cos z} \int_0^{\sqrt{16-y^2}} y\, dx\, dy\, dz$

15 A plane figure is bounded by the polar curve $r = a(1 + \cos\theta)$ between $\theta = 0$ and $\theta = \pi$, and the initial line OA. Express as a double integral the first moment of area of the figure about OA, and evaluate the integral. If the area of the figure is known to be $\dfrac{3\pi a^2}{4}$ unit^2, find the distance (h) of the centroid of the figure from OA.

16 Using double integrals, find (a) the area and (b) the second moment about OX of the plane figure bounded by the x-axis and that part of the ellipse $\dfrac{x^2}{a^2} + \dfrac{y^2}{b^2} = 1$ which lies above OX. Find also the position of the centroid.

17 The base of a solid is the plane figure in the x–y plane bounded by $x = 0$, $x = 2$, $y = x$ and $y = x^2 + 1$. The sides are vertical and the top is the surface $z = x^2 + y^2$. Calculate the volume of the solid so formed.

18 A solid consists of vertical sides standing on the plane figure enclosed by $x = 0$, $x = b$, $y = a$ and $y = c$. The top is the surface $z = xy$. Find the volume of the solid so defined.

19 Show that the area outside the circle $r = a$ and inside the circle $r = 2a\cos\theta$ is given by

$$A = 2\int_0^{\pi/3} \int_a^{2a\cos\theta} r\, dr\, d\theta$$

Evaluate the integral.

20 A rectangular block is bounded by the coordinate planes of reference and by the planes $x = 3$, $y = 4$, $z = 2$. Its density at any point is numerically equal to the square of its distance from the origin. Find the total mass of the solid.

Program 5

Differentials and line integrals

Frames
1 to 73

Learning outcomes

When you have completed this Program you will be able to:

- Understand the role of the differential of a function of two or more real variables
- Determine exact differentials in two real variables and their integrals
- Evaluate the area enclosed by a closed curve by contour integration
- Evaluate line integrals and appreciate their properties
- Evaluate line integrals around closed curves within a simply connected region
- Link line integrals to integrals along the *x*-axis
- Link line integrals to integrals along a contour given in parametric form
- Discuss the dependence of a line integral between two points on the path of integration
- Determine exact differentials in three real variables and their integrals
- Demonstrate the validity and use of Green's theorem

Differentials and line integrals

Differentials

[1]

It is convenient in various branches of the calculus to denote small increases in value of a variable by the use of *differentials*. The method is particularly useful in dealing with the effects of small finite changes and shortens the writing of calculus expressions.

We are already familiar with the diagram from which finite changes δy and δx in a function $y = f(x)$ are depicted.

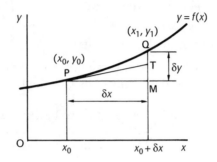

The increase in y from P to Q = MQ = $\delta y = f(x_0 + \delta x) - f(x_0)$

If PT is the tangent at P, then MQ = MT + TQ. Also $\dfrac{MT}{\delta x} = f'(x_0)$

$\therefore$ MT = $f'(x_0)\delta x$
$\therefore$ MQ = $\delta y = f'(x_0) \cdot \delta x + TQ$

and, if Q is close to P, then $\delta y \approx f'(x_0)\delta x$

We define the differentials dy and dx as finite quantities such that

$$dy = f'(x_0)\,dx$$

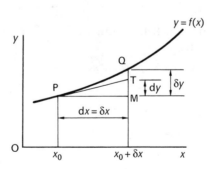

Note that the differentials dy and dx are finite quantities – not necessarily zero – and can therefore exist alone.

Note too that $dx = \delta x$.

▶

From the diagram, we can see that

δy is the increase in y as we move from P to Q along the curve.
dy is the increase in y as we move from P to T along the tangent.

As Q approaches P, the difference between δy and dy decreases to zero. The use of differentials simplifies the writing of many relationships and is based on the general statement $dy = f'(x)\,dx$.

For example

(a) $y = x^5$ then $dy = 5x^4\,dx$
(b) $y = \sin 3x$ then $dy = 3\cos 3x\,dx$
(c) $y = e^{4x}$ then $dy = 4e^{4x}\,dx$
(d) $y = \cosh 2x$ then $dy = 2\sinh 2x\,dx$

Note that when the left-hand side is a differential dy the right-hand side must also contain a differential. Remember therefore to include the 'dx' on the right-hand side.

The product and quotient rules can also be expressed in differentials.

$$\frac{d}{dx}(uv) = u\frac{dv}{dx} + v\frac{du}{dx} \quad \text{becomes} \quad d(uv) = u\,dv + v\,du$$

$$\frac{d}{dx}\left(\frac{u}{v}\right) = \frac{v\frac{du}{dx} - u\frac{dv}{dx}}{v^2} \quad \text{becomes} \quad d\left(\frac{u}{v}\right) = \frac{v\,du - u\,dv}{v^2}$$

So, if $y = e^{2x}\sin 4x$, $dy = \ldots\ldots\ldots\ldots$

and if $y = \dfrac{\cos 2t}{t^2}$ $dy = \ldots\ldots\ldots\ldots$

2

$y = e^{2x}\sin 4x$, $dy = 2e^{2x}(2\cos 4x + \sin 4x)\,dx$

$y = \dfrac{\cos 2t}{t^2}$, $dy = -\dfrac{2}{t^3}\{t\sin 2t + \cos 2t\}\,dt$

That was easy enough. Let us now consider a function of two independent variables, $z = f(x, y)$.

If $z = f(x, y)$ then $z + \delta z = f(x + \delta x, y + \delta y)$
$$\therefore \quad \delta z = f(x + \delta x, y + \delta y) - f(x, y)$$

Expanding δz in terms of δx and δy, gives

$\delta z = A\delta x + B\delta y +$ higher powers of δx and δy,

where A and B are functions of x and y.

If y remains constant, i.e. $\delta y = 0$, then

$$\delta z = A\,\delta x + \text{higher powers of } \delta x \quad \therefore \quad \frac{\delta z}{\delta x} \approx A$$

$$\therefore \text{ If } \delta x \to 0, \text{ then } A = \frac{\partial z}{\partial x}$$

▶

Differentials and line integrals

Similarly, if x remains constant, i.e. $\delta x = 0$, then

$$\delta z = B\,\delta y + \text{higher powers of } \delta y \quad \therefore \quad \frac{\delta z}{\delta y} \approx B$$

$\therefore$ If $\delta y \to 0$, then $B = \dfrac{\partial z}{\partial y}$

$\therefore \delta z = \dfrac{\partial z}{\partial x}\delta x + \dfrac{\partial z}{\partial y}\delta y + \text{higher powers of small quantities}$

$\therefore \delta z = \dfrac{\partial z}{\partial x}\delta x + \dfrac{\partial z}{\partial y}\delta y$

In terms of differentials, this result can be written

If $z = f(x, y)$, then $dz = \dfrac{\partial z}{\partial x}dx + \dfrac{\partial z}{\partial y}dy$

The result can be extended to functions of more than two independent variables.

$$\text{If } z = f(x, y, w), \quad dz = \frac{\partial z}{\partial x}dx + \frac{\partial z}{\partial y}dy + \frac{\partial z}{\partial w}dw$$

Make a note of these results in differential form as shown.

Exercise

Determine the differential dz for each of the following functions.

1. $z = x^2 + y^2$
2. $z = x^3 \sin 2y$
3. $z = (2x - 1)e^{3y}$
4. $z = x^2 + 2y^2 + 3w^2$
5. $z = x^3 y^2 w$

Finish all five and then check the results.

---

1. $dz = 2(x\,dx + y\,dy)$
2. $dz = x^2(3\sin 2y\,dx + 2x\cos 2y\,dy)$
3. $dz = e^{3y}\{2\,dx + (6x - 3)dy\}$
4. $dz = 2(x\,dx + 2y\,dy + 3w\,dw)$
5. $dz = x^2 y(3yw\,dx + 2xw\,dy + xy\,dw)$

Now move on

4 Exact differential

We have just established that if $z = f(x, y)$

$$dz = \frac{\partial z}{\partial x} dx + \frac{\partial z}{\partial y} dy$$

We now work in reverse.

Any expression $dz = P\, dx + Q\, dy$, where P and Q are functions of x and y, is an *exact differential* if it can be integrated to determine z.

$$\therefore P = \frac{\partial z}{\partial x} \quad \text{and} \quad Q = \frac{\partial z}{\partial y}$$

Now $\dfrac{\partial P}{\partial y} = \dfrac{\partial^2 z}{\partial y\, \partial x}$ and $\dfrac{\partial Q}{\partial x} = \dfrac{\partial^2 z}{\partial x\, \partial y}$ and we know that $\dfrac{\partial^2 z}{\partial y\, \partial x} = \dfrac{\partial^2 z}{\partial x\, \partial y}$.

Therefore, for dz to be an exact differential $\dfrac{\partial P}{\partial y} = \dfrac{\partial Q}{\partial x}$ and this is the test we apply.

Example 1

$dz = (3x^2 + 4y^2)\, dx + 8xy\, dy$.

If we compare the right-hand side with $P\, dx + Q\, dy$, then

$$P = 3x^2 + 4y^2 \quad \therefore \frac{\partial P}{\partial y} = 8y$$

$$Q = 8xy \quad \therefore \frac{\partial Q}{\partial x} = 8y$$

$\dfrac{\partial P}{\partial y} = \dfrac{\partial Q}{\partial x}$ $\quad \therefore dz$ is an exact differential

Similarly, we can test this one.

Example 2

$dz = (1 + 8xy)\, dx + 5x^2\, dy$.

From this we find ....

5

$$\boxed{dz \text{ is } not \text{ an exact differential}}$$

Because $dz = (1 + 8xy)\, dx + 5x^2\, dy$

$$\therefore P = 1 + 8xy \quad \therefore \frac{\partial P}{\partial y} = 8x$$

$$Q = 5x^2 \quad \therefore \frac{\partial Q}{\partial x} = 10x$$

$\dfrac{\partial P}{\partial y} \neq \dfrac{\partial Q}{\partial x}$ $\quad \therefore dz$ is not an exact differential.

Differentials and line integrals

Exercise

Determine whether each of the following is an exact differential.
1. $dz = 4x^3y^3\,dx + 3x^4y^2\,dy$
2. $dz = (4x^3y + 2xy^3)\,dx + (x^4 + 3x^2y^2)\,dy$
3. $dz = (15y^2e^{3x} + 2xy^2)\,dx + (10ye^{3x} + x^2y)\,dy$
4. $dz = (3x^2e^{2y} - 2y^2e^{3x})\,dx + (2x^3e^{2y} - 2ye^{3x})\,dy$
5. $dz = (4y^3\cos 4x + 3x^2\cos 2y)\,dx + (3y^2\sin 4x - 2x^3\sin 2y)\,dy$

| 1 Yes | 2 Yes | 3 No | 4 No | 5 Yes |

We have just tested whether certain expressions are, in fact, exact differentials – and we said previously that, by definition, an exact differential can be integrated. But how exactly do we go about it? The following examples will show.

Integration of exact differentials

$$dz = P\,dx + Q\,dy \quad \text{where} \quad P = \frac{\partial z}{\partial x} \quad \text{and} \quad Q = \frac{\partial z}{\partial y}$$

$$\therefore z = \int P\,dx \quad \text{and also} \quad z = \int Q\,dy$$

Example 1

$dz = (2xy + 6x)\,dx + (x^2 + 2y^3)\,dy$.

$$P = \frac{\partial z}{\partial x} = 2xy + 6x \quad \therefore z = \int (2xy + 6x)\,dx$$

$\therefore z = x^2y + 3x^2 + f(y)$ where $f(y)$ is an arbitrary function of y only, and is akin to the constant of integration in a normal integral.

$$\text{Also} \quad Q = \frac{\partial z}{\partial y} = x^2 + 2y^3 \quad \therefore z = \int (x^2 + 2y^3)\,dy$$

$$\therefore z = \ldots\ldots\ldots\ldots$$

114 Vector Analysis

7

$$z = x^2y + \frac{y^4}{2} + F(x) \text{ where } F(x) \text{ is an arbitrary function of } x \text{ only}$$

So the two results tell us
$$z = x^2y + 3x^2 + f(y) \tag{1}$$
and
$$z = x^2y + \frac{y^4}{2} + F(x) \tag{2}$$

For these two expressions to represent the same function, then

$$f(y) \text{ in (1) must be } \frac{y^4}{2} \text{ already in (2)}$$

and $F(x)$ in (2) must be $3x^2$ already in (1)

$$\therefore z = x^2y + 3x^2 + \frac{y^4}{2}$$

Example 2

Integrate $dz = (8e^{4x} + 2xy^2)\,dx + (4\cos 4y + 2x^2y)\,dy$.

Argue through the working in just the same way, from which we obtain

$$z = \ldots\ldots\ldots\ldots$$

8

$$z = 2e^{4x} + x^2y^2 + \sin 4y$$

Here it is. $dz = (8e^{4x} + 2xy^2)\,dx + (4\cos 4y + 2x^2y)\,dy$

$$P = \frac{\partial z}{\partial x} = 8e^{4x} + 2xy^2 \quad \therefore z = \int (8e^{4x} + 2xy^2)\,dx$$

$$\therefore z = 2e^{4x} + x^2y^2 + f(y) \tag{1}$$

$$Q = \frac{\partial z}{\partial y} = 4\cos 4y + 2x^2y \quad \therefore z = \int (4\cos 4y + 2x^2y)\,dy$$

$$\therefore z = \sin 4y + x^2y^2 + F(x) \tag{2}$$

For (1) and (2) to agree, $f(y) = \sin 4y$ and $F(x) = 2e^{4x}$

$$\therefore z = 2e^{4x} + x^2y^2 + \sin 4y$$

They are all done in the same way, so you will have no difficulty with the short exercise that follows. *On you go.*

Exercise

Integrate the following exact differentials to obtain the function z.

1. $dz = (6x^2 + 8xy^3)\,dx + (12x^2y^2 + 12y^3)\,dy$
2. $dz = (3x^2 + 2xy + y^2)\,dx + (x^2 + 2xy + 3y^2)\,dy$
3. $dz = 2(y+1)e^{2x}\,dx + (e^{2x} - 2y)\,dy$
4. $dz = (3y^2\cos 3x - 3\sin 3x)\,dx + (2y\sin 3x + 4)\,dy$
5. $dz = (\sinh y + y\sinh x)\,dx + (x\cosh y + \cosh x)\,dy$

Finish all five before checking with the next frame.

Differentials and line integrals

9

1. $z = 2x^3 + 4x^2y^3 + 3y^4$
2. $z = x^3 + x^2y + xy^2 + y^3$
3. $z = e^{2x}(1+y) - y^2$
4. $z = y^2 \sin 3x + \cos 3x + 4y$
5. $z = x \sinh y + y \cosh x$

In the last one, of course, we find that the two expressions for z agree without any further addition of $f(y)$ or $F(x)$.

We shall be meeting exact differentials again later on, but for the moment let us deal with something different. On then to the next frame

Area enclosed by a closed curve

10

One of the earliest applications of integration is finding the area of a plane figure bounded by the x-axis, the curve $y = f(x)$ and ordinates at $x = x_1$ and $x = x_2$.

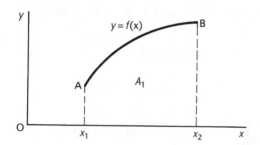

$$A_1 = \int_{x_1}^{x_2} y \, dx = \int_{x_1}^{x_2} f(x) \, dx$$

If points A and B are joined by another curve, $y = F(x)$

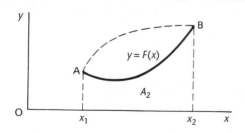

$$A_2 = \int_{x_1}^{x_2} F(x) \, dx$$

▶

Combining the two figures, we have

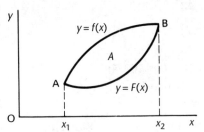

$$A = A_1 - A_2$$
$$\therefore A = \int_{x_1}^{x_2} f(x)\,dx - \int_{x_1}^{x_2} F(x)\,dx$$

It is convenient on occasions to arrange the limits so that the integration follows the path round the enclosed area in a regular order.

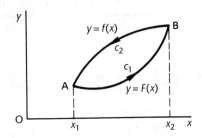

For example

$\int_{x_1}^{x_2} F(x)\,dx$ gives A_2 as before, but integrating from B to A along c_2 with $y = f(x)$, i.e. $\int_{x_2}^{x_1} f(x)\,dx$, is the integral for A_1 with the sign changed

i.e. $\int_{x_2}^{x_1} f(x)\,dx = -\int_{x_1}^{x_2} f(x)\,dx$

$\therefore$ The result $A = A_1 - A_2 = \int_{x_1}^{x_2} f(x)\,dx - \int_{x_1}^{x_2} F(x)\,dx$ becomes

$$A = \ldots\ldots\ldots\ldots$$

11

$$A = -\int_{x_1}^{x_2} F(x)\,dx - \int_{x_2}^{x_1} f(x)\,dx$$

i.e. $A = -\left\{\int_{x_1}^{x_2} F(x)\,dx + \int_{x_2}^{x_1} f(x)\,dx\right\}$

If we proceed round the boundary in an *anticlockwise manner*, the enclosed area is kept on the *left-hand side* and the resulting area is considered *positive*. If we proceed round the boundary in a *clockwise manner*, the enclosed area remains on the *right-hand side* and the resulting area is *negative*.

Differentials and line integrals

The final result above can be written in the form

$$A = -\oint y\,dx$$

where the symbol $\oint$ indicates that the integral is to be evaluated round the closed boundary in the positive (i.e. anticlockwise) direction

$$\therefore A = -\oint y\,dx = -\left\{\int_{x_1}^{x_2} F(x)\,dx + \int_{x_2}^{x_1} f(x)\,dx\right\}$$
$$\text{(along } c_1\text{)} \quad \text{(along } c_2\text{)}$$

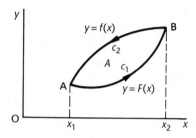

Let us apply this result to a very simple case.

Example 1

Determine the area enclosed by the graphs of $y = x^3$ and $y = 4x$ for $x \geq 0$.
First we need to know the points of intersection. These are

....

$$x = 0 \text{ and } x = 2$$

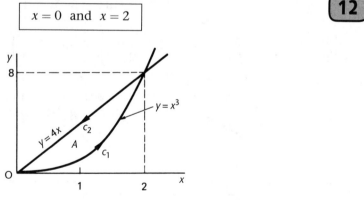

We integrate in an anticlockwise manner
c_1: $y = x^3$, limits $x = 0$ to $x = 2$
c_2: $y = 4x$, limits $x = 2$ to $x = 0$

$$A = -\oint y\,dx = \ldots\ldots\ldots$$

13

$$A = 4 \text{ square units}$$

Because

$$A = -\oint y\,dx = -\left\{\int_0^2 x^3\,dx + \int_2^0 4x\,dx\right\}$$

$$= -\left\{\left[\frac{x^4}{4}\right]_0^2 + \left[2x^2\right]_2^0\right\} = 4$$

Another example.

Example 2

Find the area of the triangle with vertices (0, 0), (5, 3) and (2, 6).

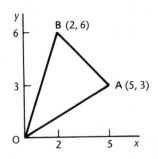

The equation of

OA is ....
BA is ....
OB is ....

14

OA is $y = \frac{3}{5}x$
BA is $y = 8 - x$
OB is $y = 3x$

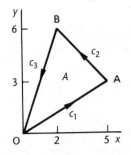

Then $A = -\oint y\,dx$

= ....

Write down the component integrals with appropriate limits.

15

$$A = -\oint y\,dx = -\left\{\int_0^5 \frac{3}{5}x\,dx + \int_5^2 (8-x)\,dx + \int_2^0 3x\,dx\right\}$$

The limits chosen must progress the integration round the boundary of the figure in an *anticlockwise* manner. Finishing off the integration, we have

$$A = \ldots\ldots\ldots$$

Differentials and line integrals　　　　　　　　　　　　　　　　　　　　　　119

16

$$A = 12 \text{ square units}$$

The actual integration is easy enough.

*The work we have just done leads us on to consider **line integrals**, so let us make a fresh start in the next frame*

Line integrals

17

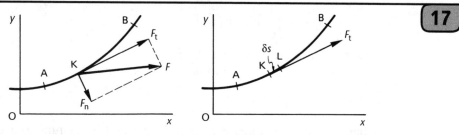

If a field exists in the x–y plane, producing a force F on a particle at K, then F can be resolved into two components

F_t along the tangent to the curve AB at K
F_n along the normal to the curve AB at K

The work done in moving the particle through a small distance δs from K to L along the curve is then approximately $F_t\, \delta s$. So the total work done in moving a particle along the curve from A to B is given by

....

18

$$\lim_{\delta s \to 0} \sum F_t\, \delta s = \int F_t\, ds \text{ from A to B}$$

This is normally written $\int_{AB} F_t\, ds$ where A and B are the end points of the curve, or as $\int_c F_t\, ds$ where the curve c connecting A and B is defined.

Such an integral thus formed is called a *line integral* since integration is carried out along the path of the particular curve c joining A and B.

$$\therefore I = \int_{AB} F_t\, ds = \int_c F_t\, ds$$

where c is the curve $y = f(x)$ between A (x_1, y_1) and B (x_2, y_2).

There is in fact an alternative form of the integral which is often useful, so let us also consider that

19 Alternative form of a line integral

It is often more convenient to integrate with respect to x or y than to take arc length as the variable.

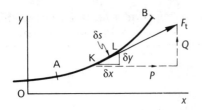

If F_t has a component

P in the x-direction

Q in the y-direction

then the work done from K to L can be stated as $P\,\delta x + Q\,\delta y$.

$$\therefore \int_{AB} F_t\,ds = \int_{AB} (P\,dx + Q\,dy)$$

where P and Q are functions of x and y.

In general then, the line integral can be expressed as

$$I = \int_c F_t\,ds = \int_c (P\,dx + Q\,dy)$$

where c is the prescribed curve and F, or P and Q, are functions of x and y.

Make a note of these results – then we will apply them to one or two examples

20 Example 1

Evaluate $\int_c (x+3y)\,dx$ from A (0, 1) to B (2, 5) along the curve $y = 1 + x^2$.

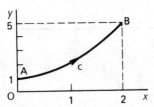

The line integral is of the form

$$\int_c (P\,dx + Q\,dy)$$

where, in this case, $Q = 0$ and c is the curve $y = 1 + x^2$.

It can be converted at once into an ordinary integral by substituting for y and applying the appropriate limits of x.

$$I = \int_c (P\,dx + Q\,dy) = \int_c (x+3y)\,dx = \int_0^2 (x + 3 + 3x^2)\,dx$$

$$= \left[\frac{x^2}{2} + 3x + x^3\right]_0^2 = 16$$

Now for another, so move on

Differentials and line integrals

Example 2

Evaluate $I = \int_c (x^2 + y)\,dx + (x - y^2)\,dy$ from A (0, 2) to B (3, 5) along the curve $y = 2 + x$.

$I = \int_c (P\,dx + Q\,dy)$

$P = x^2 + y = x^2 + 2 + x = x^2 + x + 2$

$Q = x - y^2 = x - (4 + 4x + x^2)$
$= -(x^2 + 3x + 4)$

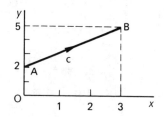

Also $y = 2 + x$ $\therefore$ $dy = dx$ and the limits are $x = 0$ to $x = 3$.

$$\therefore I = \ldots\ldots\ldots$$

$$\boxed{I = -15}$$

Because

$I = \int_0^3 \{(x^2 + x + 2)\,dx - (x^2 + 3x + 4)\,dx\}$

$\int_0^3 -(2x + 2)\,dx = \left[x^2 - 2x\right]_0^3 = -15$

Here is another.

Example 3

Evaluate $I = \int_c \{(x^2 + 2y)\,dx + xy\,dy\}$ from O (0, 0) to B (1, 4) along the curve $y = 4x^2$.

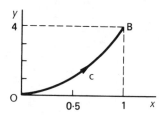

In this case, c is the curve $y = 4x^2$.

$\therefore dy = 8x\,dx$

Substitute for y in the integral and apply the limits.

Then $I = \ldots\ldots\ldots$

Finish it off: it is quite straightforward.

23

$$I = 9.4$$

Because

$$I = \int_c \{(x^2 + 2y)\,dx + xy\,dy\} \qquad y = 4x^2 \quad \therefore dy = 8x\,dx$$

Also $x^2 + 2y = x^2 + 8x^2 = 9x^2$; $\quad xy = 4x^3$

$$\therefore I = \int_0^1 \{9x^2\,dx + 32x^4\,dx\} = \int_0^1 (9x^2 + 32x^4)\,dx = 9.4$$

They are all done in very much the same way.

Move on for Example 4

24

Example 4

Evaluate $I = \int_c \{(x^2 + 2y)\,dx + xy\,dy\}$ from O (0, 0) to A (1, 0) along line $y = 0$ and then from A (1, 0) to B (1, 4) along the line $x = 1$.

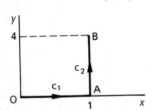

(1) OA: c_1 is the line $y = 0$ $\therefore dy = 0$. Substituting $y = 0$ and $dy = 0$ in the given integral gives

$$I_{OA} = \int_0^1 x^2\,dx = \left[\frac{x^3}{3}\right]_0^1 = \frac{1}{3}$$

(2) AB: Here c_2 is the line $x = 1$ $\quad \therefore dx = 0$

$$\therefore I_{AB} = \ldots\ldots\ldots\ldots$$

25

$$I_{AB} = 8$$

Because

$$I_{AB} = \int_0^4 \{(1 + 2y)(0) + y\,dy\}$$

$$= \int_0^4 y\,dy$$

$$= \left[\frac{y^2}{2}\right]_0^4 = 8$$

Then $I = I_{OA} + I_{AB} = \frac{1}{3} + 8 = 8\frac{1}{3}$ $\quad \therefore I = 8\frac{1}{3}$

If we now look back to Examples 3 and 4 just completed, we find that we have evaluated the same integral between the same two end points, but ....

Differentials and line integrals

along different paths of integration

If we combine the two diagrams, we have

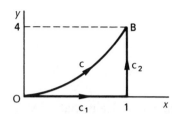

where c is the curve $y = 4x^2$ and $c_1 + c_2$ are the lines $y = 0$ and $x = 1$.

The results obtained were

$$I_c = 9\tfrac{2}{5} \text{ and } I_{c_1 + c_2} = 8\tfrac{1}{3}$$

Notice therefore that integration along two distinct paths joining the same two end points does not necessarily give the same results.

Let us pause here a moment and list the main properties of line integrals.

Properties of line integrals

1. $\displaystyle\int_c F\,ds = \int_c \{P\,dx + Q\,dy\}$

2. $\displaystyle\int_{AB} F\,ds = -\int_{BA} F\,ds$ and $\displaystyle\int_{AB}\{P\,dx + Q\,dy\} = -\int_{BA}\{P\,dx + Q\,dy\}$

   i.e. the sign of a line integral is reversed when the direction of the integration along the path is reversed.

3. (a) For a path of integration parallel to the y-axis, i.e. $x = k$,

   $$dx = 0 \quad \therefore \int_c P\,dx = 0 \quad \therefore I_c = \int_c Q\,dy$$

   (b) For a path of integration parallel to the x-axis, i.e. $y = k$,

   $$dy = 0 \quad \therefore \int_c Q\,dy = 0 \quad \therefore I_c = \int_c P\,dx$$

4. If the path of integration c joining A to B is divided into two parts AK and KB, then $I_c = I_{AB} = I_{AK} + I_{KB}$.

5. In all cases, the function $y = f(x)$ that describes the path of integration involved must be continuous and single-valued – or dealt with as in item **6** below.

6. If the function $y = f(x)$ that describes the path of integration c is not single-valued for part of its extent, the path is divided into two sections.

   $y = f_1(x)$ from A to K
   $y = f_2(x)$ from K to B

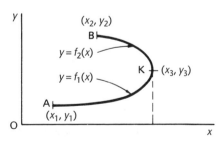

Make a note of this list for future reference and revision

28

Example

Evaluate $I = \int_c (x+y)\,dx$ from A $(0, 1)$ to B $(0, -1)$ along the semi-circle $x^2 + y^2 = 1$ for $x \geq 0$.

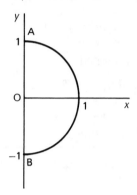

The first thing we notice is that

....

29

| the function $y = f(x)$ that describes the path of integration c is *not* single-valued |

For any value of x, $y = \pm\sqrt{1-x^2}$. Therefore, we divide c into two parts

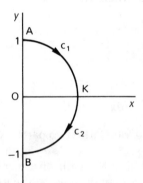

(1) $y = \sqrt{1-x^2}$ from A to K

(2) $y = -\sqrt{1-x^2}$ from K to B.

As usual, $I = \int_c (P\,dx + Q\,dy)$ and in this particular case, $Q = $ ....

30

$$Q = 0$$

$$\therefore I = \int_c P\,dx = \int_0^1 \left(x + \sqrt{1-x^2}\right) dx + \int_1^0 \left(x - \sqrt{1-x^2}\right) dx$$

$$= \int_0^1 (x + \sqrt{1-x^2} - x + \sqrt{1-x^2})\,dx = 2\int_0^1 \sqrt{1-x^2}\,dx$$

Now substitute $x = \sin\theta$ and finish it off.

$$I = \text{............}$$

$$I = \frac{\pi}{2}$$

Because

$$I = 2\int_0^1 \sqrt{1-x^2}\,dx \quad x = \sin\theta \quad \therefore dx = \cos\theta\,d\theta$$

$$\sqrt{1-x^2} = \cos\theta$$

Limits: $x = 0, \theta = 0;\ x = 1, \theta = \dfrac{\pi}{2}$

$$\therefore I = 2\int_0^{\pi/2} \cos^2\theta\,d\theta = \int_0^{\pi/2} (1 + \cos 2\theta)\,d\theta$$

$$= \left[\theta + \frac{\sin 2\theta}{2}\right]_0^{\pi/2}$$

$$= \frac{\pi}{2}$$

Now let us extend this line of development a stage further.

Regions enclosed by closed curves

A region is said to be *simply connected* if a path joining A and B can be deformed to coincide with any other line joining A and B without going outside the region.

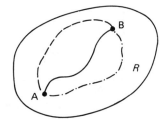

Another definition is that a region is simply connected if any closed path in the region can be contracted to a single point without leaving the region.

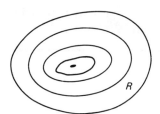

Clearly, this would not be satisfied in the case where the region R contains one or more 'holes'.

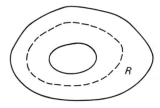

The closed curves involved in problems in this Program all relate to simply connected regions, so no difficulties will arise.

33 Line integrals round a closed curve

We have already introduced the symbol $\oint$ to indicate that an integral is to be evaluated round a closed curve in the positive (anticlockwise) direction.

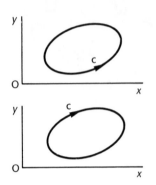

Positive direction (anticlockwise) line integral denoted by $\oint$.

Negative direction (clockwise) line integral denoted by $-\oint$.

With a closed curve, the y-values on the path c cannot be single-valued. Therefore, we divide the path into two or more parts and treat each separately.

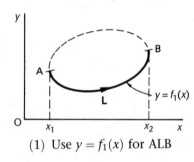   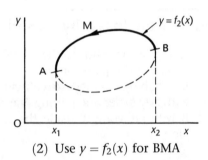

(1) Use $y = f_1(x)$ for ALB   (2) Use $y = f_2(x)$ for BMA

Unless specially required otherwise, we always proceed round the closed curve in an ....

34

anticlockwise direction

Example 1

Evaluate the line integral $I = \oint_c (x^2 \, dx - 2xy \, dy)$ where c comprises the three sides of the triangle joining O (0, 0), A (1, 0) and B (0, 1).

First draw the diagram and mark in c_1, c_2 and c_3, the proposed directions of integration. Do just that.

Differentials and line integrals

35

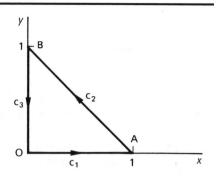

The three sections of the path of integration must be arranged in an anticlockwise manner round the figure. Now we deal with each part separately.

(a) OA: c_1 is the line $y = 0$ ∴ $dy = 0$

Then $I = \oint (x^2\, dx - 2xy\, dy)$ for this part becomes

$$I_1 = \int_0^1 x^2\, dx = \left[\frac{x^3}{3}\right]_0^1 = \frac{1}{3} \quad \therefore I_1 = \frac{1}{3}$$

(b) AB: c_2 is the line $y = 1 - x$ ∴ $dy = -dx$

$$I_2 = \ldots\ldots\ldots\ldots \quad \text{(evaluate it)}$$

36

$$\boxed{I_2 = -\tfrac{2}{3}}$$

Because c_2 is the line $y = 1 - x$ ∴ $dy = -dx$

$$I_2 = \int_1^0 \{x^2\, dx + 2x(1-x)\, dx\} = \int_1^0 (x^2 + 2x - 2x^2)\, dx$$

$$= \int_1^0 (2x - x^2)\, dx = \left[x^2 - \frac{x^3}{3}\right]_1^0 = -\frac{2}{3} \quad \therefore I_2 = -\frac{2}{3}$$

Note that anticlockwise progression is obtained by arranging the limits in the appropriate order.

Now we have to determine I_3 for BO.

(c) BO: c_3 is the line $x = 0$

$$I_3 = \ldots\ldots\ldots\ldots$$

37

$$\boxed{I_3 = 0}$$

Because for c_3, $x = 0$ ∴ $dx = 0$ ∴ $I_3 = \int 0\, dy = 0$ ∴ $I_3 = 0$

Finally, $I = I_1 + I_2 + I_3 = \tfrac{1}{3} - \tfrac{2}{3} + 0 = -\tfrac{1}{3}$ ∴ $I = -\tfrac{1}{3}$

Let us work through another example.

Example 2

Evaluate $\oint_c y\,dx$ when c is the circle $x^2 + y^2 = 4$.

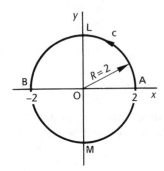

$x^2 + y^2 = 4$ ∴ $y = \pm\sqrt{4 - x^2}$

y is thus not single-valued. Therefore use $y = \sqrt{4 - x^2}$ for ALB between $x = 2$ and $x = -2$ and $y = -\sqrt{4 - x^2}$ for BMA between $x = -2$ and $x = 2$.

$$\therefore I = \int_2^{-2} \sqrt{4 - x^2}\,dx + \int_{-2}^{2} \{-\sqrt{4 - x^2}\}\,dx$$

$$= 2\int_2^{-2} \sqrt{4 - x^2}\,dx = -2\int_{-2}^{2} \sqrt{4 - x^2}\,dx$$

$$= -4\int_0^2 \sqrt{4 - x^2}\,dx$$

To evaluate this integral, substitute $x = 2\sin\theta$ and finish it off.

$$I = \ldots\ldots\ldots$$

38

$$\boxed{I = -4\pi}$$

Because

$x = 2\sin\theta$ ∴ $dx = 2\cos\theta\,d\theta$ ∴ $\sqrt{4 - x^2} = 2\cos\theta$

limits: $x = 0$, $\theta = 0$; $x = 2$, $\theta = \dfrac{\pi}{2}$

$$\therefore I = -4\int_0^{\pi/2} 2\cos\theta\, 2\cos\theta\,d\theta = -16\int_0^{\pi/2} \cos^2\theta\,d\theta$$

$$= -8\int_0^{\pi/2} (1 + \cos 2\theta)\,d\theta = -8\left[\theta + \frac{\sin 2\theta}{2}\right]_0^{\pi/2} = -4\pi$$

Now for one more

Example 3

Evaluate $I = \oint_c \{xy\,dx + (1 + y^2)\,dy\}$ where c is the boundary of the rectangle joining A (1, 0), B (3, 0), C (3, 2) and D (1, 2).

First draw the diagram and insert c_1, c_2, c_3, c_4.

That gives ....

Differentials and line integrals

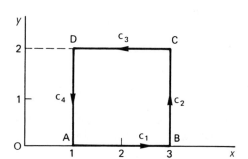

Now evaluate I_1 for AB; I_2 for BC; I_3 for CD; I_4 for DA; and finally I.

Complete the working and then check with the next frame

$$I_1 = 0; \quad I_2 = 4\tfrac{2}{3}; \quad I_3 = -8; \quad I_4 = -4\tfrac{2}{3}; \quad I = -8$$

Here is the complete working.

$$I = \oint_c \{xy\,dx + (1+y^2)\,dy\}$$

(a) AB: c_1 is $y = 0$ $\therefore dy = 0$ $\therefore I_1 = 0$

(b) BC: c_2 is $x = 3$ $\therefore dx = 0$

$$\therefore I_2 = \int_0^2 (1+y^2)\,dy = \left[y + \frac{y^3}{3}\right]_0^2 = 4\tfrac{2}{3} \quad \therefore I_2 = 4\tfrac{2}{3}$$

(c) CD: c_3 is $y = 2$ $\therefore dy = 0$

$$\therefore I_3 = \int_3^1 2x\,dx = \left[x^2\right]_3^1 = -8 \quad \therefore I_3 = -8$$

(d) DA: c_4 is $x = 1$ $\therefore dx = 0$

$$\therefore I_4 = \int_2^0 (1+y^2)\,dy = \left[y + \frac{y^3}{3}\right]_2^0 = -4\tfrac{2}{3} \quad \therefore I_4 = -4\tfrac{2}{3}$$

Finally

$$I = I_1 + I_2 + I_3 + I_4$$
$$= 0 + 4\tfrac{2}{3} - 8 - 4\tfrac{2}{3} = -8 \quad \therefore I = -8$$

Remember that, unless we are directed otherwise, we always proceed round the closed boundary in an anticlockwise manner.

On now to the next piece of work

41 Line integral with respect to arc length

We have already established that

$$I = \int_{AB} F_t \, ds = \int_{AB} \{P \, dx + Q \, dy\}$$

where F_t denoted the tangential force along the curve c at the sample point K (x, y).

The same kind of integral can, of course, relate to any function $f(x, y)$ which is a function of the position of a point on the stated curve, so that

$$I = \int_c f(x, y) \, ds.$$

This can readily be converted into an integral in terms of x. (Refer to *Engineering Mathematics (Fifth Edition)*, Program 19, Frame 30.)

$$I = \int_c f(x, y) \, ds = \int_c f(x, y) \frac{ds}{dx} dx \quad \text{where} \quad \frac{ds}{dx} = \sqrt{1 + \left(\frac{dy}{dx}\right)^2}$$

$$\therefore \int_c f(x, y) \, ds = \int_{x_1}^{x_2} f(x, y) \sqrt{1 + \left(\frac{dy}{dx}\right)^2} \, dx \tag{1}$$

Example

Evaluate $I = \int_c (4x + 3xy) \, ds$ where c is the straight line joining O (0, 0) to A (1, 2).

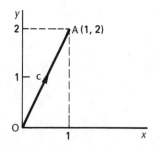

c is the line $y = 2x$ $\quad \therefore \quad \dfrac{dy}{dx} = 2$

$$\therefore \frac{ds}{dx} = \sqrt{1 + \left(\frac{dy}{dx}\right)^2} = \sqrt{5}$$

$$\therefore I = \int_{x=0}^{x=1} (4x + 3xy) \, ds = \int_0^1 (4x + 3xy)(\sqrt{5}) \, dx. \quad \text{But } y = 2x$$

$$\therefore I = \ldots\ldots\ldots\ldots$$

Differentials and line integrals

$$\boxed{I = 4\sqrt{5}}$$

Because

$$I = \int_0^1 (4x + 6x^2)(\sqrt{5})\,dx = 2\sqrt{5} \int_0^1 (2x + 3x^2)\,dx = 4\sqrt{5}$$

Try another.

The path length of the parabola defined by $y = x^2$ betwen the values $x = 0$ and $x = 2$ is given by the integral

$$I = \int_c ds = \ldots\ldots\ldots\ldots \text{ to 3 dp}$$

$$\boxed{3.393 \text{ to 3 dp}}$$

Because

$$I = \int_c ds = \int_{x=0}^{2} \sqrt{1 + \left(\frac{dy}{dx}\right)^2}\,dx$$

$$= \int_{x=0}^{2} \sqrt{1 + 2x}\,dx$$

Let $u = 1 + 2x$ so that $du = 2dx$ and so

$$I = \int_{u=1}^{5} u^{1/2} \frac{du}{2}$$

$$= \frac{1}{2}\left[\frac{2}{3}u^{3/2}\right]_1^5$$

$$= \frac{1}{3}\left(125^{1/2} - 1\right)$$

$$= 3.393 \text{ to 3 dp}$$

Parametric equations

When x and y are expressed in parametric form, e.g. $x = f(t)$, $y = g(t)$, then

$$\frac{ds}{dt} = \sqrt{\left(\frac{dx}{dt}\right)^2 + \left(\frac{dy}{dt}\right)^2} \quad \therefore \quad ds = \sqrt{\left(\frac{dx}{dt}\right)^2 + \left(\frac{dy}{dt}\right)^2}\,dt$$

and result (1) above becomes

$$I = \int_c f(x, y)\,ds = \int_{t_1}^{t_2} f(x, y)\sqrt{\left(\frac{dx}{dt}\right)^2 + \left(\frac{dy}{dt}\right)^2}\,dt \qquad (2)$$

Make a note of results (1) and (2) for future use

Vector Analysis

45

Example

Evaluate $I = \oint_c 4xy\, ds$ where c is defined as the curve $x = \sin t$, $y = \cos t$ between $t = 0$ and $t = \dfrac{\pi}{4}$.

We have $x = \sin t \quad \therefore \dfrac{dx}{dt} = \cos t$

$y = \cos t \quad \therefore \dfrac{dy}{dt} = -\sin t$

$\therefore \dfrac{ds}{dt} = \ldots\ldots\ldots\ldots$

46

$$\boxed{\dfrac{ds}{dt} = 1}$$

Because

$$\dfrac{ds}{dt} = \sqrt{\left(\dfrac{dx}{dt}\right)^2 + \left(\dfrac{dy}{dt}\right)^2} = \sqrt{\cos^2 t + \sin^2 t} = 1$$

$$\therefore I = \int_{t_1}^{t_2} f(x, y) \sqrt{\left(\dfrac{dx}{dt}\right)^2 + \left(\dfrac{dy}{dt}\right)^2}\, dt = \int_0^{\pi/4} 4 \sin t \, \cos t\, dt$$

$$= 2 \int_0^{\pi/4} \sin 2t\, dt = -2 \left[\dfrac{\cos 2t}{2}\right]_0^{\pi/4} = 1 \quad \therefore I = 1$$

Dependence of the line integral on the path of integration

We saw earlier in the Program that integration along two separate paths joining the same two end points does not necessarily give identical results.

With this in mind, let us investigate the following problem.

Example

Evaluate $I = \oint_c \{3x^2 y^2\, dx + 2x^3 y\, dy\}$ between O (0, 0) and A (2, 4)

(a) along c_1 i.e. $y = x^2$
(b) along c_2 i.e. $y = 2x$
(c) along c_3 i.e. $x = 0$ from (0, 0) to (0, 4) and $y = 4$ from (0, 4) to (2, 4).

Let us concentrate on section (a).

First we draw the figure and insert relevant information.

This gives ....

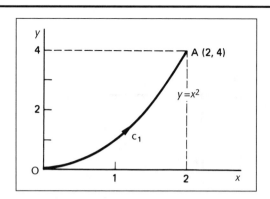

(a) $I = \int_c \{3x^2y^2\, dx + 2x^3y\, dy\}$

The path c_1 is $y = x^2$ $\quad \therefore\ dy = 2x\, dx$

$$\therefore I_1 = \int_0^2 \{3x^2x^4\, dx + 2x^3x^2 2x\, dx\} = \int_0^2 (3x^6 + 4x^4)\, dx$$

$$= \left[x^7\right]_0^2 = 128 \quad \therefore I_1 = 128$$

(b) In (b), the path of integration changes to c_2, i.e. $y = 2x$

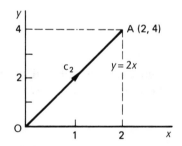

So, in this case,
$I_2 = \ldots\ldots\ldots\ldots$

$$\boxed{I_2 = 128}$$

Because with c_2, $y = 2x$ $\quad \therefore\ dy = 2\, dx$

$$\therefore I_2 = \int_0^2 \{3x^2\, 4x^2\, dx + 2x^3\, 2x2\, dx\} = \int_0^2 20x^4\, dx$$

$$= 4\left[x^5\right]_0^2 = 128 \quad \therefore I_2 = 128$$

(c) In the third case, the path c_3 is split

$x = 0$ from $(0, 0)$ to $(0, 4)$

$y = 4$ from $(0, 4)$ to $(2, 4)$

Sketch the diagram and determine I_3.

$I_3 = \ldots\ldots\ldots\ldots$

49

$$I_3 = 128$$

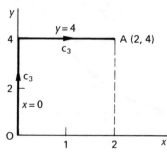

From $(0, 0)$ to $(0, 4)$ $x = 0$ $\therefore dx = 0$ $\therefore I_{3a} = 0$

From $(0, 4)$ to $(2, 4)$ $y = 4$ $\therefore dy = 0$ $\therefore I_{3b} = 48 \int_0^2 x^2 \, dx = 128$

$$\therefore I_3 = 128$$

On to the next frame

50

In the example we have just worked through, we took three different paths and in each case, the line integral produced the same result. It appears, therefore, that in this case, the value of the integral is independent of the path of integration taken.

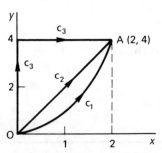

How then does this integral perhaps differ from those of previous cases?

Let us investigate

51

We have been dealing with $I = \int_c \{3x^2y^2 \, dx + 2x^3y \, dy\}$

On reflection, we see that the integrand $3x^2y^2 \, dx + 2x^3y \, dy$ is of the form $P \, dx + Q \, dy$ which we have met before and that it is, in fact, an *exact differential* of the function $z = x^3y^2$, because

$$\frac{\partial z}{\partial x} = 3x^2y^2 \quad \text{and} \quad \frac{\partial z}{\partial y} = 2x^3y$$

Provided P, Q and their first partial derivatives are finite and continuous at all points inside and on any closed curve, this always happens. If the integrand of the given integral is seen to be an *exact differential*, then the value of the line integral is *independent of the path taken and depends only on the coordinates of the two end points*

Make a note of this. It is important

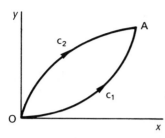

If $I = \int_c \{P\,dx + Q\,dy\}$ and $(P\,dx + Q\,dy)$ is an exact differential, then

$$I_{c_1} = I_{c_2}$$

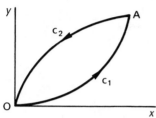

If we reverse the direction of c_2, then

$$I_{c_1} = -I_{c_2}$$

i.e. $I_{c_1} + I_{c_2} = 0$

Hence, *if $(P\,dx + Q\,dy)$ is an exact differential, then the integration taken round a closed curve is zero.*

∴ If $(P\,dx + Q\,dy)$ is an exact differential, $\oint (P\,dx + Q\,dy) = 0$

Example 1

Evaluate $I = \int_c \{3y\,dx + (3x + 2y)\,dy\}$ from A (1, 2) to B (3, 5).

No path is given, so the integrand is probably an exact differential of some function $z = f(x, y)$. In fact $\dfrac{\partial P}{\partial y} = 3 = \dfrac{\partial Q}{\partial x}$.

We have already dealt with the integration of exact differentials, so there is no difficulty. Compare with $I = \int_c \{P\,dx + Q\,dy\}$.

$P = \dfrac{\partial z}{\partial x} = 3y \qquad \therefore z = \int 3y\,dx = 3xy + f(y) \qquad (1)$

$Q = \dfrac{\partial z}{\partial y} = 3x + 2y \quad \therefore z = \int (3x + 2y)\,dy = 3xy + y^2 + F(x) \qquad (2)$

For (1) and (2) to agree

$f(y) = \ldots\ldots\ldots\ldots \quad \text{and} \quad F(x) = \ldots\ldots\ldots\ldots$

54

$$f(y) = y^2; \quad F(x) = 0$$

Hence $z = 3xy + y^2$

$$\therefore I = \int_c \{3y \, dx + (3x + 2y) \, dy\} = \int_{(1,\,2)}^{(3,\,5)} d(3xy + y^2)$$

$$= \left[3xy + y^2 \right]_{(1,\,2)}^{(3,\,5)}$$

$$= (45 + 25) - (6 + 4)$$

$$= 60$$

Example 2

Evaluate $I = \int_c \{(x^2 + ye^x) \, dx + (e^x + y) \, dy\}$ between A (0, 1) and B (1, 2).

As before, compare with $\int_c \{P \, dx + Q \, dy\}$.

$$P = \frac{\partial z}{\partial x} = x^2 + ye^x \quad \therefore z = \ldots\ldots\ldots$$

$$Q = \frac{\partial z}{\partial y} = e^x + y \quad \therefore z = \ldots\ldots\ldots$$

Continue the working and complete the evaluation.

When you have finished, check the result with the next frame

55

$$z = \frac{x^3}{3} + ye^x + f(y)$$
$$z = ye^x + \frac{y^2}{2} + F(x)$$

For these expressions to agree, $\quad f(y) = \frac{y^2}{2}; \quad F(x) = \frac{x^3}{3}$

Then $I = \left[\frac{x^3}{3} + ye^x + \frac{y^2}{2} \right]_{(0,\,1)}^{(1,\,2)}$

$$= \frac{5}{6} + 2e$$

So the main points are that, if $(P \, dx + Q \, dy)$ is an exact differential

(a) $I = \int_c (P \, dx + Q \, dy)$ is independent of the path of integration

(b) $I = \int_c (P \, dx + Q \, dy)$ is zero when c is a closed curve.

On to the next frame

Exact differentials in three independent variables

A line integral in space naturally involves three independent variables, but the method is very much like that for two independent variables.

$dw = P\,dx + Q\,dy + R\,dz$ is an exact differential of $w = f(x, y, z)$

if $\dfrac{\partial P}{\partial y} = \dfrac{\partial Q}{\partial x}$; $\dfrac{\partial P}{\partial z} = \dfrac{\partial R}{\partial x}$; $\dfrac{\partial R}{\partial y} = \dfrac{\partial Q}{\partial z}$

If the test is successful, then

(a) $\int_c (P\,dx + Q\,dy + R\,dz)$ is independent of the path of integration

(b) $\oint_c (P\,dx + Q\,dy + R\,dz)$ is zero when c is a closed curve.

Example

Verify that $dw = (3x^2yz + 6x)dx + (x^3z - 8y)dy + (x^3y + 1)dz$ is an exact differential and hence evaluate $\int_c dw$ from A (1, 2, 4) to B (2, 1, 3).

First check that dw is an exact differential by finding the partial derivatives above, when $P = 3x^2yz + 6x$; $Q = x^3z - 8y$; and $R = x^3y + 1$.

We have ....

$\dfrac{\partial P}{\partial y} = 3x^2z$; $\dfrac{\partial Q}{\partial x} = 3x^2z$ $\therefore \dfrac{\partial P}{\partial y} = \dfrac{\partial Q}{\partial x}$

$\dfrac{\partial P}{\partial z} = 3x^2y$; $\dfrac{\partial R}{\partial x} = 3x^2y$ $\therefore \dfrac{\partial P}{\partial z} = \dfrac{\partial R}{\partial x}$

$\dfrac{\partial R}{\partial y} = x^3$; $\dfrac{\partial Q}{\partial z} = x^3$ $\therefore \dfrac{\partial R}{\partial y} = \dfrac{\partial Q}{\partial z}$

$\therefore$ dw is an exact differential

Now to find w. $P = \dfrac{\partial z}{\partial x}$; $Q = \dfrac{\partial z}{\partial y}$; $R = \dfrac{\partial w}{\partial z}$

$\therefore \dfrac{\partial w}{\partial x} = 3x^2yz + 6x$ $\therefore w = \int (3x^2yz + 6x)dx$

$\qquad\qquad\qquad\qquad\qquad = x^3yz + 3x^2 + f(y, z)$

$\dfrac{\partial w}{\partial y} = x^3z - 8y$ $\therefore w = \int (x^3z - 8y)\,dy$

$\qquad\qquad\qquad\qquad\qquad = x^3zy - 4y^2 + F(x, z)$

$\dfrac{\partial w}{\partial z} = x^3y + 1$ $\therefore w = \int (x^3y + 1)\,dz$

$\qquad\qquad\qquad\qquad\qquad = x^3yz + z + g(x, y)$

For these three expressions for z to agree

$f(y, z) = $ ....; $F(x, z) = $ ....; $g(x, y) = $ ....

58

$$f(y,z) = -4y^2; \quad F(x,z) = z; \quad g(x,y) = 3x^2$$

$$\therefore w = x^3yz + 3x^2 - 4y^2 + z$$

$$\therefore I = \left[x^3yz + 3x^2 - 4y^2 + z\right]_{(1,2,4)}^{(2,1,3)}$$

$$= \ldots\ldots\ldots\ldots$$

59

$$I = 36$$

Because

$$I = \left[x^3yz + 3x^2 - 4y^2 + z\right]_{(1,2,4)}^{(2,1,3)}$$

$$= (24 + 12 - 4 + 3) - (8 + 3 - 16 + 4) = 36$$

The extension to line integrals in space is thus quite straightforward.

Finally, we have a theorem that can be very helpful on occasions and which links up with the work we have been doing.

It is important, so let us start a new section

Green's theorem in the plane

60

Let P and Q be two functions of x and y that are, along with their first partial derivatives, finite and continuous inside and on the boundary c of a region R in the x–y plane.

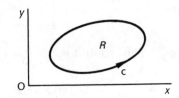

If the first partial derivatives are continuous within the region and on the boundary, then Green's theorem states that

$$\int\int_R \left(\frac{\partial P}{\partial y} - \frac{\partial Q}{\partial x}\right) dx\, dy = -\oint_c (P\, dx + Q\, dy)$$

That is, a double integral over the plane region R can be transformed into a line integral over the boundary c of the region – and the action is reversible.

Let us see how it works.

Differentials and line integrals

Example 1

Evaluate $I = \oint_c \{(2x - y)\,dx + (2y + x)\,dy\}$ around the boundary c of the ellipse $x^2 + 9y^2 = 16$.

The integral is of the form $I = \oint_c \{P\,dx + Q\,dy\}$ where

$P = 2x - y \quad \therefore \dfrac{\partial P}{\partial y} = -1$

and $Q = 2y + x \quad \therefore \dfrac{\partial Q}{\partial x} = 1$.

$\therefore I = -\int\int_R \left(\dfrac{\partial P}{\partial y} - \dfrac{\partial Q}{\partial x}\right) dx\,dy$

$= -\int\int_R (-1 - 1)\,dx\,dy$

$= 2\int\int_R dx\,dy$

But $\int\int_R dx\,dy$ over any closed region gives ....

> the area of the figure

In this case, then, $I = 2A$ where A is the area of the ellipse

$x^2 + 9y^2 = 16 \quad \text{i.e.} \quad \dfrac{x^2}{16} + \dfrac{9y^2}{16} = 1$

$\therefore a = 4;\ b = \dfrac{4}{3}$

$\therefore A = \pi ab = \dfrac{16\pi}{3}$

$\therefore I = 2A = \dfrac{32\pi}{3}$

To demonstrate the advantage of Green's theorem, let us work through the next example (a) by the method of line integrals, and (b) by applying Green's theorem.

Vector Analysis

Example 2

Evaluate $I = \oint_C \{(2x+y)\,dx + (3x-2y)\,dy\}$ taken in anticlockwise manner round the triangle with vertices at O (0, 0), A (1, 0) and B (1, 2).

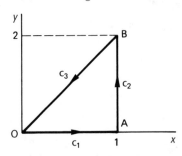

$$I = \oint_C \{(2x+y)\,dx + (3x-2y)\,dy\}$$

(a) *By the method of line integrals*

There are clearly three stages with c_1, c_2, c_3. Work through the complete evaluation to determine the value of I. It will be good revision.

When you have finished, check the result with the solution in the next frame

---

62

$$\boxed{I = 2}$$

(a) (1) c_1 is $y = 0$ ∴ $dy = 0$

$$\therefore I_1 = \int_0^1 2x\,dx = \left[x^2\right]_0^1 = 1 \quad \therefore I_1 = 1$$

(2) c_2 is $x = 1$ ∴ $dx = 0$

$$\therefore I_2 = \int_0^2 (3-2y)\,dy = \left[3y - y^2\right]_0^2 = 2 \quad \therefore I_2 = 2$$

(3) c_3 is $y = 2x$ ∴ $dy = 2\,dx$

$$\therefore I_3 = \int_1^0 \{4x\,dx + (3x-4x)2\,dx\}$$

$$= \int_1^0 2x\,dx = \left[x^2\right]_1^0 = -1 \quad \therefore I_3 = -1$$

$$I = I_1 + I_2 + I_3 = 1 + 2 + (-1) = 2 \quad \therefore I = 2$$

Now we will do the same problem by applying Green's theorem, so move on

Differentials and line integrals

(b) By Green's theorem

$$I = \oint_c \{(2x+y)\,dx + (3x-2y)\,dy\}$$

$$P = 2x+y \quad \therefore \frac{\partial P}{\partial y} = 1; \quad Q = 3x-2y \quad \therefore \frac{\partial Q}{\partial x} = 3$$

$$I = -\int_R\int \left(\frac{\partial P}{\partial y} - \frac{\partial Q}{\partial x}\right) dx\,dy$$

Finish it off. $I = \ldots\ldots\ldots$

$$\boxed{I = 2}$$

Because

$$I = -\int_R\int (1-3)\,dx\,dy$$
$$= 2\int_R\int dx\,dy = 2A$$
$$= 2 \times \text{the area of the triangle}$$
$$= 2 \times 1 = 2 \quad \therefore I = 2$$

Application of Green's theorem is not always the quickest method. It is useful, however, to have both methods available. If you have not already done so, make a note of Green's theorem.

$$\int_R\int \left(\frac{\partial P}{\partial y} - \frac{\partial Q}{\partial x}\right) dx\,dy = -\oint_c (P\,dx + Q\,dy)$$

Example 3

Evaluate the line integral $I = \oint_c \{xy\,dx + (2x-y)\,dy\}$ round the region bounded by the curves $y = x^2$ and $x = y^2$ by the use of Green's theorem.

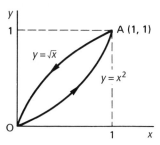

Points of intersection are O $(0, 0)$ and A $(1, 1)$. P and Q are known, so there is no difficulty.

Complete the working.

$$I = \ldots\ldots\ldots$$

66

$$I = \frac{31}{60}$$

Here is the working.

$$I = \oint_C \{xy\,dx + (2x - y)\,dy\}$$

$$\oint_C \{P\,dx + Q\,dy\} = -\int\int_R \left(\frac{\partial P}{\partial y} - \frac{\partial Q}{\partial x}\right) dx\,dy$$

$$P = xy \quad \therefore \frac{\partial P}{\partial y} = x; \quad Q = 2x - y \quad \therefore \frac{\partial Q}{\partial x} = 2$$

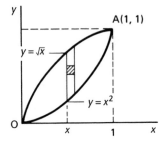

$$I = -\int\int_R (x - 2)\,dx\,dy$$

$$= -\int_0^1 \int_{y=x^2}^{y=\sqrt{x}} (x - 2)\,dy\,dx$$

$$= -\int_0^1 (x - 2)\big[y\big]_{x^2}^{\sqrt{x}} dx$$

$$\therefore I = -\int_0^1 (x - 2)(\sqrt{x} - x^2)\,dx$$

$$= -\int_0^1 (x^{3/2} - x^3 - 2x^{1/2} + 2x^2)\,dx$$

$$= -\left[\frac{2}{5}x^{5/2} - \frac{1}{4}x^4 - \frac{4}{3}x^{3/2} + \frac{2}{3}x^3\right]_0^1 = \frac{31}{60}$$

Before we finally leave this section of the work, there is one more result to note.

In the special case when $P = y$ and $Q = -x$

$$\frac{\partial P}{\partial y} = 1 \quad \text{and} \quad \frac{\partial Q}{\partial x} = -1$$

Green's theorem then states

$$\int\int_R \{1 - (-1)\}\,dx\,dy = -\oint_C (P\,dx + Q\,dy)$$

i.e. $$2\int\int_R dx\,dy = -\oint_C (y\,dx - x\,dy)$$

$$= \oint_C (x\,dy - y\,dx)$$

Therefore, the area of the closed region

$$A = \int\int_R dx\,dy = \frac{1}{2}\oint_C (x\,dy - y\,dx)$$

Note this result in your record book. Then let us see an example

Differentials and line integrals

Example 1 — 67

Determine the area of the figure enclosed by $y = 3x^2$ and $y = 6x$.

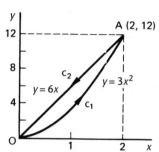

Points of intersection:

$$3x^2 = 6x \quad \therefore x = 0 \text{ or } 2$$

$$\text{Area } A = \tfrac{1}{2}\oint_c (x\,dy - y\,dx)$$

We evaluate the integral in two parts, i.e. OA along c_1
and AO along c_2

$$2A = \int_{c_1 \text{ (along OA)}} (x\,dy - y\,dx) + \int_{c_2 \text{ (along AO)}} (x\,dy - y\,dx) = I_1 + I_2$$

I_1: c_1 is $y = 3x^2$ $\quad \therefore dy = 6x\,dx$

$$\therefore I_1 = \int_0^2 (6x^2\,dx - 3x^2\,dx) = \int_0^2 3x^2\,dx = \left[x^3\right]_0^2 = 8$$

$$\therefore I_1 = 8$$

Similarly, $I_2 = \ldots\ldots\ldots\ldots$

68

$$\boxed{I_2 = 0}$$

Because

c_2 is $y = 6x$ $\quad \therefore dy = 6\,dx$

$$\therefore I_2 = \int_2^0 (6x\,dx - 6x\,dx) = 0 \quad \therefore I_2 = 0$$

$$\therefore I = I_1 + I_2 = 8 + 0 = 8 \quad \therefore A = 4 \text{ square units}$$

Finally, here is one for you to do entirely on your own.

Example 2

Determine the area bounded by the curves $y = 2x^3$, $y = x^3 + 1$ and the axis $x = 0$ for $x \geq 0$.

Complete the working and see if you agree with the working in the next frame

69

Here it is.

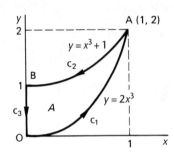

$y = 2x^3; \quad y = x^3 + 1; \quad x = 0$

Point of intersection

$2x^3 = x^3 + 1 \quad \therefore x^3 = 1 \quad \therefore x = 1$

Area $A = \dfrac{1}{2} \oint_c (x\,dy - y\,dx)$

$\therefore 2A = \oint_c (x\,dy - y\,dx)$

(a) OA: c_1 is $y = 2x^3 \quad \therefore dy = 6x^2\,dx$

$\therefore I_1 = \int_{c_1} (x\,dy - y\,dx) = \int_0^1 (6x^3\,dx - 2x^3\,dx)$

$= \int_0^1 4x^3\,dx = \left[x^4\right]_0^1 = 1 \qquad \therefore I_1 = 1$

(b) AB: c_2 is $y = x^3 + 1 \quad \therefore dy = 3x^2\,dx$

$\therefore I_2 = \int_1^0 \{3x^3\,dx - (x^3 + 1)\,dx\} = \int_1^0 (2x^3 - 1)\,dx$

$= \left[\dfrac{x^4}{2} - x\right]_1^0 = -(\tfrac{1}{2} - 1) = \tfrac{1}{2} \qquad \therefore I_2 = \tfrac{1}{2}$

(c) BO: c_3 is $x = 0 \quad \therefore dx = 0$

$I_3 = \int_{y=1}^{y=0} (x\,dy - y\,dx) = 0 \qquad \therefore I_3 = 0$

$\therefore 2A = I = I_1 + I_2 + I_3 = 1 + \tfrac{1}{2} + 0 = 1\tfrac{1}{2}$
$\therefore A = \tfrac{3}{4}$ square units

And that brings this Program to an end. We have covered some important topics, so check down the **Review summary** and the **Can You?** checklist that follow and revise any part of the text if necessary, before working through the **Test exercise**. The **Further problems** provide an opportunity for additional practice.

Differentials and line integrals

Review summary

1 *Differentials* dy *and* dx

(a)

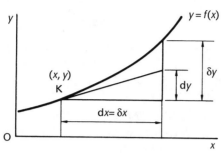

$$dy = f'(x)\, dx$$

(b) If $z = f(x, y)$, $\quad dz = \dfrac{\partial z}{\partial x} dx + \dfrac{\partial z}{\partial y} dy$

If $z = f(x, y, w)$, $\quad dz = \dfrac{\partial z}{\partial x} dx + \dfrac{\partial z}{\partial y} dy + \dfrac{\partial z}{\partial w} dw$

(c) $dz = P\, dx + Q\, dy$, where P and Q are functions of x and y, is an exact differential if $\dfrac{\partial P}{\partial y} = \dfrac{\partial Q}{\partial x}$.

2 *Line integrals – definition*

$$I = \int_c f(x, y)\, ds = \int_c (P\, dx + Q\, dy)$$

3 *Properties of line integrals*

(a) Sign of line integral is reversed when the direction of integration along the path is reversed.

(b) Path of integration parallel to y-axis, $dx = 0$ $\therefore I_c = \int_c Q\, dy$.

Path of integration parallel to x-axis, $dy = 0$ $\therefore I_c = \int_c P\, dx$.

(c) The y-values on the path of integration must be continuous and single-valued.

4 *Line of integral round a closed curve* $\oint$

Positive direction $\oint$ anticlockwise

Negative direction $\oint$ clockwise, i.e. $\oint = -\oint$.

5 *Line integral related to arc length*

$$I = \int_{AB} F \, ds = \int_{AB} (P \, dx + Q \, dy)$$

$$= \int_{x_1}^{x_2} f(x, y) \sqrt{1 + \left(\frac{dy}{dx}\right)^2} \, dx$$

With parametric equations, x and y in terms of t,

$$I = \int_c f(x, y) \, ds = \int_{t_1}^{t_2} f(x, y) \sqrt{\left(\frac{dx}{dt}\right)^2 + \left(\frac{dy}{dt}\right)^2} \, dt$$

6 *Dependence of line integral on path of integration*

In general, the value of the line integral depends on the particular path of integration.

7 *Exact differential*

If $P \, dx + Q \, dy$ is an exact differential where P, Q and their first derivatives are finite and continuous inside the simply connected region R

(a) $\dfrac{\partial P}{\partial y} = \dfrac{\partial Q}{\partial x}$

(b) $I = \int_c (P \, dx + Q \, dy)$ is independent of the path of integration where c lies entirely within R

(c) $I = \oint_c (P \, dx + Q \, dy)$ is zero when c is a closed curve lying entirely within R.

8 *Exact differentials in three variables*

If $P \, dx + Q \, dy + R \, dz$ is an exact differential where P, Q, R and their first partial derivatives are finite and continuous inside a simply connected region containing path c

(a) $\dfrac{\partial P}{\partial y} = \dfrac{\partial Q}{\partial x}$; $\dfrac{\partial P}{\partial z} = \dfrac{\partial R}{\partial x}$; $\dfrac{\partial R}{\partial y} = \dfrac{\partial Q}{\partial z}$

(b) $\int_c (P \, dx + Q \, dy + R \, dz)$ is independent of the path of integration

(c) $\oint_c (P \, dx + Q \, dy + R \, dz)$ is zero when c is a closed curve.

9 *Green's theorem*

$$\oint_c (P \, dx + Q \, dy) = -\int\int_R \left\{ \frac{\partial P}{\partial y} - \frac{\partial Q}{\partial x} \right\} dx \, dy$$

and, for a simple closed curve

$$\oint_c (x \, dy - y \, dx) = 2 \int\int_R dx \, dy = 2A$$

where A is the area of the enclosed figure.

Differentials and line integrals

✓ Can You?

Checklist 5

Check this list before and after you try the end of Program test.

On a scale of 1 to 5 how confident are you that you can: Frames

- Understand the role of the differential of a function of two or more real variables?
  Yes ☐ ☐ ☐ ☐ ☐ No   [1] to [3]

- Determine exact differentials in two real variables and their integrals?
  Yes ☐ ☐ ☐ ☐ ☐ No   [4] to [9]

- Evaluate the area enclosed by a closed curve by contour integration?
  Yes ☐ ☐ ☐ ☐ ☐ No   [10] to [16]

- Evaluate line integrals and appreciate their properties?
  Yes ☐ ☐ ☐ ☐ ☐ No   [17] to [31]

- Evaluate line integrals around closed curves within a simply connected region?
  Yes ☐ ☐ ☐ ☐ ☐ No   [32] to [40]

- Link line integrals to integrals along the *x*-axis?
  Yes ☐ ☐ ☐ ☐ ☐ No   [41] to [43]

- Link line integrals to integrals along a contour given in parametric form?
  Yes ☐ ☐ ☐ ☐ ☐ No   [44] to [46]

- Discuss the dependence of a line integral between two points on the path of integration?
  Yes ☐ ☐ ☐ ☐ ☐ No   [46] to [55]

- Determine exact differentials in three real variables and their integrals?
  Yes ☐ ☐ ☐ ☐ ☐ No   [56] to [59]

- Demonstrate the validity and use of Green's theorem?
  Yes ☐ ☐ ☐ ☐ ☐ No   [60] to [69]

Test exercise 5

1. Determine the differential dz of each of the following.
   (a) $z = x^4 \cos 3y$; (b) $z = e^{2y} \sin 4x$; (c) $z = x^2 y w^3$.

2. Determine which of the following are exact differentials and integrate where appropriate to determine z.
   (a) $dz = (3x^2 y^4 + 8x)\,dx + (4x^3 y^3 - 15y^2)\,dy$
   (b) $dz = (2x \cos 4y - 6 \sin 3x)dx - 4(x^2 \sin 4y - 2y)\,dy$
   (c) $dz = 3e^{3x}(1-y)\,dx + (e^{3x} + 3y^2)\,dy$.

3. Calculate the area of the triangle with vertices at O (0, 0), A (4, 2) and B (1, 5).

4. Evaluate the following.
   (a) $I = \int_c \{(x^2 - 3y)\,dx + xy^2\,dy\}$ from A (1, 2) to B (2, 8) along the curve $y = 2x^2$.
   (b) $I = \int_c (2x + y)\,dx$ from A (0, 1) to B (0, -1) along the semicircle $x^2 + y^2 = 1$ for $x \geq 0$.
   (c) $I = \oint_c \{(1 + xy)\,dx + (1 + x^2)\,dy\}$ where c is the boundary of the rectangle joining A (1, 0), B (4, 0), C (4, 3) and D (1, 3).
   (d) $I = \int_c 2xy\,ds$ where c is defined by the parametric equations
   $x = 4\cos\theta, \ y = 4\sin\theta$ between $\theta = 0$ and $\theta = \dfrac{\pi}{3}$.
   (e) $I = \int_c \{(8xy + y^3)\,dx + (4x^2 + 3xy^2)\,dy\}$ from A(1, 3) to B(2, 1).
   (f) $I = \oint_c \{(3x + y)\,dx + (y - 2x)\,dy\}$ round the boundary of the ellipse $x^2 + 4y^2 = 36$.

5. Apply Green's theorem to determine the area of the plane figure bounded by the curves $y = x^3$ and $y = \sqrt{x}$.

6. Verify that $dw = (2xyz + 2z - y^2)dx + (x^2 z - 2yx)dy + (x^2 y + 2x)dz$ is an exact differential and find the value of
$$\int_c dw \text{ where}$$
   (a) c is the straight line joining (0, 0, 0) to (1, 1, 1)
   (b) c is the curve of intersection of the unit sphere centred on the origin and the plane $x + y + z = 1$.

Differentials and line integrals

 Further problems 5

1. Show that $I = \int_c \{xy^2w^2\,dx + x^2yw^2\,dy + x^2y^2w\,dw\}$ is independent of the path of integration c and evaluate the integral from A $(1, 3, 2)$ to B $(2, 4, 1)$.

2. Determine whether $dz = 3x^2(x^2 + y^2)\,dx + 2y(x^3 + y^4)\,dy$ is an exact differential. If so, determine z and hence evaluate $\int_c dz$ from A $(1, 2)$ to B $(2, 1)$.

3. Evaluate the line integral $I = \oint_c \left\{\dfrac{xdy - ydx}{x^2 + y^2 + 4}\right\}$ where c is the boundary of the segment formed by the arc of the circle $x^2 + y^2 = 4$ and the chord $y = 2 - x$ for $x \geq 0$.

4. Show that
$$I = \int_c\{(3x^2\sin y + 2\sin 2x + y^3)\,dx + (x^3\cos y + 3xy^2)\,dy\}$$
is independent of the path of integration and evaluate it from A $(0,0)$ to B $\left(\dfrac{\pi}{2}, \pi\right)$.

5. Evaluate the integral $I = \int_c xy\,ds$ where c is defined by the parametric equations $x = \cos^3 t$, $y = \sin^3 t$ from $t = 0$ to $t = \dfrac{\pi}{2}$.

6. Verify that $dz = \dfrac{xdx}{x^2 - y^2} - \dfrac{ydy}{x^2 - y^2}$ for $x^2 > y^2$ is an exact differential and evaluate $z = f(x, y)$ from A $(3, 1)$ to B $(5, 3)$.

7. The parametric equations of a circle, centre $(1, 0)$ and radius 1, can be expressed as $x = 2\cos^2\theta$, $y = 2\cos\theta\sin\theta$.
   Evaluate $I = \int_c\{(x+y)\,dx + x^2\,dy\}$ along the semicircle for which $y \geq 0$ from O $(0, 0)$ to A $(2, 0)$.

8. Evaluate $\oint_c \{x^3y^2\,dx + x^2y\,dy\}$ where c is the boundary of the region enclosed by the curve $y = 1 - x^2$, $x = 0$ and $y = 0$ in the first quadrant.

9. Use Green's theorem to evaluate
$$I = \oint_c \{(4x + y)\,dx + (3x - 2y)\,dy\}$$
where c is the boundary of the trapezium with vertices A $(0, 1)$, B $(5, 1)$, C $(3, 3)$ and D $(1, 3)$.

10. Evaluate $I = \int_c\{(3x^2y^2 + 2\cos 2x - 2xy)\,dx + (2x^3y + 8y - x^2)\,dy\}$
    (a) along the curve $y = x^2 - x$ from A $(0, 0)$ to B $(2, 2)$
    (b) round the boundary of the quadrilateral joining the points $(1, 0)$, $(3, 1)$, $(2, 3)$ and $(0, 3)$

Vector Analysis

11 Verify that $dw = \dfrac{y}{z}dx + \dfrac{x}{z}dy - \dfrac{xy}{z^2}dz$ is an exact differential and find the value of

$$\int_c dw$$

where c is the straight line joining (0, 0, 1) to (1, 2, 3) for either region $z > 0$ or $z < 0$.

Program 6

Surface and volume integrals

Frames **1** to **77**

Learning outcomes

When you have completed this Program you will be able to:
- Evaluate double integrals and surface integrals
- Relate three-dimensional Cartesian coordinates to cylindrical and spherical polar forms
- Evaluate volume integrals in Cartesian coordinates and in cylindrical and spherical polar coordinates
- Use the Jacobian to convert integrals given in Cartesian coordinates into general curvilinear coordinates in two and three dimensions

152 *Vector Analysis*

Double integrals

1

Let us start off with an example with which we are already familiar.

Example 1

A solid is enclosed by the planes $z = 0$, $y = 1$, $y = 2$, $x = 0$, $x = 3$ and the surface $z = x + y^2$. We have to determine the volume of the solid so formed.

First take some care in sketching the figure, which is

....

2

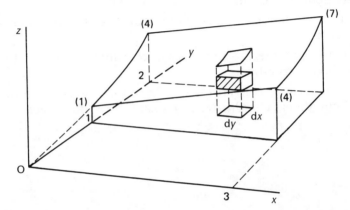

In the plane $y = 1$, $z = x + 1$, i.e. a straight line joining $(0, 1, 1)$ and $(3, 1, 4)$
In the plane $y = 2$, $z = x + 4$, i.e. a straight line joining $(0, 2, 4)$ and $(3, 2, 7)$
In the plane $x = 0$, $z = y^2$, i.e. a parabola joining $(0, 1, 1)$ and $(0, 2, 4)$
In the plane $x = 3$, $z = 3 + y^2$, i.e. a parabola joining $(3, 1, 4)$ and $(3, 2, 7)$

Consideration like this helps us to visualize the problem and the time involved is well spent.

Now we can proceed.

The element of volume $\delta v = \delta x \, \delta y \, \delta z$

Then the total volume $V = \iiint dx \, dy \, dz$ between appropriate limits in each case.

Surface and volume integrals

We could also have said that the element of area on the $z = 0$ plane

$$\delta a = \delta y \, \delta x$$

and that the volume of the column $\quad \delta v_c = z \, \delta a = z \, \delta x \, \delta y$
Then, since $z = x + y^2$, this becomes $\quad \delta v_c = (x + y^2) \, \delta x \, \delta y$
Summing in the usual way then gives

$$V = \int z \, da$$

$$= \int_R \int (x + y^2) \, dx \, dy$$

where R is the region bounded in the x–y plane.

Now we insert the appropriate limits and complete the integration

$$V = \ldots\ldots\ldots\ldots$$

$$V = 11.5 \text{ cubic units}$$

3

Because

$$V = \int_{y=1}^{y=2} \int_{x=0}^{x=3} (x + y^2) \, dx \, dy$$

$$= \int_1^2 \left[\frac{x^2}{2} + xy^2 \right]_{x=0}^{x=3} dy$$

$$= \int_1^2 \left(\frac{9}{2} + 3y^2 \right) dy$$

$$= \left[\frac{9}{2} y + y^3 \right]_1^2$$

$$= 11.5$$

$\therefore V = 11.5$ cubic units

Although we have found a volume, this is, in fact, an example of a *double integral* since the expression for z was a function of position in the x–y plane within the closed region

i.e. $\quad I = \int_R \int f(x, y) \, da$

$$= \int_R \int f(x, y) \, dy \, dx$$

In this particular case, R is the region in the x–y plane bounded by $x = 0$, $x = 3$, $y = 1$, $y = 2$.

Example 2

A triangular thin plate has the dimensions shown and a variable density ρ where $\rho = 1 + x + xy$.

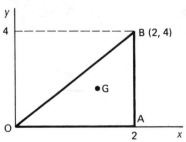

We have to determine
(a) the mass of the plate
(b) the position of its center of gravity G.

(a) Consider an element of area at the point $P(x, y)$ in the plate

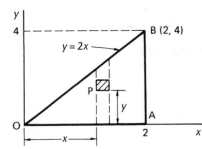

$\delta a = \delta x \, \delta y$

The mass δm of the element is then

$\delta m = \rho \, \delta x \, \delta y$

$\therefore$ Total mass $M = \int_R \int dm = \int_R \int \rho \, dx \, dy$

Now we insert the limits and complete the integration, remembering that $\rho = (1 + x + xy)$

$M = \ldots\ldots\ldots$

4

$$M = 17\frac{1}{3}$$

Because we have

$$M = \int_R \int \rho \, dx \, dy = \int_{x=0}^{x=2} \int_{y=0}^{y=2x} (1 + x + xy) \, dy \, dx$$

$$= \int_0^2 \left[y + xy + \frac{xy^2}{2} \right]_{y=0}^{y=2x} dx$$

$$= \int_0^2 \{ 2x + 2x^2 + 2x^3 \} \, dx$$

$$= \left[x^2 + \frac{2x^3}{3} + \frac{x^4}{2} \right]_0^2 = 17\frac{1}{3}$$

(b) To find the position of the center of gravity, we need to know
$\ldots\ldots\ldots$

the sum of the moments of mass about OY and OX

(1) To find $\bar{x}$, we take moments about OY.

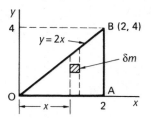

Moment of mass of element about OY

$$= x\,\delta m$$
$$= x(1 + x + xy)\,\delta x\,\delta y$$

∴ Sum of first moments $= \int_R \int (x + x^2 + x^2 y)\,dx\,dy$

$= \ldots\ldots\ldots\ldots$

$$26\frac{2}{15}$$

Because sum of first moments $= \int_{x=0}^{x=2} \int_{y=0}^{y=2x} (x + x^2 + x^2 y)\,dy\,dx$

$$= \int_0^2 \left[xy + x^2 y + \frac{x^2 y^2}{2} \right]_{y=0}^{y=2x} dx$$

$$= \int_0^2 \{2x^2 + 2x^3 + 2x^4\}\,dx$$

$$= 2\int_0^2 (x^2 + x^3 + x^4)\,dx$$

$$= 2\left[\frac{x^3}{3} + \frac{x^4}{4} + \frac{x^5}{5} \right]_0^2 = 26\frac{2}{15}$$

Now $M\bar{x}$ = sum of moments ∴ $\bar{x} = \ldots\ldots\ldots\ldots$

$$\bar{x} = 1.508$$

We found previously that $M = 17\frac{1}{3}$ ∴ $\left(17\frac{1}{3}\right)\bar{x} = 26\frac{2}{15}$

which gives $\bar{x} = 1\frac{33}{65} = 1.508$

(2) To find $\bar{y}$ we proceed in just the same way, this time taking moments about OX. Work right through it on your own.

$$\bar{y} = \ldots\ldots\ldots\ldots$$

$$\bar{y} = 1.754$$

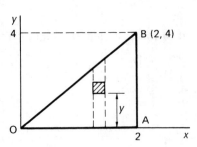

Moment of element of mass δm about OX
$$= y\,\delta m = y(1 + x + xy)\,\delta x\,\delta y$$

$\therefore$ Sum of first moments about OX $= \iint_R (y + xy + xy^2)\,dx\,dy$

$$= \int_{x=0}^{x=2} \int_{y=0}^{y=2x} (y + xy + xy^2)\,dy\,dx$$

$$= \int_0^2 \left[\frac{y^2}{2} + \frac{xy^2}{2} + \frac{xy^3}{3}\right]_{y=0}^{y=2x} dx$$

$$= \int_0^2 \left\{2x^2 + 2x^3 + \frac{8x^4}{3}\right\} dx$$

$$= \left[\frac{2x^3}{3} + \frac{x^4}{2} + \frac{8x^5}{15}\right]_0^2$$

$$= 30\frac{2}{5}$$

$\therefore M\bar{y} = 30\frac{2}{5} \quad \therefore \bar{y} = 30\frac{2}{5} \Big/ 17\frac{1}{3} = 1.754$

So we finally have:

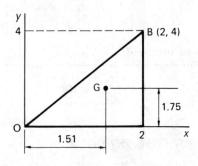

Note that this again referred to a plane figure in the x–y plane.

Now let us move on to something slightly different

Surface integrals

When the area over which we integrate is not restricted to the x–y plane, matters become rather more involved, but also more interesting.

If S is a two-sided surface in space and R is its projection on the x–y plane, then the equation of S is of the form $z = f(x, y)$ where f is a single-valued function and continuous throughout R.

Let δA denote an element of R and δS the corresponding element of area of S at the point $P(x, y, z)$ in S.

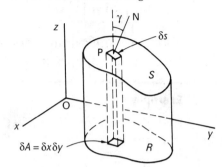

$\delta A = \delta x \delta y$

Let also $\phi(x, y, z)$ be a function of position on S (e.g. potential) and let γ denote the angle between the outward normal PN to the surface at P and the positive z-axis.

Then $\delta A \approx \delta S \cos \gamma$ i.e. $\delta S \approx \dfrac{\delta A}{\cos \gamma} = \delta A \sec \gamma$ and

$\sum \phi(x, y, z) \delta S$ is the total value of $\phi(x, y, z)$ taken over the surface S. As $\delta S \to 0$, this sum becomes the integral

$$I = \int_S \phi(x, y, z)\, dS$$

and, since $\delta S \approx \delta A \sec \gamma$, the result can be written

$$I = \int_R \int \phi(x, y, z) \sec \gamma \, dx\, dy \qquad \left(\gamma < \dfrac{\pi}{2}\right)$$

Notice that $\cos \gamma = \hat{\mathbf{n}} \cdot \mathbf{k}$, where $\mathbf{k}$ is the unit vector in the z-direction and $\hat{\mathbf{n}}$ is the unit normal to the surface at P.

With limits inserted for x and y, the integral seems straightforward, except for the factor $\sec \gamma$, which naturally varies over the surface S.

We can, in fact, show that $\sec \gamma = \sqrt{1 + \left(\dfrac{\partial z}{\partial x}\right)^2 + \left(\dfrac{\partial z}{\partial y}\right)^2}$

(see Program 8, Frame 43)

Therefore, the *surface integral* of $\phi(x, y, z)$ over the surface S is given by

(a) $I = \displaystyle\int_S \phi(x, y, z)\, dS$ \hfill (1)

or (b) $I = \displaystyle\int_R \int \phi(x, y, z) \sqrt{1 + \left(\dfrac{\partial z}{\partial x}\right)^2 + \left(\dfrac{\partial z}{\partial y}\right)^2}\, dx\, dy$ \hfill (2)

where $z = f(x, y)$

Vector Analysis

Note that, when $\phi(x, y, z) = 1$, then $I = \int_S dS$ gives the area of the surface S.

$$\therefore S = \int_S dS = \int\int_R \sqrt{1 + \left(\frac{\partial z}{\partial x}\right)^2 + \left(\frac{\partial z}{\partial y}\right)^2} \, dx \, dy \tag{3}$$

Make a note of these three important results.
Then we will apply them to a few examples.

10

Example 1

Find the area of the surface $z = \sqrt{x^2 + y^2}$ over the region bounded by $x^2 + y^2 = 1$.

$$S = \int\int_R \sqrt{1 + \left(\frac{\partial z}{\partial x}\right)^2 + \left(\frac{\partial z}{\partial y}\right)^2} \, dx \, dy$$

So we now find $\dfrac{\partial z}{\partial x}$ and $\dfrac{\partial z}{\partial y}$ and determine $\sqrt{1 + \left(\dfrac{\partial z}{\partial x}\right)^2 + \left(\dfrac{\partial z}{\partial y}\right)^2}$ which is

....

11

$$\boxed{\sqrt{2}}$$

Because

$$z = (x^2 + y^2)^{1/2} \quad \therefore \quad \frac{\partial z}{\partial x} = \frac{1}{2}(x^2 + y^2)^{-1/2} 2x = \frac{x}{\sqrt{x^2 + y^2}}$$

$$\frac{\partial z}{\partial y} = \frac{1}{2}(x^2 + y^2)^{-1/2} 2y = \frac{y}{\sqrt{x^2 + y^2}}$$

$$\therefore 1 + \left(\frac{\partial z}{\partial x}\right)^2 + \left(\frac{\partial z}{\partial y}\right)^2 = 1 + \frac{x^2 + y^2}{x^2 + y^2} = 2$$

$$\therefore \sqrt{1 + \left(\frac{\partial z}{\partial x}\right)^2 + \left(\frac{\partial z}{\partial y}\right)^2} = \sqrt{2}$$

$$\therefore S = \sqrt{2} \int\int_R dx \, dy = \sqrt{2} \times \ldots\ldots\ldots\ldots$$

Surface and volume integrals 159

the area of the region R

But R is bounded by $x^2 + y^2 = 1$, i.e. a circle, center the origin and radius 1.
∴ area $= \pi$

$$\therefore S = \sqrt{2} \int_R \int dx\,dy = \sqrt{2}\pi$$

Example 2

Find the area of the surface S of the paraboloid $z = x^2 + y^2$ cut off by the cone $z = 2\sqrt{x^2 + y^2}$.

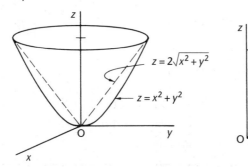

We can find the point of intersection A by considering the y–z plane, i.e. put $x = 0$.

Coordinates of A are ....

A (2, 4)

The projection of the surface S on the x–y plane is

....

the circle $x^2 + y^2 = 4$

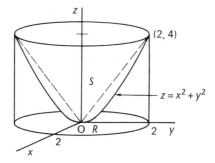

$$S = \int_R \int \sqrt{1 + \left(\frac{\partial z}{\partial x}\right)^2 + \left(\frac{\partial z}{\partial y}\right)^2}\,dx\,dy$$

For this we use the equation of the surface S. The information from the projection R on the x–y plane will later provide the limits of the two stages of integration.

For the time being, then, $S = $ ....

15

$$S = \int\int_R \sqrt{1 + 4x^2 + 4y^2}\, dx\, dy$$

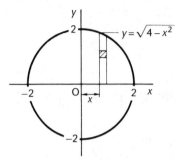

Using Cartesian coordinates, we could integrate with respect to y from $y = 0$ to $y = \sqrt{4 - x^2}$ and then with respect to x from $x = 0$ to $x = 2$. Finally, we should multiply by four to cover all four quadrants.

i.e. $S = 4\displaystyle\int_{x=0}^{x=2}\int_{y=0}^{y=\sqrt{4-x^2}} \sqrt{1 + 4x^2 + 4y^2}\, dy\, dx$

But how do we carry out the actual integration?

It becomes a lot easier if we use polar coordinates.

The same integral in polar coordinates is ....

16

$$S = \int_{\theta=0}^{\theta=2\pi}\int_{r=0}^{r=2} \sqrt{1 + 4r^2}\, r\, dr\, d\theta$$

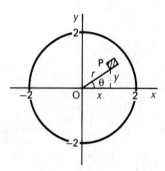

$x = r\cos\theta; \qquad y = r\sin\theta$

$x^2 + y^2 = r^2 \qquad dx\, dy = r\, dr\, d\theta$

(refer to Frame 67)

$S = \displaystyle\int_{\theta=0}^{\theta=2\pi}\int_{r=0}^{r=2} \sqrt{1 + 4r^2}\, r\, dr\, d\theta$

$\therefore S = \ldots\ldots\ldots\ldots$

Finish it off.

$$\boxed{S = 36.18 \text{ square units}}$$

Because

$$S = \int_{\theta=0}^{\theta=2\pi} \int_{r=0}^{r=2} (1+4r^2)^{1/2} r \, dr \, d\theta = \int_0^{2\pi} \left[\frac{1}{12}(1+4r^2)^{3/2}\right]_0^2 d\theta$$

$$= \frac{1}{12}\int_0^{2\pi} \{17^{3/2} - 1\} \, d\theta = 5.7577 \Big[\theta\Big]_0^{2\pi} = 36.18$$

Now on to Example 3.

Example 3

To determine the moment of inertia of a thin spherical shell of radius a about a diameter as axis. The mass per unit area of shell is ρ.

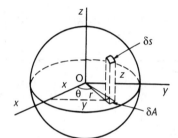

Equation of sphere

$$x^2 + y^2 + z^2 = a^2$$

Mass of element $= m = \rho \, \delta S$

$$I \approx \Sigma m r^2 \approx \Sigma \rho \, \delta S r^2$$

Let us deal with the upper hemisphere

$$\therefore I_H = \int_S \rho r^2 \, dS$$

$$= \int_R \int \rho r^2 \sqrt{1 + \left(\frac{\partial z}{\partial x}\right)^2 + \left(\frac{\partial z}{\partial y}\right)^2} \, dx \, dy$$

Now determine the partial derivatives and simplify the integral as far as possible in Cartesian coordinates.

$$I_H = \ldots\ldots\ldots\ldots$$

$$\boxed{I_H = \int_R \int \rho r^2 \frac{a}{\sqrt{a^2 - x^2 - y^2}} \, dx \, dy}$$

In this particular example, R is, of course, the region bounded by the circle $x^2 + y^2 = a^2$ in the x–y plane.

Converting to polar coordinates

$$x = r\cos\theta; \quad y = r\sin\theta; \quad dx\,dy = r\,dr\,d\theta$$

the integral becomes $I_H = \ldots\ldots\ldots\ldots$

20

$$I_H = \rho a \int_{\theta=0}^{\theta=2\pi} \int_{r=0}^{r=a} \frac{r^3}{\sqrt{a^2 - r^2}} \, dr \, d\theta$$

Because for $x^2 + y^2 = r^2$: limits of r: $r = 0$ to $r = a$
limits of θ: $\theta = 0$ to $\theta = 2\pi$

$$I_H = \int\int_R \rho r^2 \frac{a}{\sqrt{a^2 - r^2}} r \, dr \, d\theta$$

$$= \rho a \int_{\theta=0}^{\theta=2\pi} \int_{r=0}^{r=a} \frac{r^3}{\sqrt{a^2 - r^2}} \, dr \, d\theta$$

First we have to evaluate

$$I_r = \int_0^a \frac{r^3}{\sqrt{a^2 - r^2}} \, dr$$

If we substitute $u = a^2 - r^2$ then the integral is evaluated as

$$I_r = \ldots\ldots\ldots$$

21

$$I_r = \frac{2a^3}{3}$$

Because

When $u = a^2 - r^2$ then $du = -2r \, dr$ so that $r^2 = a^2 - u$ and $r \, dr = -\frac{du}{2}$. Therefore

$$I_r = \int_0^a \frac{r^3}{\sqrt{a^2 - r^2}} \, dr = \int_{r=0}^a \frac{r^2}{\sqrt{a^2 - r^2}} r \, dr$$

$$= -\int_{u=a^2}^0 \frac{a^2 - u}{\sqrt{u}} \frac{du}{2}$$

$$= -\frac{a^2}{2} \int_{u=a^2}^0 u^{-1/2} \, du + \frac{1}{2} \int_{u=a^2}^0 u^{1/2} \, du$$

$$= -\frac{a^2}{2} \left[2u^{1/2}\right]_{u=a^2}^0 + \frac{1}{2} \left[\frac{2}{3}u^{3/2}\right]_{u=a^2}^0$$

$$= a^3 - \frac{a^3}{3}$$

$$= \frac{2a^3}{3}$$

Now, to complete I_H we have

$$I_H = \rho a \int_0^{2\pi} \frac{2a^3}{3} \, d\theta$$

$$= \ldots\ldots\ldots$$

Surface and volume integrals

$$I_H = \frac{4\pi \rho a^4}{3}$$

Because

$$I_H = \rho a \int_0^{2\pi} \frac{2a^3}{3} d\theta = \frac{2a^4 \rho}{3}\left[\theta\right]_0^{2\pi} = \frac{4\pi a^4 \rho}{3}$$

Therefore, the moment of inertia for the complete spherical shell is

$$I_s = \frac{8\pi a^4 \rho}{3}$$

The total mass of the shell $M = 4\pi a^2 \rho$ $\therefore I = \frac{2Ma^2}{3}$

Now let us turn our attention towards *volume integrals* and in preparation review systems of space coordinates.

Space coordinate systems

1 *Cartesian coordinates* (x, y, z) refers to three coordinate axes OX, OY, OZ at right angles to each other. These are arranged in a *right-handed* manner, i.e. turning from OX to OY gives a right-handed screw action in the positive direction of OZ.

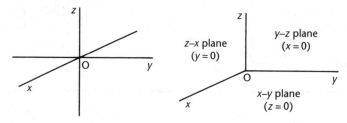

The three coordinate planes, $x = 0$, $y = 0$, $z = 0$, divide the space into eight sections called *octants*. The section containing $x \geq 0$, $y \geq 0$, $z \geq 0$ is called the *first octant*.

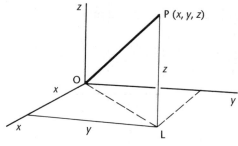

For a point P (x, y, z)
$$OL^2 = x^2 + y^2$$
$$OP^2 = x^2 + y^2 + z^2$$

Note that this is Pythagoras' theorem in three dimensions.

We are all familiar with this system of coordinates.

164 Vector Analysis

24

2 *Cylindrical coordinates* (r, θ, z) are useful where an axis of symmetry occurs.

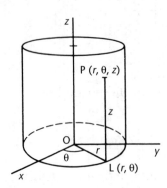

Any point P is considered as having a position on a cylinder. If L is the projection of P on the x–y plane, then (r, θ) are the usual polar coordinates of L. The cylindrical coordinates of P then merely require the addition of the z-coordinate.

$$r \geq 0$$

Relationship between Cartesian and cylindrical coordinates

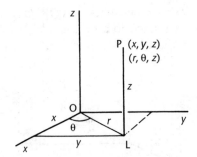

If we consider a combined figure, we can easily relate the two systems.

Expressing each of the following in terms of the alternative system,

$x = \ldots\ldots\ldots\ldots$ $r = \ldots\ldots\ldots\ldots$
$y = \ldots\ldots\ldots\ldots$ $\theta = \ldots\ldots\ldots\ldots$
$z = \ldots\ldots\ldots\ldots$ $z = \ldots\ldots\ldots\ldots$

25

$x = r \cos \theta$	$r = \sqrt{x^2 + y^2}$
$y = r \sin \theta$	$\theta = \arctan(y/x)$
$z = z$	$z = z$

So, in cylindrical coordinates, the surface defined by

(1) $r = 5$ is $\ldots\ldots\ldots\ldots$
(2) $\theta = \pi/6$ is $\ldots\ldots\ldots\ldots$
(3) $z = 4$ is $\ldots\ldots\ldots\ldots$

Surface and volume integrals

26

(1) $r = 5$ is a right cylinder, radius 5, with OZ as axis.
(2) $\theta = \pi/6$ is a plane through OZ, making an angle $\pi/6$ with OX.
(3) $z = 4$ is a plane parallel to the x–y plane cutting OZ at 4 units above the origin.

So position P (2, 3, 4) in Cartesian coordinates
= .... in cylindrical coordinates
and position Q (2.5, $\pi/3$, 6) in cylindrical coordinates
= .... in Cartesian coordinates.

27

P (2, 3, 4) = ($\sqrt{13}$, 0.983, 4) in cylindrical coordinates
Q (2.5, $\pi/3$, 6) = (1.25, 2.165, 6) in Cartesian coordinates.

3 *Spherical coordinates* (r, θ, ϕ) are appropriate where a center of symmetry occurs. The position of a point is considered as being a point on a sphere.

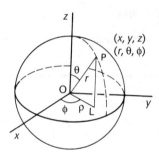

r is the distance of P from the origin and is always taken as positive

L is the projection of P on the x–y plane

θ is the angle between OP and the positive OZ axis

ϕ is the angle between OL and the OX axis

Note that (a) ϕ may be regarded as the longitude of P from OX
(b) θ may be regarded as the complement of the latitude of P

Relationship between Cartesian and spherical coordinates

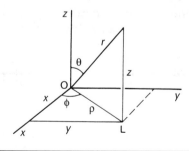

The combined figure shows the connection between the two systems, so

$x = $ .... $r = $ ....
$y = $ .... $\theta = $ ....
$z = $ .... $\phi = $ ....

28

$$x = r\sin\theta\cos\phi \qquad r = \sqrt{x^2 + y^2 + z^2}$$
$$y = r\sin\theta\sin\phi \qquad \theta = \arccos(z/r)$$
$$z = r\cos\theta \qquad \phi = \arctan(y/x)$$

For the spherical coordinates of any point in space

$$r \geq 0; \quad 0 \leq \theta \leq \pi; \quad 0 \leq \phi \leq 2\pi$$

So, converting Cartesian coordinates (2, 3, 4) to spherical coordinates gives
....

29

$$P(r, \theta, \phi) = (5.385, 0.734, 0.983)$$

Because

$$x = 2, \ y = 3, \ z = 4$$
$$\therefore r = \sqrt{x^2 + y^2 + z^2} = \sqrt{4 + 9 + 16} = \sqrt{29} = 5.385$$
$$\theta = \arccos(z/r) = \arccos(4/\sqrt{29}) = 0.734$$
$$\phi = \arctan(y/x) = \arctan 1.5 = 0.983$$

And, in reverse, spherical coordinates $(5, \pi/4, \pi/3)$ transform into Cartesian coordinates ....

30

$$P(x, y, z) = (1.768, 3.061, 3.536)$$

Because

$$x = r\sin\theta\cos\phi = 5\sin\frac{\pi}{4}\cos\frac{\pi}{3} = 5(0.707)(0.5) = 1.768$$
$$y = r\sin\theta\sin\phi = 5\sin\frac{\pi}{4}\sin\frac{\pi}{3} = 5(0.707)(0.866) = 3.061$$
$$z = r\cos\theta = 5\cos\frac{\pi}{4} = 5(0.707) = 3.536.$$

One of the main uses of cylindrical and spherical coordinates occurs in integrals dealing with volumes of solids. In preparation for this, let us consider the next important section of the work.

So move on

Element of volume in space in the three coordinate systems

1 *Cartesian coordinates*

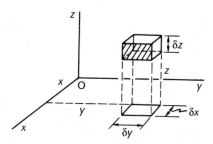

We have already used this many times.

$$\delta v = \delta x\, \delta y\, \delta z$$

2 *Cylindrical coordinates*

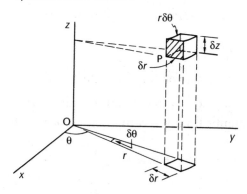
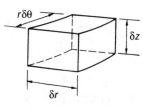

$$\delta v = r\delta\theta\, \delta r\, \delta z$$
$$\therefore\ \delta v = r\, \delta r\, \delta\theta\, \delta z$$

3 *Spherical coordinates*

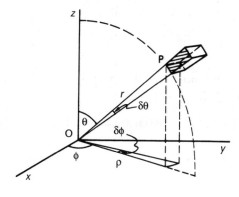
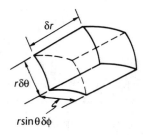

$$\delta v = \delta r\, r\delta\theta\, r\sin\theta\, \delta\phi$$
$$\therefore\ \delta v = r^2 \sin\theta\, \delta r\, \delta\theta\, \delta\phi$$

It is important to make a note of these results, since they are required when we change the variables in various types of integrals. We shall meet them again before long, so be sure of them now.

Volume integrals

32

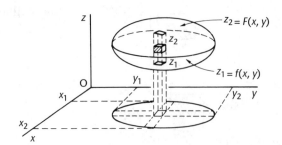

A solid is enclosed by a lower surface $z_1 = f(x, y)$ and an upper surface $z_2 = F(x, y)$.

Then, in general, using Cartesian coordinates, the element of volume is $\delta v = \delta x \, \delta y \, \delta z$.

The approximate value of the total volume V is then found

(a) by summing δv from $z = z_1$ to $z = z_2$ to obtain the volume of the column
(b) by summing all such columns from $y = y_1$ to $y = y_2$ to obtain the volume of the slice
(c) by summing all such slices from $x = x_1$ to $x = x_2$ to obtain the total volume V.

Then, when $\delta x \to 0$, $\delta y \to 0$, $\delta z \to 0$, the summation becomes an integral

$$V = \int_{x=x_1}^{x=x_2} \int_{y=y_1}^{y=y_2} \int_{z=z_1}^{z=z_2} dz \, dy \, dx$$

Example 1

Find the volume of the solid bounded by the planes $z = 0$, $x = 0$, $y = 0$, $x^2 + y^2 = 4$ and $z = 6 - xy$ for $x \geq 0$, $y \geq 0$, $z \geq 0$.

First sketch the figure, so that we can see what we are doing. Take your time over it.

Surface and volume integrals

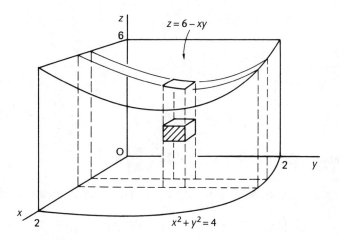

$\delta v = \delta x\, \delta y\, \delta z$

Volume of column $\approx \sum\limits_{z=0}^{z=6-xy} \delta x\, \delta y\, \delta z$

Volume of slice $\approx \sum\limits_{y=0}^{\sqrt{4-x^2}} \left\{ \sum\limits_{z=0}^{6-xy} \delta x\, \delta y\, \delta z \right\}$

Total volume $\approx \sum\limits_{x=0}^{2} \sum\limits_{y=0}^{\sqrt{4-x^2}} \sum\limits_{z=0}^{6-xy} \delta x\, \delta y\, \delta z$

If $\delta x \to 0$, $\delta y \to 0$, $\delta z \to 0$, then

$$V = \int_0^2 \int_0^{\sqrt{4-x^2}} \int_0^{6-xy} dz\, dy\, dx$$

Starting with the innermost integral

$$\int_0^{6-xy} dz = \left[z \right]_0^{6-xy}$$
$$= 6 - xy$$

Then $\displaystyle\int_0^{\sqrt{4-x^2}} (6 - xy)\, dy = \ldots\ldots\ldots\ldots$

34

$$6\sqrt{4-x^2} - \frac{x}{2}(4-x^2)$$

Because

$$\int_0^{\sqrt{4-x^2}} (6-xy)\,dy = \left[6y - \frac{xy^2}{2}\right]_{y=0}^{y=\sqrt{4-x^2}}$$

$$= 6\sqrt{4-x^2} - \frac{x}{2}(4-x^2)$$

Then finally $V = \int_0^2 \left\{6(4-x^2)^{1/2} - 2x + \frac{x^3}{2}\right\} dx$

Now we are faced with $\int (4-x^2)^{1/2}\,dx$. You may remember that this is a standard form $\int \sqrt{a^2-x^2}\,dx = \frac{1}{2}\left\{x\sqrt{a^2-x^2} + a^2 \arcsin\frac{x}{a}\right\}$.

If not, to evaluate $\int_0^2 \sqrt{4-x^2}\,dx$, put $x = 2\sin\theta$ and proceed from there.

Finish off the main integral, so that we have

$$V = \ldots\ldots\ldots$$

35

$$V = 6\pi - 2 \approx 16.8 \text{ cubic units}$$

Because we had

$$V = \int_0^2 \left\{6(4-x^2)^{1/2} - 2x + \frac{x^3}{2}\right\} dx$$

$$= 3\left[x\sqrt{4-x^2} + 4\arcsin\frac{x}{2}\right]_0^2 - \left[x^2 - \frac{x^4}{8}\right]_0^2$$

$$= 3\{4\arcsin 1 - 4\arcsin 0\} - 4 + 2$$

$$= 3\{2\pi\} - 2 = 6\pi - 2$$

$$\approx 16.8$$

Surface and volume integrals

Alternative method

We could, of course, have used cylindrical coordinates in this problem.

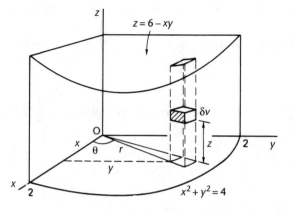

$\delta v = r\,\delta r\,\delta\theta\,\delta z$
$x = r\cos\theta;\ y = r\sin\theta$
$\therefore z = 6 - xy$
$\quad = 6 - r^2 \sin\theta\cos\theta$
$\quad = 6 - \dfrac{r^2}{2}\sin 2\theta$

$$\therefore V = \int_{r=0}^{2}\int_{\theta=0}^{\pi/2}\int_{z=0}^{6-(r^2/2)\sin 2\theta} r\,dr\,d\theta\,dz$$

$$= \int_{\theta=0}^{\pi/2}\int_{r=0}^{2}\int_{z=0}^{6-(r^2/2)\sin 2\theta} dz\,r\,dr\,d\theta$$

$$= \ldots\ldots\ldots\ldots$$

Finish it

$$V = 6\pi - 2 \text{ (as before)}$$

$$V = \int_{\theta=0}^{\pi/2}\int_{r=0}^{2}\left(6 - \dfrac{r^2}{2}\sin 2\theta\right) r\,dr\,d\theta$$

$$= \int_{\theta=0}^{\pi/2}\int_{r=0}^{2}\left(6r - \dfrac{r^3}{2}\sin 2\theta\right) dr\,d\theta$$

$$= \int_{0}^{\pi/2}\left[3r^2 - \dfrac{r^4}{8}\sin 2\theta\right]_{r=0}^{r=2} d\theta$$

$$= \int_{0}^{\pi/2}(12 - 2\sin 2\theta)\,d\theta$$

$$= \Big[12\theta + \cos 2\theta\Big]_{0}^{\pi/2}$$

$$= (6\pi - 1) - 1$$

$$\therefore V = 6\pi - 2$$

In this case, the use of cylindrical coordinates facilitates the evaluation.

Let us consider another example.

37

Example 2

To find the moment of inertia and radius of gyration of a thick hollow sphere about a diameter as axis. Outer radius $= a$; inner radius $= b$; density of material $= c$.

It is convenient to deal with one-eighth of the sphere in the first octant.

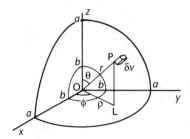

$\therefore$ Total mass of the solid $M_1 = \dfrac{1}{8} M$

$$M_1 = \dfrac{1}{8} \cdot \dfrac{4}{3} \pi (a^3 - b^3) c = \dfrac{\pi}{6}(a^3 - b^3) c$$

Using spherical coordinates, the element of volume

$$\delta v = \ldots\ldots\ldots\ldots$$

38

$$\boxed{\delta v = r^2 \sin\theta \, \delta r \, \delta\theta \, \delta\phi}$$

Also the element of mass $m = c \delta v$

Second moment of mass of the element about OZ

$$= m\rho^2 = m(r\sin\theta)^2$$
$$= c r^2 \sin\theta \, \delta r \, \delta\theta \, \delta\phi \, r^2 \sin^2\theta$$
$$= c r^4 \sin^3\theta \, \delta r \, \delta\theta \, \delta\phi$$

$\therefore$ Total second moment for the solid

$$I_1 \approx \sum_{\phi=0}^{\pi/2} \sum_{\theta=0}^{\pi/2} \sum_{r=b}^{a} c r^4 \, \delta r \, \sin^3\theta \, \delta\theta \, \delta\phi$$

Then, as usual, if $\delta r \to 0$, $\delta\theta \to 0$, $\delta\phi \to 0$, we finally obtain

$$I_1 = \int_{\phi=0}^{\pi/2} \int_{\theta=0}^{\pi/2} \int_{r=b}^{a} c r^4 \, dr \, \sin^3\theta \, d\theta \, d\phi$$

which you can evaluate without any difficulty and obtain

$$I_1 = \ldots\ldots\ldots\ldots$$

Surface and volume integrals

$$I_1 = \frac{\pi}{15}(a^5 - b^5)c$$

Because

$$I_1 = \int_0^{\pi/2} \int_0^{\pi/2} \left[c\frac{r^5}{5}\right]_b^a \sin^3\theta \, d\theta \, d\phi$$

$$= \int_0^{\pi/2} \int_0^{\pi/2} \frac{c}{5}(a^5 - b^5) \sin^3\theta \, d\theta \, d\phi$$

$$= \frac{c}{5}(a^5 - b^5) \int_0^{\pi/2} \int_0^{\pi/2} (1 - \cos^2\theta) \sin\theta \, d\theta \, d\phi$$

$$= \frac{c}{5}(a^5 - b^5) \int_0^{\pi/2} \left[-\cos\theta + \frac{\cos^3\theta}{3}\right]_0^{\pi/2} d\phi$$

$$= \frac{c}{5}(a^5 - b^5) \int_0^{\pi/2} \left(1 - \frac{1}{3}\right) d\phi$$

$$= \frac{2c}{15}(a^5 - b^5)\left[\phi\right]_0^{\pi/2} = \frac{c\pi}{15}(a^5 - b^5)$$

Therefore, the moment of inertia for the whole sphere I is

$I = 8I_1$ i.e. $I = \dfrac{8\pi}{15}(a^5 - b^5)c$

Radius of gyration (k) $Mk^2 = I$

$$\therefore k = \ldots\ldots\ldots\ldots$$

$$k = \sqrt{\frac{2}{5}\left(\frac{a^5 - b^5}{a^3 - b^3}\right)}$$

We had already calculated the total mass $M = \dfrac{4\pi}{3}(a^3 - b^3)c$ and since $I = \dfrac{8\pi}{15}(a^5 - b^5)c$ then

$$\frac{4\pi}{3}(a^3 - b^3)ck^2 = \frac{8\pi}{15}(a^5 - b^5)c$$

$$\therefore k^2 = \frac{2}{5}\left(\frac{a^5 - b^5}{a^3 - b^3}\right) \quad \therefore k = \sqrt{\frac{2}{5}\left(\frac{a^5 - b^5}{a^3 - b^3}\right)}$$

We have set the working out in considerable detail, since spherical coordinates may be a new topic. Many of the statements can be streamlined when one is familiar with the system.

Now move on for another example

41

Example 3

Find the total mass of a solid sphere of radius a, enclosed by the surface $x^2 + y^2 + z^2 = a^2$ and having variable density c where $c = 1 + r|z|$ and r is the distance of any point from the origin.

This is a case where spherical coordinates can clearly be used with advantage.

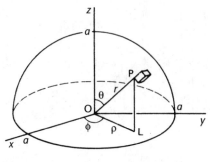

(a) ....

In the element of volume, the three dimensions are

(b) .... (c) ....

42

(a) δr (b) $r\,\delta\theta$ (c) $\rho\,\delta\phi = r\sin\theta\,\delta\phi$

so that $\delta v = $ ....

43

$$\delta v = r^2 \sin\theta\,\delta r\,\delta\theta\,\delta\phi$$

Then the mass of the element $= c\,\delta v = (1 + r|z|)\,\delta v$
and
$$z = r\cos\theta$$
$$\therefore\ m = c\,\delta v = (1 + r^2\cos\theta)\,r^2\sin\theta\,\delta r\,\delta\theta\,\delta\phi$$

Since the density uses $|z| = 1$ we must only consider the region where $\cos\theta \geq 0$ and so we consider the *upper hemisphere* only. The integral for the total mass M_1 is

$$M_1 = \text{............}$$

Write out the integral and insert the limits.

Surface and volume integrals

$$M_1 = \int_{\phi=0}^{\phi=2\pi} \int_{\theta=0}^{\theta=\pi/2} \int_{r=0}^{r=a} (1+r^2\cos\theta)r^2\sin\theta \, dr \, d\theta \, d\phi$$

i.e. $M_1 = \int_{\phi=0}^{2\pi} \int_{\theta=0}^{\pi/2} \int_{r=0}^{a} \{r^2\sin\theta \, dr \, d\theta \, d\phi + r^4\sin\theta\cos\theta \, dr \, d\theta \, d\phi\}$

$$= \qquad I_1 \qquad + \qquad I_2$$

$$I_1 = \int_0^{2\pi}\int_0^{\pi/2}\int_0^a r^2\sin\theta \, dr \, d\theta \, d\phi \text{ gives } \ldots\ldots\ldots\ldots$$

Do *not* work it out. You can doubtless recognize what the result would represent.

The volume of the hemisphere

Because the integral is simply the summation of elements of volume throughout the region of the hemisphere.

Thus, without more ado, $I_1 = \dfrac{2}{3}\pi a^3$.

Now for I_2.

$$I_2 = \int_0^{2\pi}\int_0^{\pi/2}\int_0^a r^4\sin\theta\cos\theta \, dr \, d\theta \, d\phi$$

$= \ldots\ldots\ldots\ldots$ Evaluate the triple integral.

$$I_2 = \frac{\pi a^5}{5}$$

Because

$$I_2 = \int_0^{2\pi}\int_0^{\pi/2} \frac{a^5}{5}\sin\theta\cos\theta \, d\theta \, d\phi$$

$$= \frac{a^5}{5}\int_0^{2\pi}\left[\frac{\sin^2\theta}{2}\right]_0^{\pi/2} d\phi$$

$$= \frac{a^5}{10}\int_0^{2\pi} 1 \, d\phi$$

$$= \frac{a^5}{10}\left[\phi\right]_0^{2\pi} = \frac{\pi a^5}{5}$$

$$\therefore I_2 = \frac{\pi a^5}{5}$$

So now finish it off. For the complete sphere

$$M = \ldots\ldots\ldots\ldots$$

176 Vector Analysis

47

$$M = \frac{2\pi a^3}{15}(10 + 3a^2)$$

Because

$$M_1 = I_1 + I_2 = \frac{2}{3}\pi a^3 + \frac{\pi a^5}{5} = \frac{\pi a^3}{15}(10 + 3a^2)$$

Then, for the whole sphere, $M = 2M_1 = \frac{2\pi a^3}{15}(10 + 3a^2)$

Each problem, then, is tackled in much the same way.
(a) Draw a careful sketch diagram, inserting all relevant information.
(b) Decide on the most appropriate coordinate system to use.
(c) Build up the multiple integral and insert correct limits.
(d) Evaluate the integral.

And now we can apply the general guide lines to a final problem.

Example 4

Determine the volume of the solid bounded by the planes $x = 0$, $y = 0$, $z = x$, $z = 2$ and $y = 4 - x^2$ in the first quadrant.

First we sketch the diagram.

48

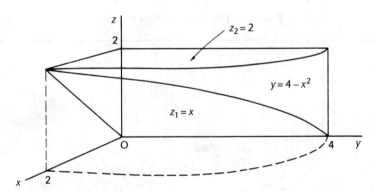

There is no axis of symmetry and no spherical center. We shall therefore use .... coordinates.

49

$$\boxed{\text{Cartesian}}$$

So off you go on your own. There are no snags.

$$V = \ldots\ldots\ldots\ldots$$

$$\boxed{V = 6\frac{2}{3} \text{ cubic units}}$$

Here is the complete solution.

$$V \approx \sum_{x=0}^{2} \sum_{y=0}^{4-x^2} \sum_{z=x}^{2} \delta x\, \delta y\, \delta z$$

$$\therefore V = \int_{x=0}^{2} \int_{y=0}^{4-x^2} \int_{z=x}^{2} dz\, dy\, dx$$

$$= \int_0^2 \int_0^{4-x^2} (2-x)\, dy\, dx$$

$$= \int_0^2 \Big[2y - xy\Big]_{y=0}^{4-x^2} dx$$

$$= \int_0^2 \{8 - 2x^2 - 4x + x^3\}\, dx$$

$$= \left[8x - \frac{2x^3}{3} - 2x^2 + \frac{x^4}{4}\right]_0^2$$

$$= 6\frac{2}{3}$$

And that is it. Now we move to the next section of work

Change of variables in multiple integrals

In Cartesian coordinates, we use the variables (x, y, z); in cylindrical coordinates, we use the variables (r, θ, z); in spherical coordinates, we use the variables (r, θ, ϕ); and we have established relationships connecting these systems of variables, permitting us to transfer from one system to another. These relationships, you will remember, were obtained geometrically in Frames 23 to 30 of this Program.

There are occasions, however, when it is expedient to make other transformations beside those we have used and it is worth looking at the problem in a rather more general manner.

This we will now do

52

First, however, let us revise a result from an earlier Program on determinants to find the area of the triangle ABC.

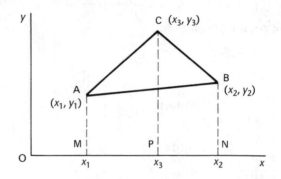

If we arrange the vertices A (x_1, y_1)
B (x_2, y_2)
C (x_3, y_3)

in an anticlockwise manner then

area triangle ABC = trapezium AMPC + trapezium CPNB

— trapezium AMNB

$= \frac{1}{2}\{(x_3 - x_1)(y_1 + y_3) + (x_2 - x_3)(y_2 + y_3) - (x_2 - x_1)(y_1 + y_2)\}$

$= \frac{1}{2}\{x_3y_1 - x_1y_1 + x_3y_3 - x_1y_3 + x_2y_2 + x_2y_3 - x_3y_2 - x_3y_3$
$\qquad - x_2y_1 - x_2y_2 + x_1y_1 + x_1y_2\}$

$= \frac{1}{2}\{(x_2y_3 - x_3y_2) + (x_3y_1 - x_1y_3) + (x_1y_2 - x_2y_1)\}$

$= \frac{1}{2}\begin{vmatrix} 1 & 1 & 1 \\ x_1 & x_2 & x_3 \\ y_1 & y_2 & y_3 \end{vmatrix}$

The determinant is positive if the points A, B, C are taken in an anticlockwise manner.

We shall need to use this result in a short while, so keep it in mind.

On to the next frame

Curvilinear coordinates

Consider the double integral $\iint_R \phi(x, y) dA$ where $dA = dx dy$ in Cartesian coordinates. Let u and v be two new independent variables defined by $u = F(x, y)$ and $v = G(x, y)$ where these equations can be simultaneously solved to obtain $x = f(u, v)$ and $y = g(u, v)$. Furthermore, these transformation equations are such that every point (x, y) is mapped to a unique point (u, v) and vice versa.

Let us see where this leads us, so on to the next frame

The equation $u = F(x, y)$ will be a family of curves depending on the particular constant value given to u in each case.

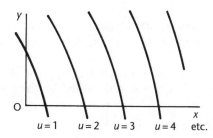

Curves $u = F(x, y)$ for different constant values of u.

Similarly, $v = G(x, y)$ will be a family of curves depending on the particular constant value assigned to v in each case.

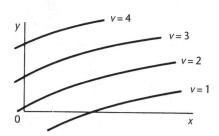

Curves $v = G(x, y)$ for different constant values of v.

Vector Analysis

These two sets of curves will therefore cover the region R and form a network, and to any point P (x_0, y_0) there will be a pair of curves $u = u_0$ (constant) and $v = v_0$ (constant) that intersect at that point.

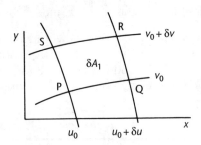

The u- and v-values relating to any particular point are known as its *curvilinear coordinates* and $x = f(u, v)$ and $y = g(u, v)$ are the *transformation equations* between the two systems.

In the Cartesian coordinates (x, y) system, the element of area $\delta A = \delta x \delta y$ and is the area bounded by the lines $x = x_0$, $x = x_0 + \delta x$, $y = y_0$, and $y = y_0 + \delta y$.

In the new system of *curvilinear coordinates* (u, v) the element of area δA_1 can be taken as that of the figure P, Q, R, S, i.e. the area bounded by the curves $u = u_0$, $u = u_0 + \delta u$, $v = v_0$ and $v = v_0 + \delta v$.

Since δA_1 is small, PQRS may be regarded as a parallelogram

i.e. $\delta A_1 \approx 2 \times$ area of triangle PQS

and this is where we make use of the result previously revised that the area of a triangle ABC with vertices (x_1, y_1), (x_2, y_2), (x_3, y_3) can be expressed in determinant form as

$$\text{Area} = \ldots\ldots\ldots\ldots$$

Surface and volume integrals

$$\text{Area} = \frac{1}{2}\begin{vmatrix} 1 & 1 & 1 \\ x_1 & x_2 & x_3 \\ y_1 & y_2 & y_3 \end{vmatrix}$$

55

Before we can apply this, we must find the Cartesian coordinates of P, Q and S in the diagram in the previous frame where we omit the subscript $_0$ on the coordinates.

If $x = f(u, v)$, then a small increase δx in x is given by

$$\delta x = \ldots\ldots\ldots\ldots$$

$$\delta x = \frac{\partial f}{\partial u}\delta u + \frac{\partial f}{\partial v}\delta v$$

56

i.e. $\delta x = \dfrac{\partial x}{\partial u}\delta u + \dfrac{\partial x}{\partial v}\delta v$

and, for $y = g(u, v)$

$$\delta y = \ldots\ldots\ldots\ldots$$

$$\delta y = \frac{\partial y}{\partial u}\delta u + \frac{\partial y}{\partial v}\delta v$$

57

Now

(a) P is the point (x, y)

(b) Q corresponds to small changes from P.

$$\delta x = \frac{\partial x}{\partial u}\delta u + \frac{\partial x}{\partial v}\delta v \quad \text{and} \quad \delta y = \frac{\partial y}{\partial u}\delta u + \frac{\partial y}{\partial v}\delta v$$

But along PQ v is constant. $\therefore \delta v = 0$.

$$\therefore \delta x = \frac{\partial x}{\partial u}\delta u \quad \text{and} \quad \delta y = \frac{\partial y}{\partial u}\delta u$$

i.e. Q is the point $\left(x + \dfrac{\partial x}{\partial u}\delta u,\ y + \dfrac{\partial y}{\partial u}\delta u\right)$.

(c) Similarly for S, since u is constant along PS $\delta u = 0$ and

$\therefore$ S is the point $\left(x + \dfrac{\partial x}{\partial v}\delta v,\ y + \dfrac{\partial y}{\partial v}\delta v\right)$

So the Cartesian coordinates of P, Q, S are

$$P\ (x, y);\ Q\ \left(x + \frac{\partial x}{\partial u}\delta u,\ y + \frac{\partial y}{\partial u}\delta u\right);\ S\ \left(x + \frac{\partial x}{\partial v}\delta v,\ y + \frac{\partial y}{\partial v}\delta v\right)$$

$\therefore$ The determinant for the area PQS is $\ldots\ldots\ldots\ldots$

58

$$\text{Area} = \frac{1}{2}\begin{vmatrix} 1 & 1 & 1 \\ x & x+\frac{\partial x}{\partial u}\delta u & x+\frac{\partial x}{\partial v}\delta v \\ y & y+\frac{\partial y}{\partial u}\delta u & y+\frac{\partial y}{\partial v}\delta v \end{vmatrix}$$

Subtracting column 1 from columns 2 and 3 gives

$$\text{Area} = \frac{1}{2}\begin{vmatrix} 1 & 0 & 0 \\ x & \frac{\partial x}{\partial u}\delta u & \frac{\partial x}{\partial v}\delta v \\ y & \frac{\partial y}{\partial u}\delta u & \frac{\partial y}{\partial v}\delta v \end{vmatrix}$$

which simplifies immediately to

....

59

$$\text{Area} = \frac{1}{2}\begin{vmatrix} \frac{\partial x}{\partial u}\delta u & \frac{\partial x}{\partial v}\delta v \\ \frac{\partial y}{\partial u}\delta u & \frac{\partial y}{\partial v}\delta v \end{vmatrix}$$

Then, taking out the factor δu from the first column and the factor δv from the second column, this becomes

$$\text{Area} = \ldots\ldots\ldots\ldots$$

60

$$\frac{1}{2}\begin{vmatrix} \frac{\partial x}{\partial u} & \frac{\partial x}{\partial v} \\ \frac{\partial y}{\partial u} & \frac{\partial y}{\partial v} \end{vmatrix}\delta u\,\delta v$$

The area of the approximate parallelogram is twice the area of the triangle.

$$\therefore \text{ Area of parallelogram} = \delta A_1 = \begin{vmatrix} \frac{\partial x}{\partial u} & \frac{\partial x}{\partial v} \\ \frac{\partial y}{\partial u} & \frac{\partial y}{\partial v} \end{vmatrix}\delta u\,\delta v$$

Expressing this in differentials

$$dA = \begin{vmatrix} \frac{\partial x}{\partial u} & \frac{\partial x}{\partial v} \\ \frac{\partial y}{\partial u} & \frac{\partial y}{\partial v} \end{vmatrix} du\,dv$$

and, for convenience, this is often written

$$dA = \frac{\partial(x, y)}{\partial(u, v)} du\,dv$$

▶

Surface and volume integrals

$\dfrac{\partial(x, y)}{\partial(u, v)}$ is called the *Jacobian of the transformation* from the Cartesian coordinates (x, y) to the curvilinear coordinates (u, v).

$$\therefore J(u, v) = \dfrac{\partial(x, y)}{\partial(u, v)} = \begin{vmatrix} \dfrac{\partial x}{\partial u} & \dfrac{\partial x}{\partial v} \\ \dfrac{\partial y}{\partial u} & \dfrac{\partial y}{\partial v} \end{vmatrix}$$

So, if the transformation equations are

$$x = u(u + v) \quad \text{and} \quad y = uv^2$$
$$J(u, v) = \ldots\ldots\ldots\ldots$$

$$\boxed{J(u, v) = uv(4u + v)}$$

61

Because

$$\dfrac{\partial x}{\partial u} = 2u + v \qquad \dfrac{\partial x}{\partial v} = u$$

$$\dfrac{\partial y}{\partial u} = v^2 \qquad \dfrac{\partial y}{\partial v} = 2uv$$

$$\therefore J(u, v) = \begin{vmatrix} 2u + v & u \\ v^2 & 2uv \end{vmatrix} = 4u^2v + 2uv^2 - uv^2$$

$$= 4u^2v + uv^2 = uv(4u + v)$$

Next frame

Sometimes the transformation equations are given the other way round. That is, where u and v are given as expressions in x and y. In such a case $J(u, v)$ can be found using the fact that

62

$$\dfrac{\partial(x, y)}{\partial(u, v)} = \dfrac{1}{\left(\dfrac{\partial(u, v)}{\partial(x, y)}\right)}$$

For example, if the transformation equations are given as $u = x^2 + y^2$ and $v = 2xy$ then

$$J(u, v) = \ldots\ldots\ldots\ldots$$

63

$$J(u, v) = \frac{1}{4\sqrt{u^2 - v^2}}$$

Because

$$\frac{\partial(u, v)}{\partial(x, y)} = \begin{vmatrix} \dfrac{\partial u}{\partial x} & \dfrac{\partial u}{\partial y} \\ \dfrac{\partial v}{\partial x} & \dfrac{\partial v}{\partial y} \end{vmatrix} = \begin{vmatrix} 2x & 2y \\ 2y & 2x \end{vmatrix} = 4x^2 - 4y^2$$

and so

$$J(u, v) = \frac{\partial(x, y)}{\partial(u, v)} = \frac{1}{\left(\dfrac{\partial(u, v)}{\partial(x, y)}\right)} = \frac{1}{4(x^2 - y^2)}$$

Now $u - v = x^2 - 2xy + y^2 = (x - y)^2$
and $u + v = x^2 + 2xy + y^2 = (x + y)^2$
and so $x^2 - y^2 = (x - y)(x + y) = \sqrt{u - v}\sqrt{u + v} = \sqrt{u^2 - v^2}$ giving

$$J(u, v) = \frac{1}{4\sqrt{u^2 - v^2}}$$

There is one further point to note in this piece of work, so move on

64

Note: In the transformation, it is possible for the order of the points P, Q, R, S to be reversed with the result that δA may give a negative result when the determinant is evaluated. To ensure a positive element of area, the result is finally written

$$dA = \left|\frac{\partial(x, y)}{\partial(u, v)}\right| du\, dv$$

where the 'modulus' lines indicate the absolute value of the Jacobian.

Therefore, to rewrite the integral $\int_R \int F(x, y)\, dx\, dy$ in terms of the new variables, u and v, where $x = f(u, v)$ and $y = g(u, v)$, we substitute for x and y in $F(x, y)$ and replace $dx\, dy$ with $\left|\dfrac{\partial(x, y)}{\partial(u, v)}\right| du\, dv$.

The integral then becomes

$$\int_R \int F\{f(u, v),\, g(u, v)\} \left|\frac{\partial(x, y)}{\partial(u, v)}\right| du\, dv$$

Make a note of this result

Surface and volume integrals

Example 1

Express $I = \int_R \int xy^2 \, dx \, dy$ in polar coordinates, making the substitutions $x = r\cos\theta$, $y = r\sin\theta$.

$$\frac{\partial x}{\partial r} = \cos\theta \qquad \frac{\partial x}{\partial \theta} = -r\sin\theta$$

$$\frac{\partial y}{\partial r} = \sin\theta \qquad \frac{\partial y}{\partial \theta} = r\cos\theta$$

$$\therefore J(r, \theta) = \ldots\ldots\ldots$$

$$\boxed{J(r, \theta) = r}$$

$$J(r, \theta) = \begin{vmatrix} \cos\theta & -r\sin\theta \\ \sin\theta & r\cos\theta \end{vmatrix} = r\cos^2\theta + r\sin^2\theta = r$$

Then $I = \int_R \int xy^2 \, dx \, dy$ becomes ....

$$\boxed{I = \int_R \int r^3 \sin^2\theta \cos\theta \, r \, dr \, d\theta}$$

Because $xy^2 = r\cos\theta \, r^2 \sin^2\theta = r^3 \sin^2\theta \cos\theta$

$$\left| \frac{\partial(x, y)}{\partial(r, \theta)} \right| dr \, d\theta = r \, dr \, d\theta$$

$$\therefore I = \int_R \int r^3 \sin^2\theta \cos\theta \, r \, dr \, d\theta = \int_R \int r^4 \sin^2\theta \cos\theta \, dr \, d\theta$$

Now this one.

Example 2

Express $I = \int_R \int (x^2 + y^2) \, dx \, dy$ in terms of u and v, given that $x = u^2 - v^2$ and $y = 2uv$.

First of all, the expression for $\dfrac{\partial(x, y)}{\partial(u, v)}$ gives ....

68

$$\boxed{4(u^2+v^2)}$$

Because

$$x = u^2 - v^2 \quad \therefore \frac{\partial x}{\partial u} = 2u \quad \frac{\partial x}{\partial v} = -2v$$

$$y = 2uv \quad \therefore \frac{\partial y}{\partial u} = 2v \quad \frac{\partial y}{\partial v} = 2u$$

$$\therefore \frac{\partial(x, y)}{\partial(u, v)} = \begin{vmatrix} \frac{\partial x}{\partial u} & \frac{\partial y}{\partial u} \\ \frac{\partial x}{\partial v} & \frac{\partial y}{\partial v} \end{vmatrix} = \begin{vmatrix} 2u & 2v \\ -2v & 2u \end{vmatrix} = 4(u^2 + v^2)$$

Also $x^2 + y^2 = (u^2 - v^2)^2 + (2uv)^2 = u^4 - 2u^2v^2 + v^4 + 4u^2v^2$
$$= u^4 + 2u^2v^2 + v^4 = (u^2 + v^2)^2$$

Then $I = \iint_R (x^2 + y^2)\,dx\,dy$ becomes $I = \ldots\ldots\ldots\ldots$

69

$$\boxed{I = 4\int_R\int (u^2+v^2)^3\,du\,dv}$$

One more.

Example 3

By substituting $x = 2uv$ and $y = u(1-v)$ where $u > 0$ and $v > 0$, express the integral $I = \int_R\int x^2 y\,dx\,dy$ in terms of u and v.

Complete it: there are no snags. $\quad I = \ldots\ldots\ldots\ldots$

Surface and volume integrals

$$I = 8\int_R\int u^4 v^2 (1-v)\,du\,dv$$

Working:

$x = 2uv \qquad \therefore \dfrac{\partial x}{\partial u} = 2v \qquad \dfrac{\partial x}{\partial v} = 2u$

$y = u - uv \qquad \dfrac{\partial y}{\partial u} = 1 - v \qquad \dfrac{\partial y}{\partial v} = -u$

$\therefore J(u,v) = \dfrac{\partial(x,y)}{\partial(u,v)} = \begin{vmatrix} \dfrac{\partial x}{\partial u} & \dfrac{\partial y}{\partial u} \\ \dfrac{\partial x}{\partial v} & \dfrac{\partial y}{\partial v} \end{vmatrix} = \begin{vmatrix} 2v & 1-v \\ 2u & -u \end{vmatrix}$

$= 2u \begin{vmatrix} v & 1-v \\ 1 & -1 \end{vmatrix} = 2u \begin{vmatrix} v & 1 \\ 1 & 0 \end{vmatrix} = -2u$

$\therefore \left| \dfrac{\partial(x,y)}{\partial(u,v)} \right| = 2u$

$x^2 y = 4u^2 v^2 (u - uv) = 4u^3 v^2 (1-v)$

$\therefore I = \int_R \int 4u^3 v^2 (1-v)\, 2u\, du\, dv$

$I = 8 \int_R \int u^4 v^2 (1-v)\, du\, dv$

Transformation in three dimensions

If we extend the previous results to convert variables (x, y, z) to (u, v, w), we proceed in just the same way.

If $x = f(u, v, w);\ y = g(u, v, w);\ z = h(u, v, w)$

Then $\quad J(u, v, w) = \dfrac{\partial(x, y, z)}{\partial(u, v, w)} = \begin{vmatrix} \dfrac{\partial x}{\partial u} & \dfrac{\partial y}{\partial u} & \dfrac{\partial z}{\partial u} \\ \dfrac{\partial x}{\partial v} & \dfrac{\partial y}{\partial v} & \dfrac{\partial z}{\partial v} \\ \dfrac{\partial x}{\partial w} & \dfrac{\partial y}{\partial w} & \dfrac{\partial z}{\partial w} \end{vmatrix}$

and the element of volume $dV = dx\,dy\,dz$ becomes

$dV = |J(u, v, w)|\,du\,dv\,dw$

Also $\iiint F(x, y, z)\,dx\,dy\,dz$ is transformed into

$\iiint G(u, v, w) \left| \dfrac{\partial(x, y, z)}{\partial(u, v, w)} \right| du\,dv\,dw$

Now for an example, so move on

188 Vector Analysis

71

Example 4

To transform a triple integral $I = \iiint F(x, y, z) \, dx \, dy \, dz$ in Cartesian coordinates to spherical coordinates by the transformation equations

$$x = r \sin \theta \cos \phi$$
$$y = r \sin \theta \sin \phi$$
$$z = r \cos \theta$$

First we need the partial derivatives, from which to build up the Jacobian. These are ....

72

$$\frac{\partial x}{\partial r} = \sin \theta \cos \phi \qquad \frac{\partial y}{\partial r} = \sin \theta \sin \phi \qquad \frac{\partial z}{\partial r} = \cos \theta$$

$$\frac{\partial x}{\partial \theta} = r \cos \theta \cos \phi \qquad \frac{\partial y}{\partial \theta} = r \cos \theta \sin \phi \qquad \frac{\partial z}{\partial \theta} = -r \sin \theta$$

$$\frac{\partial x}{\partial \phi} = -r \sin \theta \sin \phi \qquad \frac{\partial y}{\partial \phi} = r \sin \theta \cos \phi \qquad \frac{\partial z}{\partial \phi} = 0$$

$$\therefore J(r, \theta, \phi) = \begin{vmatrix} \sin \theta \cos \phi & \sin \theta \sin \phi & \cos \theta \\ r \cos \theta \cos \phi & r \cos \theta \sin \phi & -r \sin \theta \\ -r \sin \theta \sin \phi & r \sin \theta \cos \phi & 0 \end{vmatrix}$$

$$= \cos \theta \begin{vmatrix} r \cos \theta \cos \phi & r \cos \theta \sin \phi \\ -r \sin \theta \sin \phi & r \sin \theta \cos \phi \end{vmatrix}$$

$$+ r \sin \theta \begin{vmatrix} \sin \theta \cos \phi & \sin \theta \sin \phi \\ -r \sin \theta \sin \phi & r \sin \theta \cos \phi \end{vmatrix}$$

$$= \ldots \ldots \ldots$$

73

$$\boxed{r^2 \sin \theta}$$

Because

$$J(r, \theta, \phi) = r^2 \cos^2 \theta \sin \theta \begin{vmatrix} \cos \phi & \sin \phi \\ -\sin \phi & \cos \phi \end{vmatrix}$$

$$+ r^2 \sin^3 \theta \begin{vmatrix} \cos \phi & \sin \phi \\ -\sin \phi & \cos \phi \end{vmatrix}$$

$$= (r^2 \sin^3 \theta + r^2 \sin \theta \cos^2 \theta) \begin{vmatrix} \cos \phi & \sin \phi \\ -\sin \phi & \cos \phi \end{vmatrix}$$

$$= r^2 \sin \theta (\sin^2 \theta + \cos^2 \theta)(\cos^2 \phi + \sin^2 \phi) = r^2 \sin \theta$$

$$\therefore I = \iiint G(u, v, w) r^2 \sin \theta \, dr \, d\theta \, d\phi$$

which agrees, of course, with the result we had previously obtained by a geometric consideration.

Surface and volume integrals

And that is about it. Check carefully down the **Review summary** and the **Can You?** checklist that now follow, before working through the **Test exercise**. The **Further problems** give additional practice.

Review summary

1 *Surface integrals*

$$I = \int_R f(x, y)\, da = \int_R \int f(x, y)\, dy\, dx$$

2 *Surface in space*

$$I = \int_S \phi(x, y, z)\, dS = \int_R \int \phi(x, y, z) \sec\gamma\, dx\, dy \quad (\gamma < \pi/2)$$

$$= \int_R \int \phi(x, y, z) \sqrt{1 + \left(\frac{\partial z}{\partial x}\right)^2 + \left(\frac{\partial z}{\partial y}\right)^2}\, dx\, dy$$

3 *Space coordinate systems*
(a) *Cartesian coordinates* (x, y, z)

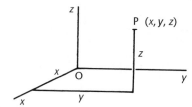

First octant:
$x \geq 0;\ y \geq 0;\ z \geq 0$

(b) *Cylindrical coordinates* $(r, \theta, z)\ r \geq 0$

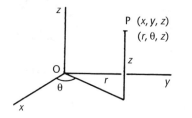

$x = r\cos\theta \qquad r = \sqrt{x^2 + y^2}$
$y = r\sin\theta \qquad \theta = \arctan(y/x)$
$z = z \qquad\qquad z = z$

(c) *Spherical coordinates* $(r, \theta, \phi)\ r \geq 0$

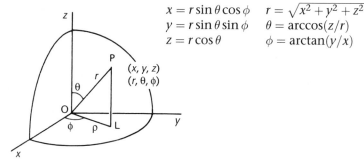

$x = r\sin\theta\cos\phi \qquad r = \sqrt{x^2 + y^2 + z^2}$
$y = r\sin\theta\sin\phi \qquad \theta = \arccos(z/r)$
$z = r\cos\theta \qquad\qquad \phi = \arctan(y/x)$

▶

4 Elements of volume
(a) Cartesian coordinates

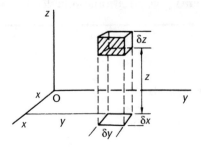

$$\delta v = \delta x \, \delta y \, \delta z$$

(b) Cylindrical coordinates $r \geq 0$

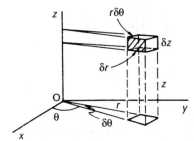
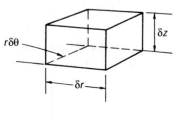

$$\delta v = r \, \delta r \, \delta \theta \, \delta z$$

(c) Spherical coordinates

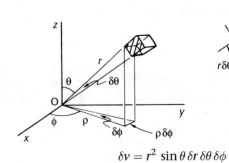
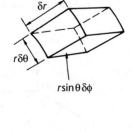

$$\delta v = r^2 \sin \theta \, \delta r \, \delta \theta \, \delta \phi$$

5 Volume integrals

$$V = \iiint dz \, dy \, dx$$

$$I = \iiint f(x, y, z) \, dz \, dy \, dx$$

Surface and volume integrals

6 **Change of variables** in multiple integrals
(a) *Double integrals* $x = f(u, v); \quad y = g(u, v)$

$$dA = \left|\frac{\partial(x, y)}{\partial(u, v)}\right| du\, dv; \quad J(u, v) = \frac{\partial(x, y)}{\partial(u, v)} = \begin{vmatrix} \frac{\partial x}{\partial u} & \frac{\partial y}{\partial u} \\ \frac{\partial x}{\partial v} & \frac{\partial y}{\partial v} \end{vmatrix}$$

$$I = \int\!\!\int_R F(x, y)\, dx\, dy = \int\!\!\int_R F\{f(u, v), g(u, v)\} \left|\frac{\partial(x, y)}{\partial(u, v)}\right| du\, dv$$

(b) *Triple integrals* $\quad x = f(u, v, w); \; y = g(u, v, w); \; z = h(u, v, w)$

$$J(u, v, w) = \frac{\partial(x, y, z)}{\partial(u, v, w)} = \begin{vmatrix} \frac{\partial x}{\partial u} & \frac{\partial y}{\partial u} & \frac{\partial z}{\partial u} \\ \frac{\partial x}{\partial v} & \frac{\partial y}{\partial v} & \frac{\partial z}{\partial v} \\ \frac{\partial x}{\partial w} & \frac{\partial y}{\partial w} & \frac{\partial z}{\partial w} \end{vmatrix}$$

Then $I = \int\!\!\int\!\!\int F(x, y, z)\, dx\, dy\, dz$

$$= \int\!\!\int\!\!\int G(u, v, w) \left|\frac{\partial(x, y, z)}{\partial(u, v, w)}\right| du\, dv\, dw$$

✓ Can You?

Checklist 6

75

Check this list before and after you try the end of Program test.

On a scale of 1 to 5, how confident are you that you can: Frames

- Evaluate double integrals and surface integrals?   1 to 22
  Yes ☐ ☐ ☐ ☐ ☐ No

- Relate three-dimensional Cartesian coordinates to cylindrical and spherical polar forms?   23 to 31
  Yes ☐ ☐ ☐ ☐ ☐ No

- Evaluate volume integrals in Cartesian coordinates and in cylindrical and spherical polar coordinates?   32 to 50
  Yes ☐ ☐ ☐ ☐ ☐ No

- Use the Jacobian to convert integrals given in Cartesian coordinates into general curvilinear coordinates in two and three dimensions?   51 to 73
  Yes ☐ ☐ ☐ ☐ ☐ No

Test exercise 6

1. Determine the area of the surface $z = \sqrt{x^2 + y^2}$ over the region bounded by $x^2 + y^2 = 4$.

2. Evaluate the surface integral $I = \int_S \phi \, dS$ where $\phi = \dfrac{1}{\sqrt{x^2 + y^2}}$ over the surface of the sphere $x^2 + y^2 + z^2 = a^2$ in the first octant.

3. (a) Transform the Cartesian coordinates
    (1) $(4, 2, 3)$ to cylindrical coordinates (r, θ, z).
    (2) $(3, 1, 5)$ to spherical coordinates (r, θ, ϕ).
   (b) Express in Cartesian coordinates (x, y, z)
    (1) the cylindrical coordinates $(5, \pi/4, 3)$
    (2) the spherical coordinates $(4, \pi/6, 2)$.

4. Determine the volume of the solid bounded by the plane $z = 0$ and the surfaces $x^2 + y^2 = 4$ and $z = x^2 + y^2 + 1$.

5. Determine the total mass of a solid hemisphere bounded by the plane $z = 0$ and the surface $x^2 + y^2 + z^2 = a^2$ $(z \geq 0)$ if the density at any point is given by $\rho = 1 - z$ $(z < a)$.

6. (a) Express the integral $I = \int\int_R (x - y) \, dx \, dy$ in terms of u and v, where $x = u(1 + v)$ and $y = u - v$.
   (b) Express the triple integral $I = \int\int\int \left(\dfrac{x + z}{y}\right) dx \, dy \, dz$ in terms of u, v, w using the transformation equations
   $x = u + v + w; \quad y = v^2 w; \quad z = u - w.$

Further problems 6

1. Evaluate the surface integral $I = \int_S (x^2 + y^2) \, dS$ over the surface of the cone $z^2 = 4(x^2 + y^2)$ between $z = 0$ and $z = 4$.

2. Find the position of the center of gravity of that part of a thin spherical shell $x^2 + y^2 + z^2 = a^2$ which exists in the first octant.

3. Determine the surface area of the plane $6x + 3y + 4z = 60$ cut off by $x = 0$, $y = 0$, $x = 5$, $y = 8$.

4. Find the surface area of the plane $3x + 2y + 3z = 12$ cut off by the planes $x = 0$, $y = 0$, and the cylinder $x^2 + y^2 = 16$ for $x \geq 0$, $y \geq 0$.

Surface and volume integrals

5 Determine the area of the paraboloid $z = 2(x^2 + y^2)$ cut off by the cone $z = \sqrt{x^2 + y^2}$.

6 Find the area of the cone $z^2 = 4(x^2 + y^2)$ which is inside the paraboloid $z = 2(x^2 + y^2)$.

7 Cylinders $x^2 + y^2 = a^2$ and $x^2 + z^2 = a^2$ intersect. Determine the total external surface area of the common portion.

8 Determine the surface area of the sphere $x^2 + y^2 + z^2 = a^2$ cut off by the cylinder $x^2 + y^2 = ax$.

9 A cylinder of radius b, with the z-axis as its axis of symmetry, is removed from a sphere of radius a, $a > b$, with center at the origin. Calculate the total curved surface area of the ring so formed, including the inner cylindrical surface.

10 Find the volume enclosed by the cylinder $x^2 + y^2 = 9$ and the planes $z = 0$ and $z = 5 - x$.

11 Determine the volume of the solid bounded by the surfaces $y = x^2$, $x = y^2$, $z = 2$ and $x + y + z = 4$.

12 Find the volume of the solid bounded by the plane $z = 0$, the cylinder $x^2 + y^2 = a^2$ and the surface $z = x^2 + y^2$.

13 A solid is bounded by the planes $x = 0$, $y = 0$, $z = 2$, $z = x$ and the surface $x^2 + y^2 = 4$. Determine the volume of the solid.

14 Find the position of the center of gravity of the part of the solid sphere $x^2 + y^2 + z^2 = a^2$ in the first octant.

15 A solid is bounded by the cone $z = 2\sqrt{x^2 + y^2}$, $z \geq 0$, and the sphere $x^2 + y^2 + (z - a)^2 = 2a^2$. Determine the volume of the solid so formed.

16 Determine the volume enclosed by the ellipsoid $\dfrac{x^2}{a^2} + \dfrac{y^2}{b^2} + \dfrac{z^2}{c^2} = 1$.

17 Find the volume of the solid in the first octant bounded by the planes $x = 0$, $y = 0$, $z = 0$, $z = x + y$ and the surface $x^2 + y^2 = a^2$.

18 Express the integral $\iint (x^2 + y^2)\, dx\, dy$ in terms of u and v, using the transformations $u = x + y$, $v = x - y$.

19 Determine an expression for the element of volume $dx\, dy\, dz$ in terms of u, v, w using the transformations $x = u(1 - v)$, $y = uv$, $z = uvw$.

20 A solid sphere of radius a has variable density c at any point (x, y, z) given by $c = k(a - z)$ where k is a constant. Determine the position of the center of gravity of the sphere.

21 Calculate $\iint x^2 y^2\, dx\, dy$ over the triangular region in the x-y plane with vertices $(0, 0)$, $(1, 1)$, $(1, 2)$.

22 Evaluate the integral $I = \int_0^2 \int_{\sqrt{y(2-y)}}^{\sqrt{4-y^2}} \frac{y}{x^2+y^2} \, dx \, dy$ by transforming to polar coordinates.

23 Evaluate $I = \int_0^1 \int_0^y \frac{xy^2}{\sqrt{x^2+y^2}} \, dx \, dy$.

24 Find the volume bounded by the cylinder $x^2 + y^2 = a^2$, the plane $z = 0$ and the surface $z = x^2 + y^2$. Convert to polar coordinates and show that
$$V = \frac{\pi a^4}{2}.$$

25 By changing the order of integration in the integral
$$I = \int_0^a \int_x^a \frac{y^2 \, dy \, dx}{\sqrt{x^2+y^2}}$$
show that $I = \frac{1}{3}a^3 \ln(1+\sqrt{2})$.

Program 7

Vectors

Frames 1 to 79

Learning outcomes

When you have completed this Program you will be able to:
- Define a vector
- Represent a vector by a directed straight line
- Add vectors
- Write a vector in terms of component vectors
- Write a vector in terms of component unit vectors
- Set up a coordinate system for representing vectors
- Obtain the direction cosines of a vector
- Calculate the scalar product of two vectors
- Calculate the vector product of two vectors
- Determine the angle between two vectors
- Evaluate the direction ratios of a vector
- Obtain the scalar and vector triple products and appreciate their geometric significance

Introduction: scalar and vector quantities

1

Physical quantities can be divided into two main groups, scalar quantities and vector quantities.

(a) A *scalar quantity* is one that is defined completely by a single number with appropriate units, e.g. length, area, volume, mass, time, etc. Once the units are stated, the quantity is denoted entirely by its size or *magnitude*.

(b) A *vector quantity* is defined completely when we know not only its magnitude (with units) but also the direction in which it operates, e.g. force, velocity, acceleration. A vector quantity necessarily involves *direction* as well as magnitude.

So (a) a speed of 10 km/h is a scalar quantity, but
 (b) a velocity of 10 km/h due north is a .... quantity.

2

> vector

A force F acting at a point P is a vector quantity, since to define it completely we must give:

(a) its magnitude, and also
(b) its ....

3

> direction

So that:
(a) A temperature of 100°C is a .... quantity.
(b) An acceleration of 9.8 m/s² vertically downwards is a .... quantity.
(c) The weight of a 7 kg mass is a .... quantity.
(d) The sum of £500 is a .... quantity.
(e) A north-easterly wind of 20 knots is a .... quantity.

4

> (a) scalar   (b) vector   (c) vector   (d) scalar   (e) vector

Since, in (b), (c) and (e) the complete description of the quantity includes not only its magnitude, but also its ....

5

> direction

Move on to Frame 6

Vector representation

A vector quantity can be represented graphically by a line, drawn so that:

(a) the *length* of the line denotes the magnitude of the quantity, according to some stated vector scale

(b) the *direction* of the line denotes the direction in which the vector quantity acts. The sense of the direction is indicated by an arrowhead.

e.g. A horizontal force of 35 N acting to the right, would be indicated by a line ⟶ and if the chosen vector scale were 1 cm ≡ 10 N, the line would be .... cm long.

The vector quantity AB is referred to as

$\overline{AB}$ or **a**

The magnitude of the vector quantity is written $|\overline{AB}|$, or $|\mathbf{a}|$, or simply AB or a.

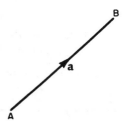

Note that $\overline{BA}$ would represent a vector quantity of the same magnitude but with opposite sense.

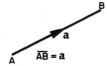 

On to Frame 8

Two equal vectors

If two vectors, **a** and **b**, are said to be equal, they have the same magnitude and the same direction.

If **a** = **b**, then

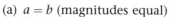

(a) $a = b$ (magnitudes equal)

(b) the direction of **a** = direction of **b**, i.e. the two vectors are parallel and in the same sense.

Similarly, if two vectors **a** and **b** are such that **b** = −**a**, what can we say about:

(a) their magnitudes,
(b) their directions?

9

(a) Magnitudes are equal
(b) The vectors are parallel but opposite in sense

i.e. if **b** = −**a**, then

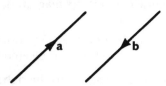

10 Types of vectors

(a) A *position vector* $\overline{AB}$ occurs when the point A is fixed.
(b) A *line vector* is such that it can slide along its line of action, e.g. a mechanical force acting on a body.
(c) A *free vector* is not restricted in any way. It is completely defined by its magnitude and direction and can be drawn as any one of a set of equal-length parallel lines.

Most of the vectors we shall consider will be free vectors

So on now to Frame 11

11 Addition of vectors

The sum of two vectors, $\overline{AB}$ and $\overline{BC}$, is defined as the single or equivalent or resultant vector $\overline{AC}$

i.e. $\overline{AB} + \overline{BC} = \overline{AC}$

or $\mathbf{a} + \mathbf{b} = \mathbf{c}$

To find the sum of two vectors **a** and **b** then, we draw them as a chain, starting the second where the first ends: the sum **c** is given by the single vector joining the start of the first to the end of the second.

e.g. if **p** ≡ a force of 40 N, acting in the direction due east

q ≡ a force of 30 N, acting in the direction due north

then the magnitude of the vector sum *r* of these will forces will be ....

12

$r = 50$ N

Because

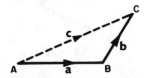

$r^2 = p^2 + q^2$
$= 1600 + 900 = 2500$
$r = \sqrt{2500} = 50$ N

Vectors

The sum of a number of vectors a + b + c + d + ...

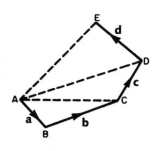

(a) Draw the vectors as a chain.
(b) Then:
$$\mathbf{a} + \mathbf{b} = \overline{AC}$$
$$\overline{AC} + \mathbf{c} = \overline{AD}$$
$$\therefore \mathbf{a} + \mathbf{b} + \mathbf{c} = \overline{AD}$$
$$\overline{AD} + \mathbf{d} = \overline{AE}$$
$$\therefore \mathbf{a} + \mathbf{b} + \mathbf{c} + \mathbf{d} = \overline{AE}$$

i.e. the sum of all vectors, **a, b, c, d**, is given by the single vector joining the start of the first to the end of the last – in this case, $\overline{AE}$. This follows directly from our previous definition of the sum of two vectors.

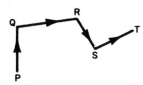

Similarly:
$$\overline{PQ} + \overline{QR} + \overline{RS} + \overline{ST} = \ldots\ldots\ldots$$

| $\overline{PT}$ |

| 13 |

Now suppose that in another case, we draw the vector diagram to find the sum of **a, b, c, d, e**, and discover that the resulting diagram is, in fact, a closed figure.

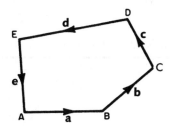

What is the sum of the vectors **a, b, c, d, e** in this case?

Think carefully and when you have decided, move on to Frame 14

| Sum of the vectors = **0** |

| 14 |

Because we said in the previous case, that the vector sum was given by the single equivalent vector joining the beginning of the first vector to the end of the last.

But, if the vector diagram is a closed figure, the end of the last vector coincides with the beginning of the first, so that the resultant sum is a vector with no magnitude.

▶

Now for some examples:

Find the vector sum $\overline{AB} + \overline{BC} + \overline{CD} + \overline{DE} + \overline{EF}$.

Without drawing a diagram, we can see that the vectors are arranged in a chain, each beginning where the previous one left off. The sum is therefore given by the vector joining the beginning of the first vector to the end of the last.

$$\therefore \text{Sum} = \overline{AF}$$

In the same way:

$$\overline{AK} + \overline{KL} + \overline{LP} + \overline{PQ} = \ldots\ldots\ldots\ldots$$

15

$$\boxed{\overline{AQ}}$$

Right. Now what about this one?

Find the sum of $\overline{AB} - \overline{CB} + \overline{CD} - \overline{ED}$

We must beware of the negative vectors. Remember that $-\overline{CB} = \overline{BC}$, i.e. the same magnitude and direction but in the opposite sense.
Also $-\overline{ED} = \overline{DE}$

$$\therefore \overline{AB} - \overline{CB} + \overline{CD} - \overline{ED} = \overline{AB} + \overline{BC} + \overline{CD} + \overline{DE}$$
$$= \overline{AE}$$

Now you do this one:

Find the vector sum $\overline{AB} + \overline{BC} - \overline{DC} - \overline{AD}$

When you have the result, move on to Frame 16

16

Because

$$\overline{AB} + \overline{BC} - \overline{DC} - \overline{AD} = \overline{AB} + \overline{BC} + \overline{CD} + \overline{DA}$$

and the lettering indicates that the end of the last vector coincides with the beginning of the first. The vector diagram is thus a closed figure and therefore the sum of the vectors is **0**.

Now here are some for you to do:

(a) $\overline{PQ} + \overline{QR} + \overline{RS} + \overline{ST} = \ldots\ldots\ldots\ldots$
(b) $\overline{AC} + \overline{CL} - \overline{ML} = \ldots\ldots\ldots\ldots$
(c) $\overline{GH} + \overline{HJ} + \overline{JK} + \overline{KL} + \overline{LG} = \ldots\ldots\ldots\ldots$
(d) $\overline{AB} + \overline{BC} + \overline{CD} + \overline{DB} = \ldots\ldots\ldots\ldots$

When you have finished all four, check with the results in the next frame

Vectors

Here are the results:

(a) $\overline{PQ} + \overline{QR} + \overline{RS} + \overline{ST} = \overline{PT}$

(b) $\overline{AC} + \overline{CL} - \overline{ML} = \overline{AC} + \overline{CL} + \overline{LM} = \overline{AM}$

(c) $\overline{GH} + \overline{HJ} + \overline{JK} + \overline{KL} + \overline{LG} = \mathbf{0}$

[Since the end of the last vector coincides with the beginning of the first.]

(d) $\overline{AB} + \overline{BC} + \overline{CD} + \overline{DB} = \overline{AB}$

The last three vectors form a closed figure and therefore the sum of these three vectors is zero, leaving only $\overline{AB}$ to be considered.

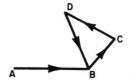

Now on to Frame 18

Components of a given vector

Just as $\overline{AB} + \overline{BC} + \overline{CD} + \overline{DE}$ can be replaced by $\overline{AE}$, so any single vector $\overline{PT}$ can be replaced by any number of component vectors so long as they form a chain in the vector diagram, beginning at P and ending at T.

e.g.

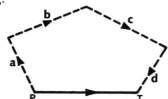

$$\overline{PT} = \mathbf{a} + \mathbf{b} + \mathbf{c} + \mathbf{d}$$

Example 1

ABCD is a quadrilateral, with G and H the mid-points of DA and BC respectively. Show that $\overline{AB} + \overline{DC} = 2\overline{GH}$.

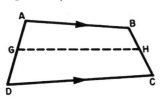

We can replace vector $\overline{AB}$ by any chain of vectors so long as they start at A and end at B e.g. we could say

$$\overline{AB} = \overline{AG} + \overline{GH} + \overline{HB}$$

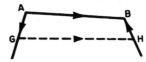

Similarly, we could say

$$\overline{DC} = \ldots\ldots\ldots\ldots$$

19

$$\overline{DC} = \overline{DG} + \overline{GH} + \overline{HC}$$

So we have:

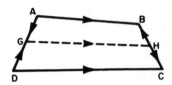

$$\overline{AB} = \overline{AG} + \overline{GH} + \overline{HB}$$
$$\overline{DC} = \overline{DG} + \overline{GH} + \overline{HC}$$

$$\therefore \overline{AB} + \overline{DC} = \overline{AG} + \overline{GH} + \overline{HB} + \overline{DG} + \overline{GH} + \overline{HC}$$
$$= 2\overline{GH} + (\overline{AG} + \overline{DG}) + (\overline{HB} + \overline{HC})$$

Now, G is the mid-point of AD. Therefore, vectors $\overline{AG}$ and $\overline{DG}$ are equal in length but opposite in sense.

$$\therefore \overline{DG} = -\overline{AG}$$

Similarly $\overline{HC} = -\overline{HB}$

$$\therefore \overline{AB} + \overline{DC} = 2\overline{GH} + (\overline{AG} - \overline{AG}) + (\overline{HB} - \overline{HB})$$
$$= 2\overline{GH}$$

Next frame

20

Example 2

Points L, M, N are mid-points of the sides AB, BC, CA of the triangle ABC. Show that:

(a) $\overline{AB} + \overline{BC} + \overline{CA} = \mathbf{0}$
(b) $2\overline{AB} + 3\overline{BC} + \overline{CA} = 2\overline{LC}$
(c) $\overline{AM} + \overline{BN} + \overline{CL} = \mathbf{0}$

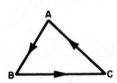

(a) We can dispose of the first part straight away without any trouble. We can see from the vector diagram that $\overline{AB} + \overline{BC} + \overline{CA} = \mathbf{0}$ since these three vectors form a ....

21

closed figure

Now for part (b):

To show that $2\overline{AB} + 3\overline{BC} + \overline{CA} = 2\overline{LC}$

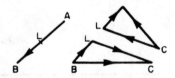

Vectors

From the figure:
$$\overline{AB} = 2\overline{AL}; \quad \overline{BC} = \overline{BL} + \overline{LC}; \quad \overline{CA} = \overline{CL} + \overline{LA}$$
$$\therefore 2\overline{AB} + 3\overline{BC} + \overline{CA} = 4\overline{AL} + 3\overline{BL} + 3\overline{LC} + \overline{CL} + \overline{LA}$$
Now $\overline{BL} = -\overline{AL}; \quad \overline{CL} = -\overline{LC}; \quad \overline{LA} = -\overline{AL}$

Substituting these in the previous line, gives

$2\overline{AB} + 3\overline{BC} + \overline{CA} = \ldots\ldots\ldots\ldots$

$$\boxed{2\overline{LC}}$$

Because
$$2\overline{AB} + 3\overline{BC} + \overline{CA} = 4\overline{AL} + 3\overline{BL} + 3\overline{LC} + \overline{CL} + \overline{LA}$$
$$= 4\overline{AL} - 3\overline{AL} + 3\overline{LC} - \overline{LC} - \overline{AL}$$
$$= 4\overline{AL} - 4\overline{AL} + 3\overline{LC} - \overline{LC}$$
$$= 2\overline{LC}$$

Now part (c):

To prove that $\overline{AM} + \overline{BN} + \overline{CL} = \mathbf{0}$

From the figure in Frame 21, we can say:
$$\overline{AM} = \overline{AB} + \overline{BM}$$
$$\overline{BN} = \overline{BC} + \overline{CN}$$
Similarly $\overline{CL} = \ldots\ldots\ldots\ldots$

$$\boxed{\overline{CL} = \overline{CA} + \overline{AL}}$$

So $\overline{AM} + \overline{BN} + \overline{CL} = \overline{AB} + \overline{BM} + \overline{BC} + \overline{CN} + \overline{CA} + \overline{AL}$
$$= (\overline{AB} + \overline{BC} + \overline{CA}) + (\overline{BM} + \overline{CN} + \overline{AL})$$
$$= (\overline{AB} + \overline{BC} + \overline{CA}) + \frac{1}{2}(\overline{BC} + \overline{CA} + \overline{AB})$$
$$= \ldots\ldots\ldots\ldots$$

Finish it off

$$\boxed{\overline{AM} + \overline{BN} + \overline{CL} = \mathbf{0}}$$

Because $\overline{AM} + \overline{BN} + \overline{CL} = (\overline{AB} + \overline{BC} + \overline{CA}) + \frac{1}{2}(\overline{BC} + \overline{CA} + \overline{AB})$

Now $\overline{AB} + \overline{BC} + \overline{CA}$ is a closed figure $\therefore$ Vector sum $= \mathbf{0}$
and $\overline{BC} + \overline{CA} + \overline{AB}$ is a closed figure $\therefore$ Vector sum $= \mathbf{0}$
$\therefore \overline{AM} + \overline{BN} + \overline{CL} = \mathbf{0}$

Here is another.

Example 3

ABCD is a quadrilateral in which P and Q are the mid-points of the diagonals AC and BD respectively.

Show that $\overline{AB} + \overline{AD} + \overline{CB} + \overline{CD} = 4\overline{PQ}$

First, just draw the figure.

Then move on to Frame 25

25

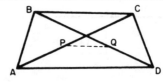

To prove that $\overline{AB} + \overline{AD} + \overline{CB} + \overline{CD} = 4\overline{PQ}$

Taking the vectors on the left-hand side, one at a time, we can write:

$\overline{AB} = \overline{AP} + \overline{PQ} + \overline{QB}$
$\overline{AD} = \overline{AP} + \overline{PQ} + \overline{QD}$
$\overline{CB} = \ldots\ldots\ldots$
$\overline{CD} = \ldots\ldots\ldots$

26

$$\boxed{\overline{CB} = \overline{CP} + \overline{PQ} + \overline{QB}; \quad \overline{CD} = \overline{CP} + \overline{PQ} + \overline{QD}}$$

Adding all four lines together, we have:

$\overline{AB} + \overline{AD} + \overline{CB} + \overline{CD} = 4\overline{PQ} + 2\overline{AP} + 2\overline{CP} + 2\overline{QB} + 2\overline{QD}$
$\qquad\qquad\qquad\qquad\qquad = 4\overline{PQ} + 2(\overline{AP} + \overline{CP}) + 2(\overline{QB} + \overline{QD})$

Now what can we say about $(\overline{AP} + \overline{CP})$?

27

$$\boxed{\overline{AP} + \overline{CP} = \mathbf{0}}$$

Because P is the mid-point of AC $\therefore$ AP = PC

$\therefore \overline{CP} = -\overline{PC} = -\overline{AP}$

$\therefore \overline{AP} + \overline{CP} = \overline{AP} - \overline{AP} = \mathbf{0}$.

In the same way, $(\overline{QB} + \overline{QD}) = \ldots\ldots\ldots$

28

$$\boxed{\overline{QB} + \overline{QD} = \mathbf{0}}$$

Since Q is the mid-point of BD $\therefore \overline{QD} = -\overline{QB}$

$\therefore \overline{QB} + \overline{QD} = \overline{QB} - \overline{QB} = \mathbf{0}$

$\therefore \overline{AB} + \overline{AD} + \overline{CB} + \overline{CD} = 4\overline{PQ} + \mathbf{0} + \mathbf{0} = 4\overline{PQ}$

Here is one more.

Example 4

Prove by vectors that the line joining the mid-points of two sides of a triangle is parallel to the third side and half its length.

Let D and E by the mid-points of AB and AC respectively.

We have $\quad \overline{DE} = \overline{DA} + \overline{AE}$

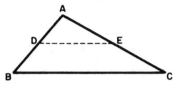

Now express $\overline{DA}$ and $\overline{AE}$ in terms of $\overline{BA}$ and $\overline{AC}$ respectively and see if you can get the required results.

Then on to Frame 30

Here is the working. Check through it.

$$\overline{DE} = \overline{DA} + \overline{AE}$$
$$= \frac{1}{2}\overline{BA} + \frac{1}{2}\overline{AC} = \frac{1}{2}(\overline{BA} + \overline{AC})$$
$$\overline{DE} = \frac{1}{2}\overline{BC}$$

∴ $\overline{DE}$ is half the magnitude (length) of $\overline{BC}$ and acts in the same direction.

i.e. DE and BC are parallel.

Now for the next section of the work: move on to Frame 31

Components of a vector in terms of unit vectors

The vector $\overline{OP}$ is defined by its magnitude (r) and its direction (θ). It could also be defined by its two components in the OX and OY directions.

i.e. $\overline{OP}$ is equivalent to a vector **a** in the OX direction + a vector **b** in the OY direction.

i.e. $\overline{OP} = \mathbf{a}$ (along OX) $+ \mathbf{b}$ (along OY)

If we now define **i** to be a *unit vector* in the OX direction,

then $\quad \mathbf{a} = a\mathbf{i}$

Similarly, if we define **j** to be a *unit vector* in the OY direction,

then $\quad \mathbf{b} = b\mathbf{j}$

So that the vector OP can be written as:

$\mathbf{r} = a\mathbf{i} + b\mathbf{j}$

where **i** and **j** are unit vectors in the OX and OY directions.

32

Let $z_1 = 2i + 4j$ and $z_2 = 5i + 2j$

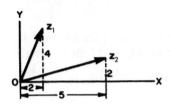

To find $z_1 + z_2$, draw the two vectors in a chain.

$$z_1 + z_2 = \overline{OB} = (2+5)i + (4+2)j = 7i + 6j$$

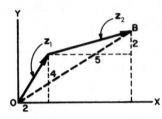

i.e. total up the vector components along OX, and total up the vector components along OY

Of course, we can do this without a diagram:

If $z_1 = 3i + 2j$ and $z_2 = 4i + 3j$

$$\begin{aligned} z_1 + z_2 &= 3i + 2j + 4i + 3j \\ &= 7i + 5j \end{aligned}$$

And in much the same way, $z_2 - z_1 = \ldots\ldots\ldots\ldots$

33

$$\boxed{z_2 - z_1 = i + j}$$

Because
$$\begin{aligned} z_2 - z_1 &= (4i + 3j) - (3i + 2j) \\ &= 4i + 3j - 3i - 2j \\ &= 1i + 1j \\ &= i + j \end{aligned}$$

Similarly, if $z_1 = 5i - 2j$; $z_2 = 3i + 3j$; $z_3 = 4i - 1j$

then (a) $z_1 + z_2 + z_3 = \ldots\ldots\ldots\ldots$
and (b) $z_1 - z_2 - z_3 = \ldots\ldots\ldots\ldots$

When you have the results, move on to Frame 34

Vectors

34

$$\boxed{\text{(a) } 12\mathbf{i} \qquad \text{(b) } -2\mathbf{i} - 4\mathbf{j}}$$

Here is the working:

(a) $\mathbf{z}_1 + \mathbf{z}_2 + \mathbf{z}_3 = 5\mathbf{i} - 2\mathbf{j} + 3\mathbf{i} + 3\mathbf{j} + 4\mathbf{i} - 1\mathbf{j}$
$= (5 + 3 + 4)\mathbf{i} + (3 - 2 - 1)\mathbf{j} = 12\mathbf{i}$

(b) $\mathbf{z}_1 - \mathbf{z}_2 - \mathbf{z}_3 = (5\mathbf{i} - 2\mathbf{j}) - (3\mathbf{i} + 3\mathbf{j}) - (4\mathbf{i} - 1\mathbf{j})$
$= (5 - 3 - 4)\mathbf{i} + (-2 - 3 + 1)\mathbf{j} = -2\mathbf{i} - 4\mathbf{j}$

Now this one.

If $\overline{OA} = 3\mathbf{i} + 5\mathbf{j}$ and $\overline{OB} = 5\mathbf{i} - 2\mathbf{j}$, find $\overline{AB}$.

As usual, a diagram will help. Here it is:

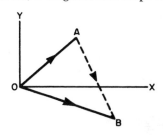

First of all, from the diagram, write down a relationship between the vectors. Then express them in terms of the unit vectors.

$\overline{AB} = \ldots\ldots\ldots\ldots$

35

$$\boxed{\overline{AB} = 2\mathbf{i} - 7\mathbf{j}}$$

Because we have

$\overline{OA} + \overline{AB} = \overline{OB}$ (from diagram)

$\therefore \overline{AB} = \overline{OB} - \overline{OA}$
$= (5\mathbf{i} - 2\mathbf{j}) - (3\mathbf{i} + 5\mathbf{j}) = 2\mathbf{i} - 7\mathbf{j}$

On to Frame 36

Vectors in space

36

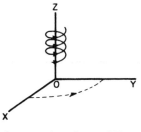

The axes of reference are defined by the 'right-hand' rule.

OX, OY, OZ form a right-handed set if rotation from OX to OY takes a right-handed corkscrew action along the positive direction of OZ.

Similarly, rotation from OY to OZ gives right-hand corkscrew action along the positive direction of $\ldots\ldots\ldots\ldots$

37

$$\boxed{OX}$$

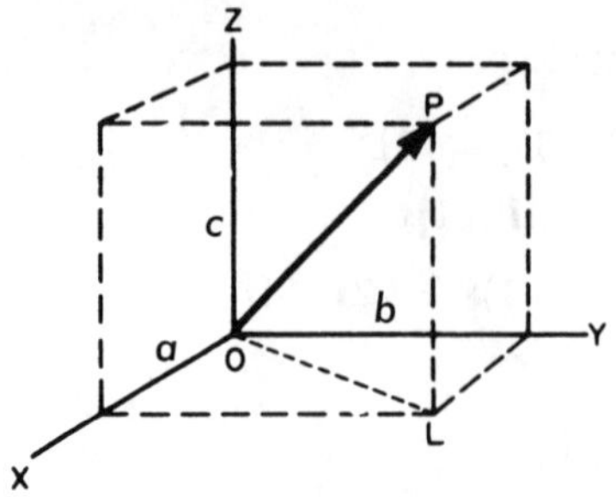

Vector $\overline{OP}$ is defined by its components

a along OX
b along OY
c along OZ

Let $\mathbf{i}$ = unit vector in OX direction
$\mathbf{j}$ = unit vector in OY direction
$\mathbf{k}$ = unit vector in OZ direction

Then $\quad\overline{OP} = a\mathbf{i} + b\mathbf{j} + c\mathbf{k}$

Also $\quad OL^2 = a^2 + b^2$ and $OP^2 = OL^2 + c^2$

$\quad\quad OP^2 = a^2 + b^2 + c^2$

So, if $\mathbf{r} = a\mathbf{i} + b\mathbf{j} + c\mathbf{k}$, then $r = \sqrt{a^2 + b^2 + c^2}$

This gives us an easy way of finding the magnitude of a vector expressed in terms of the unit vectors.

Now you can do this one:

If $\overline{PQ} = 4\mathbf{i} + 3\mathbf{j} + 2\mathbf{k}$, then $|\overline{PQ}| = \ldots\ldots\ldots\ldots$

38

$$\boxed{|\overline{PQ}| = \sqrt{29} = 5.385}$$

Because

$\overline{PQ} = 4\mathbf{i} + 3\mathbf{j} + 2\mathbf{k}$
$|\overline{PQ}| = \sqrt{4^2 + 3^2 + 2^2}$
$\quad\quad = \sqrt{16 + 9 + 4}$
$\quad\quad = \sqrt{29}$
$\quad\quad = 5.385$

Now move on to Frame 39

Direction cosines

39

The direction of a vector in three dimensions is determined by the angles which the vector makes with the three axes of reference.

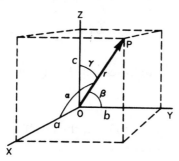

Let $\overline{OP} = \mathbf{r} = a\mathbf{i} + b\mathbf{j} + c\mathbf{k}$

Then

$\dfrac{a}{r} = \cos \alpha \qquad \therefore a = r \cos \alpha$

$\dfrac{b}{r} = \cos \beta \qquad b = r \cos \beta$

$\dfrac{c}{r} = \cos \gamma \qquad c = r \cos \gamma$

Also $a^2 + b^2 + c^2 = r^2$

$\therefore r^2 \cos^2 \alpha + r^2 \cos^2 \beta + r^2 \cos^2 \gamma = r^2$

$\therefore \cos^2 \alpha + \cos^2 \beta + \cos^2 \gamma = 1$

If $l = \cos \alpha$

$m = \cos \beta$

$n = \cos \gamma$ then $l^2 + m^2 + n^2 = 1$

Note: $[l, m, n]$ written in square brackets are called the *direction cosines* of the vector $\overline{OP}$ and are the values of the cosines of the angles which the vector makes with the three axes of reference.

So for the vector $\mathbf{r} = a\mathbf{i} + b\mathbf{j} + c\mathbf{k}$

$$l = \dfrac{a}{r}; \quad m = \dfrac{b}{r}; \quad n = \dfrac{c}{r}; \quad \text{and, of course } r = \sqrt{a^2 + b^2 + c^2}$$

So, with that in mind, find the direction cosines $[l, m, n]$ of the vector

$\mathbf{r} = 3\mathbf{i} - 2\mathbf{j} + 6\mathbf{k}$

Then to Frame 40

40

$\mathbf{r} = 3\mathbf{i} - 2\mathbf{j} + 6\mathbf{k}$

$\therefore a = 3, \ b = -2, \ c = 6, \ r = \sqrt{9 + 4 + 36}$

$\therefore r = \sqrt{49} = 7$

$\therefore l = \dfrac{3}{7}; \ m = -\dfrac{2}{7}; \ n = \dfrac{6}{7}$

Just as easy as that!

On to the next frame

Scalar product of two vectors

41

If **a** and **b** are two vectors, the *scalar product* of **a** and **b** is defined as the *scalar* (number) $ab\cos\theta$ where a and b are the magnitudes of the vectors **a** and **b** and θ is the angle between them.

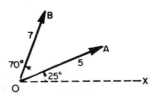

The scalar product is denoted by **a** · **b** (often called the 'dot product' for obvious reasons).

$$\therefore \mathbf{a} \cdot \mathbf{b} = ab\cos\theta$$
$$= a \times \text{projection of } \mathbf{b} \text{ on } \mathbf{a}$$
$$= b \times \text{projection of } \mathbf{a} \text{ on } \mathbf{b}$$

In both cases the result is a *scalar* quantity.

For example:

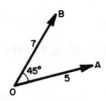

$\overline{OA} \cdot \overline{OB} = \ldots\ldots\ldots\ldots$

42

$$\boxed{\overline{OA} \cdot \overline{OB} = \frac{35\sqrt{2}}{2}}$$

Because we have:

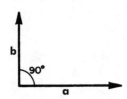

$$\overline{OA} \cdot \overline{OB} = OA \cdot OB \cdot \cos\theta$$
$$= 5 \cdot 7 \cdot \cos 45°$$
$$= 35 \cdot \frac{1}{\sqrt{2}} = \frac{35\sqrt{2}}{2}$$

Now what about this case:

The scalar product of **a** and **b** = **a.b** = $\ldots\ldots\ldots\ldots$

Vectors

$$\boxed{0}$$

Because in this case $\mathbf{a} \cdot \mathbf{b} = ab \cos 90° = ab0 = 0$. So the scalar product of any two vectors at right-angles to each other is always zero.

And in this case now, with two vectors in the same direction, $\theta = 0°$

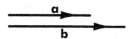

so $\mathbf{a} \cdot \mathbf{b} = \ldots\ldots\ldots\ldots$

$$\boxed{ab}$$

Because $\mathbf{a} \cdot \mathbf{b} = ab \cos 0° = ab \cdot 1 = ab$

Now suppose our two vectors are expressed in terms of the unit vectors $\mathbf{i}$, $\mathbf{j}$ and $\mathbf{k}$.

Let $\quad \mathbf{a} = a_1\mathbf{i} + a_2\mathbf{j} + a_3\mathbf{k}$
and $\quad \mathbf{b} = b_1\mathbf{i} + b_2\mathbf{j} + b_3\mathbf{k}$

Then $\mathbf{a} \cdot \mathbf{b} = (a_1\mathbf{i} + a_2\mathbf{j} + a_3\mathbf{k}) \cdot (b_1\mathbf{i} + b_2\mathbf{j} + b_3\mathbf{k})$
$= a_1b_1\mathbf{i} \cdot \mathbf{i} + a_1b_2\mathbf{i} \cdot \mathbf{j} + a_1b_3\mathbf{i} \cdot \mathbf{k} + a_2b_1\mathbf{j} \cdot \mathbf{i} + a_2b_2\mathbf{j} \cdot \mathbf{j} + a_2b_3\mathbf{j} \cdot \mathbf{k}$
$\quad + a_3b_1\mathbf{k} \cdot \mathbf{i} + a_3b_2\mathbf{k} \cdot \mathbf{j} + a_3b_3\mathbf{k} \cdot \mathbf{k}$

This can now be simplified.

Because $\mathbf{i} \cdot \mathbf{i} = (1)(1)(\cos 0°) = 1$

$\therefore \mathbf{i} \cdot \mathbf{i} = 1; \quad \mathbf{j} \cdot \mathbf{j} = 1; \quad \mathbf{k} \cdot \mathbf{k} = 1 \quad\quad\quad (a)$

Also $\mathbf{i} \cdot \mathbf{j} = (1)(1)(\cos 90°) = 0$

$\therefore \mathbf{i} \cdot \mathbf{j} = 0; \quad \mathbf{j} \cdot \mathbf{k} = 0; \quad \mathbf{k} \cdot \mathbf{i} = 0 \quad\quad\quad (b)$

So, using the results (a) and (b), we get:

$$\mathbf{a} \cdot \mathbf{b} = \ldots\ldots\ldots\ldots$$

$$\boxed{\mathbf{a} \cdot \mathbf{b} = a_1b_1 + a_2b_2 + a_3b_3}$$

Because

$\mathbf{a} \cdot \mathbf{b} = a_1b_1 \cdot 1 + a_1b_2 \cdot 0 + a_1b_3 \cdot 0 + a_2b_1 \cdot 0 + a_2b_2 \cdot 1 + a_2b_3 \cdot 0$
$\quad + a_3b_1 \cdot 0 + a_3b_2 \cdot 0 + a_3b_3 \cdot 1$
$\therefore \mathbf{a} \cdot \mathbf{b} = a_1b_1 + a_2b_2 + a_3b_3$

i.e. we just sum the products of the coefficients of the unit vectors along the corresponding axes.

▶

For example:

If $\mathbf{a} = 2\mathbf{i} + 3\mathbf{j} + 5\mathbf{k}$ and $\mathbf{b} = 4\mathbf{i} + 1\mathbf{j} + 6\mathbf{k}$

then $\mathbf{a} \cdot \mathbf{b} = 2 \times 4 + 3 \times 1 + 5 \times 6$

$\qquad = 8 + 3 + 30$

$\qquad = 41 \qquad \therefore \mathbf{a} \cdot \mathbf{b} = 41$

One for you: If $\mathbf{p} = 3\mathbf{i} - 2\mathbf{j} + 1\mathbf{k}$; $\mathbf{q} = 2\mathbf{i} + 3\mathbf{j} - 4\mathbf{k}$

then $\mathbf{p} \cdot \mathbf{q} = \ldots\ldots\ldots\ldots$

46

$$\boxed{-4}$$

Because

$\mathbf{p} \cdot \mathbf{q} = 3 \times 2 + (-2) \times 3 + 1 \times (-4)$

$\qquad = 6 - 6 - 4$

$\qquad = -4 \qquad\qquad \therefore \mathbf{p} \cdot \mathbf{q} = -4$

Now on to Frame 47

Vector product of two vectors

47

The vector product of $\mathbf{a}$ and $\mathbf{b}$ is written $\mathbf{a} \times \mathbf{b}$ (often called the 'cross product') and is defined as a *vector* having magnitude $ab \sin \theta$ where θ is the angle between the two given vectors. The product vector acts in a direction perpendicular to both $\mathbf{a}$ and $\mathbf{b}$ in such a sense that $\mathbf{a}$, $\mathbf{b}$ and $\mathbf{a} \times \mathbf{b}$ form a right-handed set – in that order.

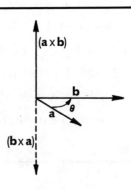

$|\mathbf{a} \times \mathbf{b}| = ab \sin \theta$

Note that $\mathbf{b} \times \mathbf{a}$ reverses the direction of rotation and the product vector would now act downwards, i.e.

$\mathbf{b} \times \mathbf{a} = -(\mathbf{a} \times \mathbf{b})$

If $\theta = 0°$, then $|\mathbf{a} \times \mathbf{b}| = \ldots\ldots\ldots\ldots$

and if $\theta = 90°$, then $|\mathbf{a} \times \mathbf{b}| = \ldots\ldots\ldots\ldots$

Vectors

$$\boxed{\begin{array}{l}\theta = 0°, \quad |\mathbf{a} \times \mathbf{b}| = 0 \\ \theta = 90°, \quad |\mathbf{a} \times \mathbf{b}| = ab\end{array}}$$

48

If $\mathbf{a}$ and $\mathbf{b}$ are given in terms of the unit vectors $\mathbf{i}$, $\mathbf{j}$ and $\mathbf{k}$:

$\mathbf{a} = a_1\mathbf{i} + a_2\mathbf{j} + a_3\mathbf{k}$ and $\mathbf{b} = b_1\mathbf{i} + b_2\mathbf{j} + b_3\mathbf{k}$

Then:

$$\mathbf{a} \times \mathbf{b} = a_1b_1\mathbf{i} \times \mathbf{i} + a_1b_2\mathbf{i} \times \mathbf{j} + a_1b_3\mathbf{i} \times \mathbf{k} + a_2b_1\mathbf{j} \times \mathbf{i} + a_2b_2\mathbf{j} \times \mathbf{j}$$
$$+ a_2b_3\mathbf{j} \times \mathbf{k} + a_3b_1\mathbf{k} \times \mathbf{i} + a_3b_2\mathbf{k} \times \mathbf{j} + a_3b_3\mathbf{k} \times \mathbf{k}$$

But $|\mathbf{i} \times \mathbf{i}| = (1)(1)(\sin 0°) = 0 \quad \therefore \quad \mathbf{i} \times \mathbf{i} = \mathbf{j} \times \mathbf{j} = \mathbf{k} \times \mathbf{k} = 0$ (a)

Also $|\mathbf{i} \times \mathbf{j}| = (1)(1)(\sin 90°) = 1$ and $\mathbf{i} \times \mathbf{j}$ is in the direction of $\mathbf{k}$, i.e. $\mathbf{i} \times \mathbf{j} = \mathbf{k}$ (same magnitude and same direction). Therefore:

$\mathbf{i} \times \mathbf{j} = \mathbf{k}$
$\mathbf{j} \times \mathbf{k} = \mathbf{i}$
$\mathbf{k} \times \mathbf{i} = \mathbf{j}$ (b)

And remember too that therefore:

$\mathbf{i} \times \mathbf{j} = -(\mathbf{j} \times \mathbf{i})$
$\mathbf{j} \times \mathbf{k} = -(\mathbf{k} \times \mathbf{j})$
$\mathbf{k} \times \mathbf{i} = -(\mathbf{i} \times \mathbf{k})$ since the sense of rotation is reversed

Now with the results of (a) and (b), and this last reminder, you can simplify the expression for $\mathbf{a} \times \mathbf{b}$.

Remove the zero terms and tidy up what is left.

....

Then on to Frame 49

$$\boxed{\mathbf{a} \times \mathbf{b} = (a_2b_3 - a_3b_2)\mathbf{i} - (a_1b_3 - a_3b_1)\mathbf{j} + (a_1b_2 - a_2b_1)\mathbf{k}}$$

49

Because

$$\mathbf{a} \times \mathbf{b} = a_1b_1\mathbf{0} + a_1b_2\mathbf{k} + a_1b_3(-\mathbf{j}) + a_2b_1(-\mathbf{k}) + a_2b_2\mathbf{0} + a_2b_3\mathbf{i}$$
$$+ a_3b_1\mathbf{j} + a_3b_2(-\mathbf{i}) + a_3b_3\mathbf{0}$$
$$\mathbf{a} \times \mathbf{b} = (a_2b_3 - a_3b_2)\mathbf{i} - (a_1b_3 - a_3b_1)\mathbf{j} + (a_1b_2 - a_2b_1)\mathbf{k}$$

and you may recognize this as the pattern of a determinant where the first row is made up of the vectors $\mathbf{i}$, $\mathbf{j}$ and $\mathbf{k}$.

▶

214 Vector Analysis

So now we have that:

If $\mathbf{a} = a_1\mathbf{i} + a_2\mathbf{j} + a_3\mathbf{k}$ and $\mathbf{b} = b_1\mathbf{i} + b_2\mathbf{j} + b_3\mathbf{k}$ then:

$$\mathbf{a} \times \mathbf{b} = \begin{vmatrix} \mathbf{i} & \mathbf{j} & \mathbf{k} \\ a_1 & a_2 & a_3 \\ b_1 & b_2 & b_3 \end{vmatrix} = (a_2b_3 - a_3b_2)\mathbf{i} - (a_1b_3 - a_3b_1)\mathbf{j} + (a_1b_2 - a_2b_1)\mathbf{k}$$

and that is the easiest way to write out the vector product of two vectors.

Notes: (a) The top row consists of the unit vectors in order $\mathbf{i}, \mathbf{j}, \mathbf{k}$.
  (b) The second row consists of the coefficients of $\mathbf{a}$.
  (c) The third row consists of the coefficients of $\mathbf{b}$.

For example, if $\mathbf{p} = 2\mathbf{i} + 4\mathbf{j} + 3\mathbf{k}$ and $\mathbf{q} = \mathbf{i} + 5\mathbf{j} - 2\mathbf{k}$, first write down the determinant that represents the vector product $\mathbf{p} \times \mathbf{q}$.

50

$$\mathbf{p} \times \mathbf{q} = \begin{vmatrix} \mathbf{i} & \mathbf{j} & \mathbf{k} \\ 2 & 4 & 3 \\ 1 & 5 & -2 \end{vmatrix} \quad \begin{array}{l} \text{Unit vectors} \\ \text{Coefficients of } \mathbf{p} \\ \text{Coefficients of } \mathbf{q} \end{array}$$

And now, expanding the determinant, we get:

$$\mathbf{p} \times \mathbf{q} = \ldots\ldots\ldots$$

51

$$\boxed{\mathbf{p} \times \mathbf{q} = -23\mathbf{i} + 7\mathbf{j} + 6\mathbf{k}}$$

Because

$$\mathbf{p} \times \mathbf{q} = \begin{vmatrix} \mathbf{i} & \mathbf{j} & \mathbf{k} \\ 2 & 4 & 3 \\ 1 & 5 & -2 \end{vmatrix} = \mathbf{i}\begin{vmatrix} 4 & 3 \\ 5 & -2 \end{vmatrix} - \mathbf{j}\begin{vmatrix} 2 & 3 \\ 1 & -2 \end{vmatrix} + \mathbf{k}\begin{vmatrix} 2 & 4 \\ 1 & 5 \end{vmatrix}$$

$$= \mathbf{i}(-8 - 15) - \mathbf{j}(-4 - 3) + \mathbf{k}(10 - 4)$$
$$= -23\mathbf{i} + 7\mathbf{j} + 6\mathbf{k}$$

So, by way of revision:

(a) *Scalar product* ('dot product')

  $\mathbf{a} \cdot \mathbf{b} = ab\cos\theta$ a scalar quantity

(b) *Vector product* ('cross product')

  $\mathbf{a} \times \mathbf{b} =$ vector of magnitude $ab\sin\theta$, acting in a direction to make $\mathbf{a}, \mathbf{b}$ and $\mathbf{a} \times \mathbf{b}$ a right-handed set. Also:

$$\mathbf{a} \times \mathbf{b} = \begin{vmatrix} \mathbf{i} & \mathbf{j} & \mathbf{k} \\ a_1 & a_2 & a_3 \\ b_1 & b_2 & b_3 \end{vmatrix} = (a_2b_3 - a_3b_2)\mathbf{i} - (a_1b_3 - a_3b_1)\mathbf{j} + (a_1b_2 - a_2b_1)\mathbf{k}$$

And here is one final example on this point.

Find the vector product of $\mathbf{p}$ and $\mathbf{q}$ where:

$\mathbf{p} = 3\mathbf{i} - 4\mathbf{j} + 2\mathbf{k}$ and $\mathbf{q} = 2\mathbf{i} + 5\mathbf{j} - \mathbf{k}$

Vectors

$$\boxed{\mathbf{p} \times \mathbf{q} = -6\mathbf{i} + 7\mathbf{j} + 23\mathbf{k}}$$

52

Because

$$\mathbf{p} \times \mathbf{q} = \begin{vmatrix} \mathbf{i} & \mathbf{j} & \mathbf{k} \\ 3 & -4 & 2 \\ 2 & 5 & -1 \end{vmatrix}$$

$$= \mathbf{i} \begin{vmatrix} -4 & 2 \\ 5 & -1 \end{vmatrix} - \mathbf{j} \begin{vmatrix} 3 & 2 \\ 2 & -1 \end{vmatrix} + \mathbf{k} \begin{vmatrix} 3 & -4 \\ 2 & 5 \end{vmatrix}$$

$$= \mathbf{i}(4 - 10) - \mathbf{j}(-3 - 4) + \mathbf{k}(15 + 8)$$

$$= -6\mathbf{i} + 7\mathbf{j} + 23\mathbf{k}$$

Remember that the order in which the vectors appear in the vector product is important. It is a simple matter to verify that:

$$\mathbf{q} \times \mathbf{p} = 6\mathbf{i} - 7\mathbf{j} - 23\mathbf{k} = -(\mathbf{p} \times \mathbf{q})$$

On to Frame 53

Angle between two vectors

53

Let **a** be one vector with direction cosines $[l, m, n]$
Let **b** be the other vector with direction cosines $[l', m', n']$

We have to find the angle between these two vectors.

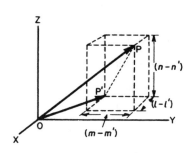

Let $\overline{OP}$ and $\overline{OP'}$ be *unit* vectors parallel to **a** and **b** respectively. Then P has coordinates (l, m, n) and P' has coordinates (l', m', n').

Then

$$(PP')^2 = (l - l')^2 + (m - m')^2 + (n - n')^2$$
$$= l^2 - 2 \cdot l \cdot l' + l'^2 + m^2 - 2 \cdot m \cdot m' + m'^2 + n^2 - 2n \cdot n' + n'^2$$
$$= (l^2 + m^2 + n^2) + (l'^2 + m'^2 + n'^2) - 2(ll' + mm' + nn')$$

But $(l^2 + m^2 + n^2) = 1$ and $(l'^2 + m'^2 + n'^2) = 1$ as was proved earlier.

$$\therefore (PP')^2 = 2 - 2(ll' + mm' + nn') \quad \text{(a)}$$

Also, by the cosine rule:

$$(PP')^2 = OP^2 + OP'^2 - 2 \cdot OP \cdot OP' \cdot \cos\theta$$
$$= 1 + 1 - 2 \cdot 1 \cdot 1 \cdot \cos\theta$$
$$= 2 - 2\cos\theta \quad \text{(b)}$$

$\left\{\begin{array}{l} \overline{OP} \text{ and } \overline{OP'} \text{ are} \\ \text{unit vectors} \end{array}\right\}$

So from (a) and (b), we have:

$$(PP')^2 = 2 - 2(ll' + mm' + nn')$$
and $(PP')^2 = 2 - 2\cos\theta$
$\therefore \quad \cos\theta = \ldots\ldots\ldots\ldots$

54

$$\boxed{\cos\theta = ll' + mm' + nn'}$$

That is, just sum the products of the corresponding direction cosines of the two given vectors.

So, if $[l, m, n] = [0.54, 0.83, -0.14]$
and $[l', m', n'] = [0.25, 0.60, 0.76]$

the angle between the vectors is $\theta = \ldots\ldots\ldots\ldots$

55

$$\boxed{\theta = 58°13'}$$

Because, we have:

$\cos\theta = ll'$ $\qquad + \quad mm' \qquad + \quad nn'$
$\quad = (0.54)(0.25) \; + \; (0.83)(0.60) \; + \; (-0.14)(0.76)$
$\quad = 0.1350 \qquad\quad + \; 0.4980 \qquad\quad - \; 0.1064$
$\quad = 0.6330 \qquad\quad - \; 0.1064 \qquad\qquad\qquad\qquad\qquad = 0.5266$

$\theta = 58°13'$

Note: For *parallel* vectors, $\theta = 0°$ $\therefore$ $ll' + mm' + nn' = 1$

For *perpendicular* vectors, $\theta = 90°$, $\therefore$ $ll' + mm' + nn' = 0$

Now an example for you to work:

Find the angle between the vectors

$$\mathbf{p} = 2\mathbf{i} + 3\mathbf{j} + 4\mathbf{k} \text{ and } \mathbf{q} = 4\mathbf{i} - 3\mathbf{j} + 2\mathbf{k}$$

First of all, find the direction cosines of **p**. So do that.

Vectors

56

$$l = \frac{2}{\sqrt{29}} \quad m = \frac{3}{\sqrt{29}} \quad n = \frac{4}{\sqrt{29}}$$

Because

$$p = |\mathbf{p}| = \sqrt{2^2 + 3^2 + 4^2} = \sqrt{4 + 9 + 16} = \sqrt{29}$$

$$\therefore l = \frac{a}{p} = \frac{2}{\sqrt{29}}$$

$$m = \frac{b}{p} = \frac{3}{\sqrt{29}}$$

$$n = \frac{c}{p} = \frac{4}{\sqrt{29}}$$

$$\therefore [l, m, n] = \left[\frac{2}{\sqrt{29}}, \frac{3}{\sqrt{29}}, \frac{4}{\sqrt{29}}\right]$$

Now find the direction cosines $[l', m', n']$ of $\overline{Q}$ in just the same way.

When you have done that move on to the next frame

57

$$l' = \frac{4}{\sqrt{29}} \quad m' = \frac{-3}{\sqrt{29}} \quad n' = \frac{2}{\sqrt{29}}$$

Because

$$q = |\mathbf{q}| = \sqrt{4^2 + 3^2 + 2^2} = \sqrt{16 + 9 + 4} = \sqrt{29}$$

$$\therefore [l', m', n'] = \left[\frac{4}{\sqrt{29}}, \frac{-3}{\sqrt{29}}, \frac{2}{\sqrt{29}}\right]$$

We already know that, for $\overline{P}$:

$$[l, m, n] = \left[\frac{2}{\sqrt{29}}, \frac{3}{\sqrt{29}}, \frac{4}{\sqrt{29}}\right]$$

So, using $\cos \theta = ll' + mm' + nn'$, you can finish it off and find the angle θ. Off you go.

58

$$\theta = 76°2'$$

Because

$$\cos \theta = \frac{2}{\sqrt{29}} \cdot \frac{4}{\sqrt{29}} + \frac{3}{\sqrt{29}} \cdot \frac{(-3)}{\sqrt{29}} + \frac{4}{\sqrt{29}} \cdot \frac{2}{\sqrt{29}}$$

$$= \frac{8}{29} - \frac{9}{29} + \frac{8}{29}$$

$$= \frac{7}{29} = 0.2414 \quad \therefore \theta = 76°2'$$

Now on to Frame 59

Direction ratios

59

If $\overline{OP} = a\mathbf{i} + b\mathbf{j} + c\mathbf{k}$, we know that:

$$|\overline{OP}| = r = \sqrt{a^2 + b^2 + c^2}$$

and that the direction cosines of $\overline{OP}$ are given by:

$$l = \frac{a}{r}, \quad m = \frac{b}{r}, \quad n = \frac{c}{r}$$

We can see that the components, a, b, c, are proportional to the direction cosines, l, m, n, respectively and they are sometimes referred to as the *direction ratios* of the vector $\overline{OP}$.

Note: The direction ratios can be converted into the direction cosines by dividing each of them by r (the magnitude of the vector).

Move on to Frame 60

Triple products

60

We now deal with the various products that we form with three vectors.

Scalar product of three vectors

If $\mathbf{A}, \mathbf{B}, \mathbf{C}$ are three vectors, the scalar formed by the product $\mathbf{A} \cdot (\mathbf{B} \times \mathbf{C})$ is called the scalar triple product.

If $\mathbf{A} = a_x\mathbf{i} + a_y\mathbf{j} + a_z\mathbf{k}; \quad \mathbf{B} = b_x\mathbf{i} + b_y\mathbf{j} + b_z\mathbf{k}; \quad \mathbf{C} = c_x\mathbf{i} + c_y\mathbf{j} + c_z\mathbf{k};$

then $\quad \mathbf{B} \times \mathbf{C} = \begin{vmatrix} \mathbf{i} & \mathbf{j} & \mathbf{k} \\ b_x & b_y & b_z \\ c_x & c_y & c_z \end{vmatrix}$

$$\therefore \mathbf{A} \cdot (\mathbf{B} \times \mathbf{C}) = (a_x\mathbf{i} + a_y\mathbf{j} + a_z\mathbf{k}) \cdot \begin{vmatrix} \mathbf{i} & \mathbf{j} & \mathbf{k} \\ b_x & b_y & b_z \\ c_x & c_y & c_z \end{vmatrix}$$

Multiplying the top row by the external bracket and remembering that

$$\mathbf{i} \cdot \mathbf{j} = \mathbf{j} \cdot \mathbf{k} = \mathbf{k} \cdot \mathbf{i} = 0 \quad \text{and} \quad \mathbf{i} \cdot \mathbf{i} = \mathbf{j} \cdot \mathbf{j} = \mathbf{k} \cdot \mathbf{k} = 1$$

we have $\quad \mathbf{A} \cdot (\mathbf{B} \times \mathbf{C}) = \begin{vmatrix} a_x & a_y & a_z \\ b_x & b_y & b_z \\ c_x & c_y & c_z \end{vmatrix}$

▶

Vectors

Example

If $\mathbf{A} = 2\mathbf{i} - 3\mathbf{j} + 4\mathbf{k}$; $\mathbf{B} = \mathbf{i} - 2\mathbf{j} - 3\mathbf{k}$; $\mathbf{C} = 2\mathbf{i} + \mathbf{j} + 2\mathbf{k}$;

then $\mathbf{A} \cdot (\mathbf{B} \times \mathbf{C}) = \begin{vmatrix} 2 & -3 & 4 \\ 1 & -2 & -3 \\ 2 & 1 & 2 \end{vmatrix}$

$= \ldots\ldots\ldots\ldots$

$$\boxed{\mathbf{A} \cdot (\mathbf{B} \times \mathbf{C}) = 42}$$

61

Because

$\mathbf{A} \cdot (\mathbf{B} \times \mathbf{C}) = \begin{vmatrix} 2 & -3 & 4 \\ 1 & -2 & -3 \\ 2 & 1 & 2 \end{vmatrix}$

$= 2(-4 + 3) + 3(2 + 6) + 4(1 + 4) = 42$

As simple as that.

Properties of scalar triple products

62

(a) $\quad \mathbf{B} \cdot (\mathbf{C} \times \mathbf{A}) = \begin{vmatrix} b_x & b_y & b_z \\ c_x & c_y & c_z \\ a_x & a_y & a_z \end{vmatrix} = -\begin{vmatrix} a_x & a_y & a_z \\ c_x & c_y & c_z \\ b_x & b_y & b_z \end{vmatrix}$

since interchanging two rows in a determinant reverses the sign. If we now interchange rows 2 and 3 and again change the sign, we have

$$\mathbf{B} \cdot (\mathbf{C} \times \mathbf{A}) = \begin{vmatrix} a_x & a_y & a_z \\ b_x & b_y & b_z \\ c_x & c_y & c_z \end{vmatrix} = \mathbf{A} \cdot (\mathbf{B} \times \mathbf{C})$$

$\therefore \mathbf{A} \cdot (\mathbf{B} \times \mathbf{C}) = \mathbf{B} \cdot (\mathbf{C} \times \mathbf{A}) = \mathbf{C} \cdot (\mathbf{A} \times \mathbf{B})$

i.e. the scalar triple product is unchanged by a cyclic change of the vectors involved.

(b) $\quad \mathbf{B} \cdot (\mathbf{A} \times \mathbf{C}) = \begin{vmatrix} b_x & b_y & b_z \\ a_x & a_y & a_z \\ c_x & c_y & c_z \end{vmatrix} = -\begin{vmatrix} a_x & a_y & a_z \\ b_x & b_y & b_z \\ c_x & c_y & c_z \end{vmatrix}$

$\therefore \mathbf{B} \cdot (\mathbf{A} \times \mathbf{C}) = -\mathbf{A} \cdot (\mathbf{B} \times \mathbf{C})$

i.e. a change of vectors not in cyclic order, changes the sign of the scalar triple product.

(c) $\quad \mathbf{A} \cdot (\mathbf{B} \times \mathbf{A}) = \begin{vmatrix} a_x & a_y & a_z \\ b_x & b_y & b_z \\ a_x & a_y & a_z \end{vmatrix} = 0$ since two rows are identical.

$\therefore \mathbf{A} \cdot (\mathbf{B} \times \mathbf{A}) = \mathbf{B} \cdot (\mathbf{C} \times \mathbf{B}) = \mathbf{C} \cdot (\mathbf{A} \times \mathbf{C}) = 0$

▶

Example

If $\mathbf{A} = \mathbf{i} + 2\mathbf{j} + 3\mathbf{k}$; $\mathbf{B} = 2\mathbf{i} - 3\mathbf{j} + \mathbf{k}$; $\mathbf{C} = 3\mathbf{i} + \mathbf{j} - 2\mathbf{k}$

$\mathbf{A} \cdot (\mathbf{B} \times \mathbf{C}) = \ldots\ldots\ldots\ldots$ $\mathbf{C} \cdot (\mathbf{B} \times \mathbf{A}) = \ldots\ldots\ldots\ldots$

63

$$\mathbf{A} \cdot (\mathbf{B} \times \mathbf{C}) = 52; \quad \mathbf{C} \cdot (\mathbf{A} \times \mathbf{B}) = -52$$

Because

$$\mathbf{A} \cdot (\mathbf{B} \times \mathbf{C}) = \begin{vmatrix} 1 & 2 & 3 \\ 2 & -3 & 1 \\ 3 & 1 & -2 \end{vmatrix} = 1(6-1) - 2(-4-3) + 3(2+9) = 52$$

$\mathbf{C} \cdot (\mathbf{B} \times \mathbf{A})$ is not a cyclic change from the above. Therefore

$\mathbf{C} \cdot (\mathbf{B} \times \mathbf{A}) = -\mathbf{A} \cdot (\mathbf{B} \times \mathbf{C}) = -52$

Coplanar vectors

The magnitude of the scalar triple product $|\mathbf{A} \cdot (\mathbf{B} \times \mathbf{C})|$ is equal to the volume of the parallelepiped with three adjacent sides defined by $\mathbf{A}$, $\mathbf{B}$ and $\mathbf{C}$.

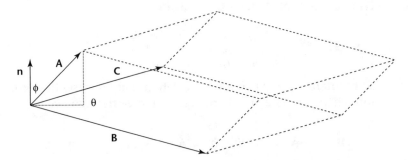

The scalar triple product $\mathbf{A} \cdot (\mathbf{B} \times \mathbf{C}) = \mathbf{A} \cdot (BC \sin \theta \, \mathbf{n}) = ABC \sin \theta \cos \phi$ where $\mathbf{n}$ is a unit vector perpendicular to the plane containing $\mathbf{B}$ and $\mathbf{C}$, θ is the angle between $\mathbf{B}$ and $\mathbf{C}$ and ϕ is the angle between $\mathbf{A}$ and $\mathbf{n}$. Therefore

$$|\mathbf{A} \cdot (\mathbf{B} \times \mathbf{C})| = ABC |\sin \theta \cos \phi|$$

Notice that in the figure both θ and ϕ are drawn as acute but in the general case this may not be so. Now, $BC|\sin \theta|$ is the area of the parallelogram defined by $\mathbf{B}$ and $\mathbf{C}$. The altitude of the parallelepiped is $A|\cos \phi|$ and so $ABC|\sin \theta \cos \phi|$ is the *volume* of the parallelepiped with three adjacent sides defined by $\mathbf{A}$, $\mathbf{B}$ and $\mathbf{C}$.

Consequently if $\mathbf{A} \cdot (\mathbf{B} \times \mathbf{C}) = 0$ then the volume of the parallelepiped is zero and the three vectors $\mathbf{A}$, $\mathbf{B}$ and $\mathbf{C}$ are coplanar.

Example 1

Show that $\mathbf{A} = \mathbf{i} + 2\mathbf{j} - 3\mathbf{k}$; $\mathbf{B} = 2\mathbf{i} - \mathbf{j} + 2\mathbf{k}$; and $\mathbf{C} = 3\mathbf{i} + \mathbf{j} - \mathbf{k}$ are coplanar.

We just evaluate $\mathbf{A} \cdot (\mathbf{B} \times \mathbf{C}) = \ldots\ldots\ldots\ldots$ and apply the test.

$$\mathbf{A} \cdot (\mathbf{B} \times \mathbf{C}) = 0$$

64

Because

$$\mathbf{A} \cdot (\mathbf{B} \times \mathbf{C}) = \begin{vmatrix} 1 & 2 & -3 \\ 2 & -1 & 2 \\ 3 & 1 & -1 \end{vmatrix} = 1(1-2) - 2(-2-6) - 3(2+3) = 0.$$

Therefore **A**, **B**, **C** are coplanar.

Example 2

If $\mathbf{A} = 2\mathbf{i} - \mathbf{j} + 3\mathbf{k}$; $\mathbf{B} = 3\mathbf{i} + 2\mathbf{j} + \mathbf{k}$; $\mathbf{C} = \mathbf{i} + p\mathbf{j} + 4\mathbf{k}$ are coplanar, find the value of p.

The method is clear enough. We merely set up and evaluate the determinant and solve the equation $\mathbf{A} \cdot (\mathbf{B} \times \mathbf{C}) = 0$.

$$p = \ldots\ldots\ldots\ldots$$

$$p = -3$$

65

Because

$$\mathbf{A} \cdot (\mathbf{B} \times \mathbf{C}) = 0 \quad \therefore \begin{vmatrix} 2 & -1 & 3 \\ 3 & 2 & 1 \\ 1 & p & 4 \end{vmatrix} = 0$$

$\therefore 2(8-p) + 1(12-1) + 3(3p-2) = 0 \quad \therefore 7p = -21 \quad \therefore p = -3$

One more.

Example 3

Determine whether the three vectors $\mathbf{A} = 3\mathbf{i} + 2\mathbf{j} - \mathbf{k}$; $\mathbf{B} = 2\mathbf{i} - \mathbf{j} + 3\mathbf{k}$; $\mathbf{C} = \mathbf{i} - 2\mathbf{j} + 2\mathbf{k}$ are coplanar.

Work through it on your own. The result shows that

$$\ldots\ldots\ldots\ldots$$

$$\text{\textbf{A}, \textbf{B}, \textbf{C} are not coplanar}$$

66

Because

in this case $\mathbf{A} \cdot (\mathbf{B} \times \mathbf{C}) = \begin{vmatrix} 3 & 2 & -1 \\ 2 & -1 & 3 \\ 1 & -2 & 2 \end{vmatrix} = 13$

$\therefore \mathbf{A} \cdot (\mathbf{B} \times \mathbf{C}) \neq 0 \quad \therefore$ **A**, **B**, **C** are not coplanar.

Now on to something different

67 Vector triple product of three vectors

If **A**, **B** and **C** are three vectors, then

$$\left. \begin{array}{c} \mathbf{A} \times (\mathbf{B} \times \mathbf{C}) \\ \text{and} \quad (\mathbf{A} \times \mathbf{B}) \times \mathbf{C} \end{array} \right\} \text{ are called the vector triple products.}$$

Consider $\mathbf{A} \times (\mathbf{B} \times \mathbf{C})$ where $\mathbf{A} = a_x\mathbf{i} + a_y\mathbf{j} + a_z\mathbf{k}$; $\mathbf{B} = b_x\mathbf{i} + b_y\mathbf{j} + b_z\mathbf{k}$ and $\mathbf{C} = c_x\mathbf{i} + c_y\mathbf{j} + c_z\mathbf{k}$.

Then $(\mathbf{B} \times \mathbf{C})$ is a vector perpendicular to the plane of **B** and **C** and $\mathbf{A} \times (\mathbf{B} \times \mathbf{C})$ is a vector perpendicular to the plane containing **A** and $(\mathbf{B} \times \mathbf{C})$, i.e. coplanar with **B** and **C**.

Note that, similarly, $(\mathbf{A} \times \mathbf{B}) \times \mathbf{C}$ is coplanar with **A** and **B** and so in general $\mathbf{A} \times (\mathbf{B} \times \mathbf{C}) \neq (\mathbf{A} \times \mathbf{B}) \times \mathbf{C}$.

Now

$$(\mathbf{B} \times \mathbf{C}) = \begin{vmatrix} \mathbf{i} & \mathbf{j} & \mathbf{k} \\ b_x & b_y & b_z \\ c_x & c_y & c_z \end{vmatrix} = \mathbf{i}\begin{vmatrix} b_y & b_z \\ c_y & c_z \end{vmatrix} - \mathbf{j}\begin{vmatrix} b_x & b_z \\ c_x & c_z \end{vmatrix} + \mathbf{k}\begin{vmatrix} b_x & b_y \\ c_x & c_y \end{vmatrix}$$

Then $\mathbf{A} \times (\mathbf{B} \times \mathbf{C}) = \begin{vmatrix} \mathbf{i} & \mathbf{j} & \mathbf{k} \\ a_x & a_y & a_z \\ \begin{vmatrix} b_y & b_z \\ c_y & c_z \end{vmatrix} & -\begin{vmatrix} b_x & b_z \\ c_x & c_z \end{vmatrix} & \begin{vmatrix} b_x & b_y \\ c_x & c_y \end{vmatrix} \end{vmatrix}$

$$= \begin{vmatrix} \mathbf{i} & \mathbf{j} & \mathbf{k} \\ a_x & a_y & a_z \\ \begin{vmatrix} b_y & b_z \\ c_y & c_z \end{vmatrix} & \begin{vmatrix} b_z & b_x \\ c_z & c_x \end{vmatrix} & \begin{vmatrix} b_x & b_y \\ c_x & c_y \end{vmatrix} \end{vmatrix}$$

In symbolic form, further expansion of the determinant becomes somewhat tedious. However a numerical example will clarify the method.

Make a note of the definition above and then go on to the next frame

68

Example 1

If $\mathbf{A} = 2\mathbf{i} - 3\mathbf{j} + \mathbf{k}$; $\mathbf{B} = \mathbf{i} + 2\mathbf{j} - \mathbf{k}$; $\mathbf{C} = 3\mathbf{i} + \mathbf{j} + 3\mathbf{k}$; determine the vector triple product $\mathbf{A} \times (\mathbf{B} \times \mathbf{C})$.

We start off with $\mathbf{B} \times \mathbf{C} = \ldots\ldots\ldots\ldots$

Vectors

$$\boxed{\mathbf{B} \times \mathbf{C} = 7\mathbf{i} - 6\mathbf{j} - 5\mathbf{k}}$$

69

Because

$$\mathbf{B} \times \mathbf{C} = \begin{vmatrix} \mathbf{i} & \mathbf{j} & \mathbf{k} \\ 1 & 2 & -1 \\ 3 & 1 & 3 \end{vmatrix} = \mathbf{i}(6+1) - \mathbf{j}(3+3) + \mathbf{k}(1-6) \\ = 7\mathbf{i} - 6\mathbf{j} - 5\mathbf{k}$$

Then $\mathbf{A} \times (\mathbf{B} \times \mathbf{C}) = \ldots\ldots\ldots\ldots$

$$\boxed{\mathbf{A} \times (\mathbf{B} \times \mathbf{C}) = 21\mathbf{i} + 17\mathbf{j} + 9\mathbf{k}}$$

70

Because

$$\mathbf{A} \times (\mathbf{B} \times \mathbf{C}) = \begin{vmatrix} \mathbf{i} & \mathbf{j} & \mathbf{k} \\ 2 & -3 & 1 \\ 7 & -6 & 5 \end{vmatrix}$$
$$= \mathbf{i}(15+6) - \mathbf{j}(-10-7) + \mathbf{k}(-12+21)$$
$$= 21\mathbf{i} + 17\mathbf{j} + 9\mathbf{k}$$

That is fundamental enough. There is, however, an even easier way of determining a vector triple product. It can be proved that

$$\mathbf{A} \times (\mathbf{B} \times \mathbf{C}) = (\mathbf{A} \cdot \mathbf{C})\mathbf{B} - (\mathbf{A} \cdot \mathbf{B})\mathbf{C}$$
and $(\mathbf{A} \times \mathbf{B}) \times \mathbf{C} = (\mathbf{C} \cdot \mathbf{A})\mathbf{B} - (\mathbf{C} \cdot \mathbf{B})\mathbf{A}$ (11)

Make a careful note of the expressions: then we will apply the method to the example we have just completed.

$\mathbf{A} = 2\mathbf{i} - 3\mathbf{j} + \mathbf{k}; \quad \mathbf{B} = \mathbf{i} + 2\mathbf{j} - \mathbf{k}; \quad \mathbf{C} = 3\mathbf{i} + \mathbf{j} + 3\mathbf{k}$ and we have

71

$$\mathbf{A} \times (\mathbf{B} \times \mathbf{C}) = (\mathbf{A} \cdot \mathbf{C})\mathbf{B} - (\mathbf{A} \cdot \mathbf{B})\mathbf{C}$$
$$= (6 - 3 + 3)(\mathbf{i} + 2\mathbf{j} - \mathbf{k}) - (2 - 6 - 1)(3\mathbf{i} + \mathbf{j} + 3\mathbf{k})$$
$$= 6(\mathbf{i} + 2\mathbf{j} - \mathbf{k}) + 5(3\mathbf{i} + \mathbf{j} + 3\mathbf{k})$$
$$= 21\mathbf{i} + 17\mathbf{j} + 9\mathbf{k}$$

which is, of course, the result we achieved before.

Here is another.

Example 2

If $\mathbf{A} = 3\mathbf{i} + 2\mathbf{j} - 2\mathbf{k}; \quad \mathbf{B} = 4\mathbf{i} - \mathbf{j} + 3\mathbf{k}; \quad \mathbf{C} = 2\mathbf{i} - 3\mathbf{j} + \mathbf{k}$ determine $(\mathbf{A} \times \mathbf{B}) \times \mathbf{C}$ using the relationship $(\mathbf{A} \times \mathbf{B}) \times \mathbf{C} = (\mathbf{C} \cdot \mathbf{A})\mathbf{B} - (\mathbf{C} \cdot \mathbf{B})\mathbf{A}$.

$$(\mathbf{A} \times \mathbf{B}) \times \mathbf{C} = \ldots\ldots\ldots\ldots$$

72

$$\boxed{-50\mathbf{i} - 26\mathbf{j} + 22\mathbf{k}}$$

Because

$$(\mathbf{A} \times \mathbf{B}) \times \mathbf{C} = (\mathbf{C} \cdot \mathbf{A})\mathbf{B} - (\mathbf{C} \cdot \mathbf{B})\mathbf{A}$$
$$= (6 - 6 - 2)(4\mathbf{i} - \mathbf{j} + 3\mathbf{k}) - (8 + 3 + 3)(3\mathbf{i} + 2\mathbf{j} - 2\mathbf{k})$$
$$= -2(4\mathbf{i} - \mathbf{j} + 3\mathbf{k}) - 14(3\mathbf{i} + 2\mathbf{j} - 2\mathbf{k})$$
$$= -50\mathbf{i} - 26\mathbf{j} + 22\mathbf{k}$$

Now one more.

Example 3

If $\mathbf{A} = \mathbf{i} + 3\mathbf{j} + 2\mathbf{k}$; $\mathbf{B} = 2\mathbf{i} + 5\mathbf{j} - \mathbf{k}$; $\mathbf{C} = \mathbf{i} + 2\mathbf{j} + 3\mathbf{k}$

$$\mathbf{A} \times (\mathbf{B} \times \mathbf{C}) = \ldots\ldots\ldots$$
$$(\mathbf{A} \times \mathbf{B}) \times \mathbf{C} = \ldots\ldots\ldots$$

Finish them both.

73

$$\boxed{\begin{array}{l}\mathbf{A} \times (\mathbf{B} \times \mathbf{C}) = 11\mathbf{i} + 35\mathbf{j} - 58\mathbf{k} \\ (\mathbf{A} \times \mathbf{B}) \times \mathbf{C} = 17\mathbf{i} + 38\mathbf{j} - 31\mathbf{k}\end{array}}$$

Because

$$\mathbf{A} \times (\mathbf{B} \times \mathbf{C}) = (\mathbf{A} \cdot \mathbf{C})\mathbf{B} - (\mathbf{A} \cdot \mathbf{B})\mathbf{C}$$
$$= (1 + 6 + 6)(2\mathbf{i} + 5\mathbf{j} - \mathbf{k}) - (2 + 15 - 2)(\mathbf{i} + 2\mathbf{j} + 3\mathbf{k})$$
$$= 13(2\mathbf{i} + 5\mathbf{j} - \mathbf{k}) - 15(\mathbf{i} + 2\mathbf{j} + 3\mathbf{k})$$
$$= 11\mathbf{i} + 35\mathbf{j} - 58\mathbf{k}$$

and

$$(\mathbf{A} \times \mathbf{B}) \times \mathbf{C} = (\mathbf{C} \cdot \mathbf{A})\mathbf{B} - (\mathbf{C} \cdot \mathbf{B})\mathbf{A}$$
$$= (1 + 6 + 6)(2\mathbf{i} + 5\mathbf{j} - \mathbf{k}) - (2 + 10 - 3)(\mathbf{i} + 3\mathbf{j} + 2\mathbf{k})$$
$$= 13(2\mathbf{i} + 5\mathbf{j} - \mathbf{k}) - 9(\mathbf{i} + 3\mathbf{j} + 2\mathbf{k}) = 17\mathbf{i} + 38\mathbf{j} - 31\mathbf{k}$$

These two results clearly confirm that

$$\mathbf{A} \times (\mathbf{B} \times \mathbf{C}) \neq (\mathbf{A} \times \mathbf{B}) \times \mathbf{C} \quad \text{so beware!}$$

Before we proceed, note the following concerning the unit vectors.

(a) $(\mathbf{i} \times \mathbf{j}) = \mathbf{k}$
$\therefore \mathbf{i} \times (\mathbf{i} \times \mathbf{j}) = \mathbf{i} \times \mathbf{k} = -\mathbf{j}$
$\therefore \mathbf{i} \times (\mathbf{i} \times \mathbf{j}) = -\mathbf{j}$

(b) $(\mathbf{i} \times \mathbf{i}) \times \mathbf{j} = (0) \times \mathbf{j} = 0$
$\therefore (\mathbf{i} \times \mathbf{i}) \times \mathbf{j} = 0$

and once again, we see that
$\mathbf{i} \times (\mathbf{i} \times \mathbf{j}) \neq (\mathbf{i} \times \mathbf{i}) \times \mathbf{j}$

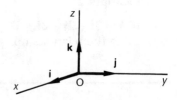

On to the next

Vectors

74

Finally, by way of revision:

Example 4

If $\mathbf{A} = 5\mathbf{i} - 2\mathbf{j} + 3\mathbf{k}$; $\mathbf{B} = 3\mathbf{i} + \mathbf{j} - 2\mathbf{k}$; $\mathbf{C} = \mathbf{i} - 3\mathbf{j} + 4\mathbf{k}$; determine
(a) the scalar triple product $\mathbf{A} \cdot (\mathbf{B} \times \mathbf{C})$
(b) the vector triple products (1) $\mathbf{A} \times (\mathbf{B} \times \mathbf{C})$
　　　　　　　　　　　　　　(2) $(\mathbf{A} \times \mathbf{B}) \times \mathbf{C}$.

Finish all these and then check with the next frame

75

(a) $\mathbf{A} \cdot (\mathbf{B} \times \mathbf{C}) = -12$
(b) (1) $\mathbf{A} \times (\mathbf{B} \times \mathbf{C}) = 62\mathbf{i} + 44\mathbf{j} - 74\mathbf{k}$
　　(2) $(\mathbf{A} \times \mathbf{B}) \times \mathbf{C} = 109\mathbf{i} + 7\mathbf{j} - 22\mathbf{k}$

Here is the working.

(a) $\mathbf{A} \cdot (\mathbf{B} \times \mathbf{C}) = \begin{vmatrix} 5 & -2 & 3 \\ 3 & 1 & -2 \\ 1 & -3 & 4 \end{vmatrix}$

$= 5(4 - 6) + 2(12 + 2) + 3(-9 - 1) = -12$

(b) (1) $\mathbf{A} \times (\mathbf{B} \times \mathbf{C}) = (\mathbf{A} \cdot \mathbf{C})\mathbf{B} - (\mathbf{A} \cdot \mathbf{B})\mathbf{C}$
$= (5 + 6 + 12)(3\mathbf{i} + \mathbf{j} - 2\mathbf{k})$
$\quad - (15 - 2 - 6)(\mathbf{i} - 3\mathbf{j} + 4\mathbf{k})$
$= 23(3\mathbf{i} + \mathbf{j} - 2\mathbf{k}) - 7(\mathbf{i} - 3\mathbf{j} + 4\mathbf{k})$
$= 62\mathbf{i} + 44\mathbf{j} - 74\mathbf{k}$

(2) $(\mathbf{A} \times \mathbf{B}) \times \mathbf{C} = (\mathbf{C} \cdot \mathbf{A})\mathbf{B} - (\mathbf{C} \cdot \mathbf{B})\mathbf{A}$
$= 23(3\mathbf{i} + \mathbf{j} - 2\mathbf{k}) - (-8)(5\mathbf{i} - 2\mathbf{j} + 3\mathbf{k})$
$= 109\mathbf{i} + 7\mathbf{j} - 22\mathbf{k}$

Review summary

76

1. A *scalar* quantity has magnitude only; a *vector* quantity has both magnitude and direction.

2. The axes of reference, OX, OY, OZ, are chosen so that they form a right-handed set. The symbols **i, j, k** denote *unit vectors* in the directions OX, OY, OZ, respectively.

   If $\overline{OP} = a\mathbf{i} + b\mathbf{j} + c\mathbf{k}$, then $|\overline{OP}| = r = \sqrt{a^2 + b^2 + c^2}$

3. The *direction cosines* $[l, m, n]$ are the cosines of the angles between the vector and the axes OX, OY, OZ respectively.

   For any vector: $l = \dfrac{a}{r}$, $m = \dfrac{b}{r}$, $n = \dfrac{c}{r}$; and $l^2 + m^2 + n^2 = 1$

▶

4 *Scalar product* ('dot product')

   $\mathbf{a} \cdot \mathbf{b} = ab \cos \theta$ where θ is the angle between $\mathbf{a}$ and $\mathbf{b}$.
   If $\mathbf{a} = a_1\mathbf{i} + a_2\mathbf{j} + a_3\mathbf{k}$ and $\mathbf{b} = b_1\mathbf{i} + b_2\mathbf{j} + b_3\mathbf{k}$
   then $\mathbf{a} \cdot \mathbf{b} = a_1b_1 + a_2b_2 + a_3b_3$

5 *Vector product* ('cross product')
   $\mathbf{a} \times \mathbf{b} = (ab \sin \theta)$ in direction perpendicular to $\mathbf{a}$ and $\mathbf{b}$, so that $\mathbf{a}$, $\mathbf{b}$ and $(\mathbf{a} \times \mathbf{b})$ form a right-handed set.

   Also $\mathbf{a} \times \mathbf{b} = \begin{vmatrix} \mathbf{i} & \mathbf{j} & \mathbf{k} \\ a_1 & a_2 & a_3 \\ b_1 & b_2 & b_3 \end{vmatrix}$

6 *Angle between two vectors*

   $\cos \theta = ll' + mm' + nn'$
   For perpendicular vectors, $ll' + mm' + nn' = 0$

If $\mathbf{A} = a_x\mathbf{i} + a_y\mathbf{j} + a_z\mathbf{k}$; $\mathbf{B} = b_x\mathbf{i} + b_y\mathbf{j} + b_z\mathbf{k}$; $\mathbf{C} = c_x\mathbf{i} + c_y\mathbf{j} + c_z\mathbf{k}$; then we have the following relationships.

7 *Scalar product* (dot product) $\mathbf{A} \cdot \mathbf{B} = AB \cos \theta$
   $\mathbf{A} \cdot \mathbf{B} = \mathbf{B} \cdot \mathbf{A}$ and $\mathbf{A} \cdot (\mathbf{B} + \mathbf{C}) = \mathbf{A} \cdot \mathbf{B} + \mathbf{A} \cdot \mathbf{C}$
   If $\mathbf{A} \cdot \mathbf{B} = 0$ and $\mathbf{A}, \mathbf{B} \neq 0$ then $\mathbf{A} \perp \mathbf{B}$.

8 *Vector product* (cross product) $\mathbf{A} \times \mathbf{B} = (AB \sin \theta)\mathbf{n}$
   $\mathbf{n}$ = unit normal vector where $\mathbf{A}$, $\mathbf{B}$, $\mathbf{n}$ form a right-handed set.

   $\mathbf{A} \times \mathbf{B} = \begin{vmatrix} \mathbf{i} & \mathbf{j} & \mathbf{k} \\ a_x & a_y & a_z \\ b_x & b_y & b_z \end{vmatrix}$

   $\mathbf{A} \times \mathbf{B} = -(\mathbf{B} \times \mathbf{A})$ and $\mathbf{A} \times (\mathbf{B} + \mathbf{C}) = \mathbf{A} \times \mathbf{B} + \mathbf{A} \times \mathbf{C}$

9 *Unit vectors*
   (a) $\mathbf{i} \cdot \mathbf{i} = \mathbf{j} \cdot \mathbf{j} = \mathbf{k} \cdot \mathbf{k} = 1$
       $\mathbf{i} \cdot \mathbf{j} = \mathbf{j} \cdot \mathbf{k} = \mathbf{k} \cdot \mathbf{i} = 0$.
   (b) $\mathbf{i} \times \mathbf{i} = \mathbf{j} \times \mathbf{j} = \mathbf{k} \times \mathbf{k} = 0$
       $\mathbf{i} \times \mathbf{j} = \mathbf{k}$, $\mathbf{j} \times \mathbf{k} = \mathbf{i}$, $\mathbf{k} \times \mathbf{i} = \mathbf{j}$.

10 *Scalar triple product* $\mathbf{A} \cdot (\mathbf{B} \times \mathbf{C})$

   $\mathbf{A} \cdot (\mathbf{B} \times \mathbf{C}) = \begin{vmatrix} a_x & a_y & a_z \\ b_x & b_y & b_z \\ c_x & c_y & c_z \end{vmatrix}$

   $\mathbf{A} \cdot (\mathbf{B} \times \mathbf{C}) = \mathbf{B} \cdot (\mathbf{C} \times \mathbf{A}) = \mathbf{C} \cdot (\mathbf{A} \times \mathbf{B})$

   Unchanged by cyclic change of vectors.
   Sign reversed by non-cyclic change of vectors.

11 *Coplanar vectors* $\mathbf{A} \cdot (\mathbf{B} \times \mathbf{C}) = 0$.

▶

Vectors

12 *Vector triple product* $\mathbf{A} \times (\mathbf{B} \times \mathbf{C})$ and $(\mathbf{A} \times \mathbf{B}) \times \mathbf{C}$

$$\mathbf{A} \times (\mathbf{B} \times \mathbf{C}) = (\mathbf{A} \cdot \mathbf{C})\mathbf{B} - (\mathbf{A} \cdot \mathbf{B})\mathbf{C}$$

and $(\mathbf{A} \times \mathbf{B}) \times \mathbf{C} = (\mathbf{C} \cdot \mathbf{A})\mathbf{B} - (\mathbf{C} \cdot \mathbf{B})\mathbf{A}$.

Now you are ready for the **Can You?** checklist and **Test exercise**.

So off you go

✓ Can You?

Checklist 7

[77]

Check this list before and after you try the end of Program test.

On a scale of 1 to 5 how confident are you that you can: Frames

- Define a vector?   [1] to [5]
  Yes ☐ ☐ ☐ ☐ ☐ No
- Represent a vector by a directed straight line?   [6] to [10]
  Yes ☐ ☐ ☐ ☐ ☐ No
- Add vectors?   [11] to [17]
  Yes ☐ ☐ ☐ ☐ ☐ No
- Write a vector in terms of component vectors?   [18] to [30]
  Yes ☐ ☐ ☐ ☐ ☐ No
- Write a vector in terms of component unit vectors?   [31] to [35]
  Yes ☐ ☐ ☐ ☐ ☐ No
- Set up a coordinate system for representing vectors?   [36] to [38]
  Yes ☐ ☐ ☐ ☐ ☐ No
- Obtain the direction cosines of a vector?   [39] to [40]
  Yes ☐ ☐ ☐ ☐ ☐ No
- Calculate the scalar product of two vectors?   [41] to [46]
  Yes ☐ ☐ ☐ ☐ ☐ No
- Calculate the vector product of two vectors?   [47] to [52]
  Yes ☐ ☐ ☐ ☐ ☐ No
- Determine the angle between two vectors?   [53] to [58]
  Yes ☐ ☐ ☐ ☐ ☐ No
- Evaluate the direction ratios of a vector?   [59]
  Yes ☐ ☐ ☐ ☐ ☐ No
- Obtain the scalar and vector triple products and appreciate their geometric significance?   [60] to [75]
  Yes ☐ ☐ ☐ ☐ ☐ No

Test exercise 7

Take your time: the problems are all straightforward so avoid careless slips. Diagrams often help where appropriate.

1. If $\overline{OA} = 4\mathbf{i} + 3\mathbf{j}$, $\overline{OB} = 6\mathbf{i} - 2\mathbf{j}$, $\overline{OC} = 2\mathbf{i} - \mathbf{j}$, find $\overline{AB}$, $\overline{BC}$ and $\overline{CA}$, and deduce the lengths of the sides of the triangle ABC.

2. Find the direction cosines of the vector joining the two points (4, 2, 2) and (7, 6, 14).

3. If $\mathbf{a} = 2\mathbf{i} + 2\mathbf{j} - \mathbf{k}$ and $\mathbf{b} = 3\mathbf{i} - 6\mathbf{j} + 2\mathbf{k}$, find (a) $\mathbf{a} \cdot \mathbf{b}$ and (b) $\mathbf{a} \times \mathbf{b}$.

4. If $\mathbf{a} = 5\mathbf{i} + 4\mathbf{j} + 2\mathbf{k}$, $\mathbf{b} = 4\mathbf{i} - 5\mathbf{j} + 3\mathbf{k}$ and $\mathbf{c} = 2\mathbf{i} - \mathbf{j} - 2\mathbf{k}$, where $\mathbf{i}$, $\mathbf{j}$, $\mathbf{k}$ are the unit vectors, determine:
   (a) the value of $\mathbf{a} \cdot \mathbf{b}$ and the angle between the vectors $\mathbf{a}$ and $\mathbf{b}$
   (b) the magnitude and the direction cosines of the product vector ($\mathbf{a} \times \mathbf{b}$) and also the angle which this product vector makes with the vector $\mathbf{c}$.

5. Find (a) the scalar product and (b) the vector product of the vectors $\mathbf{A} = 3\mathbf{i} - 2\mathbf{j} + 4\mathbf{k}$ and $\mathbf{B} = \mathbf{i} + 5\mathbf{j} - 2\mathbf{k}$.

6. If $\mathbf{A} = 2\mathbf{i} + 3\mathbf{j} - 5\mathbf{k}$; $\mathbf{B} = 3\mathbf{i} + \mathbf{j} + 2\mathbf{k}$; $\mathbf{C} = \mathbf{i} - \mathbf{j} + 3\mathbf{k}$; determine
   (a) the scalar triple product $\mathbf{A} \cdot (\mathbf{B} \times \mathbf{C})$
   (b) the vector triple product $\mathbf{A} \times (\mathbf{B} \times \mathbf{C})$.

7. Determine whether the three vectors $\mathbf{A} = 2\mathbf{i} + 3\mathbf{j} + \mathbf{k}$; $\mathbf{B} = \mathbf{i} - 2\mathbf{j} + 2\mathbf{k}$; $\mathbf{C} = 3\mathbf{i} + \mathbf{j} + 3\mathbf{k}$ are coplanar.

Further problems 7

1. The centroid of the triangle OAB is denoted by G. If O is the origin and $\overline{OA} = 4\mathbf{i} + 3\mathbf{j}$, $\overline{OB} = 6\mathbf{i} - \mathbf{j}$, find $\overline{OG}$ in terms of the unit vectors, $\mathbf{i}$ and $\mathbf{j}$.

2. Find the direction cosines of the vectors whose direction ratios are (3, 4, 5) and (1, 2, −3). Hence find the angle between the two vectors.

3. Find the modulus and the direction cosines of each of the vectors $3\mathbf{i} + 7\mathbf{j} - 4\mathbf{k}$, $\mathbf{i} - 5\mathbf{j} - 8\mathbf{k}$ and $6\mathbf{i} - 2\mathbf{j} + 12\mathbf{k}$. Find also the modulus and the direction cosines of their sum.

4. If $\mathbf{a} = 2\mathbf{i} + 4\mathbf{j} - 3\mathbf{k}$ and $\mathbf{b} = \mathbf{i} + 3\mathbf{j} + 2\mathbf{k}$, determine the scalar and vector products, and the angle between the two given vectors.

5. If $\overline{OA} = 2\mathbf{i} + 3\mathbf{j} - \mathbf{k}$ and $\overline{OB} = \mathbf{i} - 2\mathbf{j} + 3\mathbf{k}$, determine:
   (a) the value of $\overline{OA} \cdot \overline{OB}$
   (b) the product $\overline{OA} \times \overline{OB}$ in terms of the unit vectors
   (c) the cosine of the angle between $\overline{OA}$ and $\overline{OB}$

6. Find the cosine of the angle between the vectors $2\mathbf{i} + 3\mathbf{j} - \mathbf{k}$ and $3\mathbf{i} - 5\mathbf{j} + 2\mathbf{k}$.

7. Find the scalar product ($\mathbf{a} \cdot \mathbf{b}$) and the vector product ($\mathbf{a} \times \mathbf{b}$), when
   (a) $\mathbf{a} = \mathbf{i} + 2\mathbf{j} - \mathbf{k}$, $\mathbf{b} = 2\mathbf{i} + 3\mathbf{j} + \mathbf{k}$ (b) $\mathbf{a} = 2\mathbf{i} + 3\mathbf{j} + 4\mathbf{k}$, $\mathbf{b} = 5\mathbf{i} - 2\mathbf{j} + \mathbf{k}$

Vectors

8 Find the unit vector perpendicular to each of the vectors $2\mathbf{i} - \mathbf{j} + \mathbf{k}$ and $3\mathbf{i} + 4\mathbf{j} - \mathbf{k}$, where $\mathbf{i}, \mathbf{j}, \mathbf{k}$ are the mutually perpendicular unit vectors. Calculate the sine of the angle between the two vectors.

9 If A is the point $(1, -1, 2)$, B is the point $(-1, 2, 2)$ and C is the point $(4, 3, 0)$, find the direction cosines of $\overline{BA}$ and $\overline{BC}$, and hence show that the angle ABC $= 69°14'$.

10 If $\mathbf{a} = 3\mathbf{i} - \mathbf{j} + 2\mathbf{k}$, $\mathbf{b} = \mathbf{i} + 3\mathbf{j} - 2\mathbf{k}$, determine the magnitude and direction cosines of the product vector $(\mathbf{a} \times \mathbf{b})$ and show that it is perpendicular to a vector $\mathbf{c} = 9\mathbf{i} + 2\mathbf{j} + 2\mathbf{k}$.

11 $\mathbf{a}$ and $\mathbf{b}$ are vectors defined by $\mathbf{a} = 8\mathbf{i} + 2\mathbf{j} - 3\mathbf{k}$ and $\mathbf{b} = 3\mathbf{i} - 6\mathbf{j} + 4\mathbf{k}$, where $\mathbf{i}, \mathbf{j}, \mathbf{k}$ are mutually perpendicular unit vectors.

(a) Calculate $\mathbf{a} \cdot \mathbf{b}$ and show that $\mathbf{a}$ and $\mathbf{b}$ are perpendicular to each other.

(b) Find the magnitude and the direction cosines of the product vector $\mathbf{a} \times \mathbf{b}$.

12 If the position vectors of P and Q are $\mathbf{i} + 3\mathbf{j} - 7\mathbf{k}$ and $5\mathbf{i} - 2\mathbf{j} + 4\mathbf{k}$ respectively, find $\overline{PQ}$ and determine its direction cosines.

13 If position vectors, $\overline{OA}, \overline{OB}, \overline{OC}$, are defined by $\overline{OA} = 2\mathbf{i} - \mathbf{j} + 3\mathbf{k}$, $\overline{OB} = 3\mathbf{i} + 2\mathbf{j} - 4\mathbf{k}$, $\overline{OC} = -\mathbf{i} + 3\mathbf{j} - 2\mathbf{k}$, determine:

(a) the vector $\overline{AB}$

(b) the vector $\overline{BC}$

(c) the vector product $\overline{AB} \times \overline{BC}$

(d) the unit vector perpendicular to the plane ABC.

14 If $\mathbf{A} = 2\mathbf{i} + 3\mathbf{j} - 4\mathbf{k}$; $\mathbf{B} = 3\mathbf{i} + 5\mathbf{j} + 2\mathbf{k}$; $\mathbf{C} = \mathbf{i} - 2\mathbf{j} + 3\mathbf{k}$; determine $\mathbf{A} \cdot (\mathbf{B} \times \mathbf{C})$.

15 If $\mathbf{A} = 2\mathbf{i} + \mathbf{j} - 3\mathbf{k}$; $\mathbf{B} = \mathbf{i} - 2\mathbf{j} + 2\mathbf{k}$; $\mathbf{C} = 3\mathbf{i} + 2\mathbf{j} - \mathbf{k}$; find $\mathbf{A} \times (\mathbf{B} \times \mathbf{C})$.

16 If $\mathbf{A} = \mathbf{i} - 2\mathbf{j} + 3\mathbf{k}$; $\mathbf{B} = 2\mathbf{i} + \mathbf{j} - 2\mathbf{k}$; $\mathbf{C} = 3\mathbf{i} + 2\mathbf{j} + \mathbf{k}$; find

(a) $\mathbf{A} \times (\mathbf{B} \times \mathbf{C})$; (b) $(\mathbf{A} \times \mathbf{B}) \times \mathbf{C}$.

17 Find the scalar triple product of

(a) $\mathbf{A} = \mathbf{i} + 2\mathbf{j} - 3\mathbf{k}$; $\mathbf{B} = 2\mathbf{i} - \mathbf{j} + 4\mathbf{k}$; $\mathbf{C} = 3\mathbf{i} + \mathbf{j} - 2\mathbf{k}$.

(b) $\mathbf{A} = 2\mathbf{i} - 3\mathbf{j} + \mathbf{k}$; $\mathbf{B} = 3\mathbf{i} + \mathbf{j} + 2\mathbf{k}$; $\mathbf{C} = \mathbf{i} + 4\mathbf{j} - 2\mathbf{k}$.

(c) $\mathbf{A} = -2\mathbf{i} + 3\mathbf{j} - 2\mathbf{k}$; $\mathbf{B} = 3\mathbf{i} - \mathbf{j} + 3\mathbf{k}$; $\mathbf{C} = 2\mathbf{i} - 5\mathbf{j} + \mathbf{k}$.

18 Find the vector triple product $\mathbf{A} \times (\mathbf{B} \times \mathbf{C})$ of the following.

(a) $\mathbf{A} = 3\mathbf{i} + \mathbf{j} - 2\mathbf{k}$; $\mathbf{B} = 2\mathbf{i} + 4\mathbf{j} + 3\mathbf{k}$; $\mathbf{C} = \mathbf{i} - 2\mathbf{j} + \mathbf{k}$.

(b) $\mathbf{A} = 2\mathbf{i} - \mathbf{j} + 3\mathbf{k}$; $\mathbf{B} = \mathbf{i} + 4\mathbf{j} - 5\mathbf{k}$; $\mathbf{C} = 3\mathbf{i} - 2\mathbf{j} + \mathbf{k}$.

(c) $\mathbf{A} = 4\mathbf{i} + 2\mathbf{j} - 3\mathbf{k}$; $\mathbf{B} = 2\mathbf{i} - 3\mathbf{j} + 2\mathbf{k}$; $\mathbf{C} = 3\mathbf{i} - 3\mathbf{j} + \mathbf{k}$.

19 Determine the value of p such that the three vectors $\mathbf{A}, \mathbf{B}, \mathbf{C}$ are coplanar when $\mathbf{A} = 2\mathbf{i} + \mathbf{j} + 4\mathbf{k}$; $\mathbf{B} = 3\mathbf{i} + 2\mathbf{j} + p\mathbf{k}$; $\mathbf{C} = \mathbf{i} + 4\mathbf{j} + 2\mathbf{k}$.

20 If $\mathbf{A} = p\mathbf{i} - 6\mathbf{j} - 3\mathbf{k}$; $\mathbf{B} = 4\mathbf{i} + 3\mathbf{j} - \mathbf{k}$; $\mathbf{C} = \mathbf{i} - 3\mathbf{j} + 2\mathbf{k}$

(a) find the values of p for which

(1) $\mathbf{A}$ and $\mathbf{B}$ are perpendicular to each other

(2) $\mathbf{A}, \mathbf{B}$ and $\mathbf{C}$ are coplanar.

(b) determine a unit vector perpendicular to both $\mathbf{A}$ and $\mathbf{B}$ when $p = 2$.

Program 8

Vector differentiation

Frames
1 to 68

Learning outcomes

When you have completed this Program you will be able to:
- Differentiate a vector field and derive a unit vector tangential to the vector field at a point
- Integrate a vector field
- Obtain the gradient of a scalar field, the directional derivative and a unit normal to a surface
- Obtain the divergence of a vector field and recognise a solenoidal vector field
- Obtain the curl of a vector field
- Obtain combinations of div, grad and curl acting on scalar and vector fields as appropriate

Differentiation of vectors

In many practical problems, we often deal with vectors that change with time, e.g. velocity, acceleration, etc. If a vector **A** depends on a scalar variable t, then **A** can be represented as $\mathbf{A}(t)$ and **A** is then said to be a function of t.

If $\mathbf{A} = a_x\mathbf{i} + a_y\mathbf{j} + a_z\mathbf{k}$ then a_x, a_y, a_z will also be dependent on the parameter t.

i.e. $\mathbf{A}(t) = a_x(t)\mathbf{i} + a_y(t)\mathbf{j} + a_z(t)\mathbf{k}$

Differentiating with respect to t gives ....

$$\frac{d}{dt}\{\mathbf{A}(t)\} = \mathbf{i}\frac{d}{dt}\{a_x(t)\} + \mathbf{j}\frac{d}{dt}\{a_y(t)\} + \mathbf{k}\frac{d}{dt}\{a_z(t)\}$$

In short $\dfrac{d\mathbf{A}}{dt} = \mathbf{i}\dfrac{da_x}{dt} + \mathbf{j}\dfrac{da_y}{dt} + \mathbf{k}\dfrac{da_z}{dt}$.

The independent scalar variable is not, of course, restricted to t. In general, if u is the parameter, then

$$\frac{d\mathbf{A}}{du} = \ldots\ldots\ldots\ldots$$

$$\frac{d\mathbf{A}}{du} = \mathbf{i}\frac{da_x}{du} + \mathbf{j}\frac{da_y}{du} + \mathbf{k}\frac{da_z}{du}$$

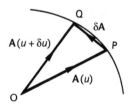

If a position vector $\overline{OP}$ moves to $\overline{OQ}$ when u becomes $u + \delta u$, then as $\delta u \to 0$, the direction of the chord $\overline{PQ}$ becomes that of the tangent to the curve at **P**, i.e. the direction of $\dfrac{d\mathbf{A}}{du}$ is along the tangent to the locus of **P**.

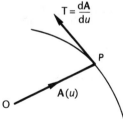

Example 1

If $\mathbf{A} = (3u^2 + 4)\mathbf{i} + (2u - 5)\mathbf{j} + 4u^3\mathbf{k}$, then

$$\frac{d\mathbf{A}}{du} = \ldots\ldots\ldots\ldots$$

4

$$\frac{d\mathbf{A}}{du} = 6u\mathbf{i} + 2\mathbf{j} + 12u^2\mathbf{k}$$

If we differentiate this again, we get $\frac{d^2\mathbf{A}}{du^2} = 6\mathbf{i} + 24u\mathbf{k}$

When $u = 2$, $\frac{d\mathbf{A}}{du} = 12\mathbf{i} + 2\mathbf{j} + 48\mathbf{k}$ and $\frac{d^2\mathbf{A}}{du^2} = 6\mathbf{i} + 48\mathbf{k}$

Then $\left|\frac{d\mathbf{A}}{du}\right| = \ldots\ldots\ldots$ and $\left|\frac{d^2\mathbf{A}}{du^2}\right| = \ldots\ldots\ldots$

5

$$\left|\frac{d\mathbf{A}}{du}\right| = 49.52; \quad \left|\frac{d^2\mathbf{A}}{du^2}\right| = 48.37$$

Because

$$\left|\frac{d\mathbf{A}}{du}\right| = \{12^2 + 2^2 + 48^2\}^{1/2} = \{2452\}^{1/2} = 49.52$$

and $\left|\frac{d^2\mathbf{A}}{du^2}\right| = \{6^2 + 48^2\}^{1/2} = \{2340\}^{1/2} = 48.37$

Example 2

If $\mathbf{F} = \mathbf{i}\sin 2t + \mathbf{j}e^{3t} + \mathbf{k}(t^3 - 4t)$, then when $t = 1$

$$\frac{d\mathbf{F}}{dt} = \ldots\ldots\ldots; \quad \frac{d^2\mathbf{F}}{dt^2} = \ldots\ldots\ldots$$

6

$$\frac{d\mathbf{F}}{dt} = 2\cos 2\mathbf{i} + 3e^3\mathbf{j} - \mathbf{k}$$

$$\frac{d^2\mathbf{F}}{dt^2} = -4\sin 2\mathbf{i} + 9e^3\mathbf{j} + 6\mathbf{k}$$

From these, we could if required find the magnitudes of $\frac{d\mathbf{F}}{dt}$ and $\frac{d^2\mathbf{F}}{dt^2}$.

$$\left|\frac{d\mathbf{F}}{dt}\right| = \ldots\ldots\ldots; \quad \left|\frac{d^2\mathbf{F}}{dt^2}\right| = \ldots\ldots\ldots$$

Vector differentiation

$$\left|\frac{d\mathbf{F}}{dt}\right| = 60.27; \quad \left|\frac{d^2\mathbf{F}}{dt^2}\right| = 180.9$$

Because

$$\left|\frac{d\mathbf{F}}{dt}\right| = \{(2\cos 2)^2 + 9e^6 + 1\}^{1/2}$$

$$= \{0.6927 + 3631 + 1\}^{1/2} = 60.27$$

and $\left|\frac{d^2\mathbf{F}}{dt^2}\right| = \{(-4\sin 2)^2 + 81e^6 + 36\}^{1/2}$

$$= \{13.23 + 32{,}678 + 36\}^{1/2} = 180.9$$

One more example.

Example 3

If $\mathbf{A} = (u+3)\mathbf{i} - (2+u^2)\mathbf{j} + 2u^3\mathbf{k}$, determine

(a) $\dfrac{d\mathbf{A}}{du}$ (b) $\dfrac{d^2\mathbf{A}}{du^2}$ (c) $\left|\dfrac{d\mathbf{A}}{du}\right|$ (d) $\left|\dfrac{d^2\mathbf{A}}{du^2}\right|$ at $u = 3$.

Work through all sections and then check with the next frame

Here is the working. $\mathbf{A} = (u+3)\mathbf{i} - (2+u^2)\mathbf{j} + 2u^3\mathbf{k}$

(a) $\dfrac{d\mathbf{A}}{du} = \mathbf{i} - 2u\mathbf{j} + 6u^2\mathbf{k}$ At $u = 3$, $\dfrac{d\mathbf{A}}{du} = \mathbf{i} - 6\mathbf{j} + 54\mathbf{k}$

(b) $\dfrac{d^2\mathbf{A}}{du^2} = -2\mathbf{j} + 12u\mathbf{k}$ At $u = 3$, $\dfrac{d^2\mathbf{A}}{du^2} = -2\mathbf{j} + 36\mathbf{k}$

(c) $\left|\dfrac{d\mathbf{A}}{du}\right| = \{1 + 36 + 2916\}^{1/2} = (2953)^{1/2} = 54.34$

(d) $\left|\dfrac{d^2\mathbf{A}}{du^2}\right| = \{4 + 1296\}^{1/2} = (1300)^{1/2} = 36.06$

The next example is of a rather different kind, so move on

Example 4

A particle moves in space so that at time t its position is stated as $x = 2t + 3$, $y = t^2 + 3t$, $z = t^3 + 2t^2$. We are required to find the components of its velocity and acceleration in the direction of the vector $2\mathbf{i} + 3\mathbf{j} + 4\mathbf{k}$ when $t = 1$.

First we can write the position as a vector $\mathbf{r}$

$$\mathbf{r} = (2t+3)\mathbf{i} + (t^2+3t)\mathbf{j} + (t^3+2t^2)\mathbf{k}$$

Then, at $t = 1$

$$\frac{d\mathbf{r}}{dt} = \ldots\ldots\ldots\ldots; \quad \frac{d^2\mathbf{r}}{dt^2} = \ldots\ldots\ldots\ldots$$

10

$$\frac{d\mathbf{r}}{dt} = 2\mathbf{i} + 5\mathbf{j} + 7\mathbf{k}; \quad \frac{d^2\mathbf{r}}{dt^2} = 2\mathbf{j} + 10\mathbf{k}$$

Because

$$\frac{d\mathbf{r}}{dt} = 2\mathbf{i} + (2t+3)\mathbf{j} + (3t^2+4t)\mathbf{k}$$

$$\therefore \text{At } t=1, \quad \frac{d\mathbf{r}}{dt} = 2\mathbf{i} + 5\mathbf{j} + 7\mathbf{k}$$

and $\quad \dfrac{d^2\mathbf{r}}{dt^2} = 2\mathbf{j} + (6t+4)\mathbf{k}$

$$\therefore \text{At } t=1, \quad \frac{d^2\mathbf{r}}{dt^2} = 2\mathbf{j} + 10\mathbf{k}$$

Now, a unit vector parallel to $2\mathbf{i} + 3\mathbf{j} + 4\mathbf{k}$ is ....

11

$$\frac{2\mathbf{i} + 3\mathbf{j} + 4\mathbf{k}}{\sqrt{4+9+16}} = \frac{1}{\sqrt{29}}(2\mathbf{i} + 3\mathbf{j} + 4\mathbf{k})$$

Denote this unit vector by **I**. Then the component of $\dfrac{d\mathbf{r}}{dt}$ in the direction of **I**

$$= \frac{d\mathbf{r}}{dt}\cos\theta$$

$$= \frac{d\mathbf{r}}{dt}\cdot\mathbf{I}$$

$$= \frac{1}{\sqrt{29}}(2\mathbf{i} + 5\mathbf{j} + 7\mathbf{k})\cdot(2\mathbf{i} + 3\mathbf{j} + 4\mathbf{k})$$

$$= \ldots\ldots\ldots$$

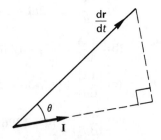

12

$$\boxed{8.73}$$

Because

$$\frac{1}{\sqrt{29}}(2\mathbf{i} + 5\mathbf{j} + 7\mathbf{k})\cdot(2\mathbf{i} + 3\mathbf{j} + 4\mathbf{k}) = \frac{1}{\sqrt{29}}(4 + 15 + 28)$$

$$= \frac{47}{\sqrt{29}}$$

$$= 8.73$$

Similarly, the component of $\dfrac{d^2\mathbf{r}}{dt^2}$ in the direction of **I** is

....

Vector differentiation

$$\boxed{8.54}$$

Because

$$\frac{d^2\mathbf{r}}{dt^2}\cos\theta = \frac{d^2\mathbf{r}}{dt^2}\cdot \mathbf{I}$$

$$= \frac{1}{\sqrt{29}}(2\mathbf{j} + 10\mathbf{k})\cdot(2\mathbf{i} + 3\mathbf{j} + 4\mathbf{k})$$

$$= \frac{1}{\sqrt{29}}(6 + 40)$$

$$= \frac{46}{\sqrt{29}}$$

$$= 8.54$$

Differentiation of sums and products of vectors

If $\mathbf{A} = \mathbf{A}(u)$ and $\mathbf{B} = \mathbf{B}(u)$, then

(a) $\dfrac{d}{du}\{c\mathbf{A}\} = c\dfrac{d\mathbf{A}}{du}$

(b) $\dfrac{d}{du}\{\mathbf{A} + \mathbf{B}\} = \dfrac{d\mathbf{A}}{du} + \dfrac{d\mathbf{B}}{du}$

(c) $\dfrac{d}{du}\{\mathbf{A}\cdot\mathbf{B}\} = \mathbf{A}\cdot\dfrac{d\mathbf{B}}{du} + \dfrac{d\mathbf{A}}{du}\cdot\mathbf{B}$

(d) $\dfrac{d}{du}\{\mathbf{A}\times\mathbf{B}\} = \mathbf{A}\times\dfrac{d\mathbf{B}}{du} + \dfrac{d\mathbf{A}}{du}\times\mathbf{B}$.

These are very much like the normal rules of differentiation. However, if $\mathbf{A}(u)\cdot\mathbf{A}(u) = a_x^2 + a_y^2 + a_z^2 = |\mathbf{A}|^2 = A^2$ is a constant then

$$\frac{d}{du}\{\mathbf{A}(u)\cdot\mathbf{A}(u)\} = \mathbf{A}(u)\cdot\frac{d}{du}\{\mathbf{A}(u)\} + \mathbf{A}(u)\cdot\frac{d}{du}\{\mathbf{A}(u)\}$$

$$= 2\mathbf{A}(u)\cdot\frac{d}{du}\{\mathbf{A}(u)\} = \frac{d}{du}\{\mathbf{A}^2\} = 0$$

Assuming that $\mathbf{A}(u) \neq 0$, then since $\mathbf{A}(u)\cdot\dfrac{d}{du}\{\mathbf{A}(u)\} = 0$ it follows that $\mathbf{A}(u)$ and $\dfrac{d}{du}\{\mathbf{A}(u)\}$ are perpendicular vectors because

....

14

$$\mathbf{A}(u) \cdot \frac{d}{du}\{\mathbf{A}(u)\} = |\mathbf{A}(u)| \left|\frac{d}{du}\{\mathbf{A}(u)\}\right| \cos\theta = 0$$

$$\therefore \cos\theta = 0 \quad \therefore \theta = \frac{\pi}{2}$$

Now let us deal with unit tangent vectors.

Unit tangent vectors

We have already established in Frame 3 of this Program that if $\overline{OP}$ is a position vector $\mathbf{A}(u)$ in space, then the direction of the vector denoting $\frac{d}{du}\{\mathbf{A}(u)\}$ is

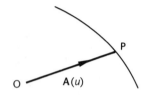

....

15

parallel to the tangent to the curve at P

Then the unit tangent vector $\mathbf{T}$ at P can be found from

$$\mathbf{T} = \frac{\frac{d}{du}\{\mathbf{A}(u)\}}{\left|\frac{d}{du}\{\mathbf{A}(u)\}\right|}$$

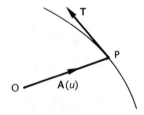

In simpler notation, this becomes:

If $\mathbf{r} = a_x\mathbf{i} + a_y\mathbf{j} + a_z\mathbf{k}$ then the unit tangent vector $\mathbf{T}$ is given by

$$\mathbf{T} = \frac{d\mathbf{r}/du}{|d\mathbf{r}/du|}$$

Example 1

Determine the unit tangent vector at the point (2, 4, 7) for the curve with parametric equations $x = 2u;\ y = u^2 + 3;\ z = 2u^2 + 5$.

First we see that the point (2, 4, 7) corresponds to $u = 1$.

The vector equation of the curve is

$$\mathbf{r} = a_x\mathbf{i} + a_y\mathbf{j} + a_z\mathbf{k} = 2u\mathbf{i} + (u^2 + 3)\mathbf{j} + (2u^2 + 5)\mathbf{k}$$

$$\therefore \frac{d\mathbf{r}}{du} = \ldots\ldots\ldots$$

Vector differentiation

16

$$\frac{d\mathbf{r}}{du} = 2\mathbf{i} + 2u\mathbf{j} + 4u\mathbf{k}$$

and at $u = 1$, $\dfrac{d\mathbf{r}}{du} = 2\mathbf{i} + 2\mathbf{j} + 4\mathbf{k}$

Hence $\left|\dfrac{d\mathbf{r}}{du}\right| = \ldots\ldots\ldots\ldots$ and $\mathbf{T} = \ldots\ldots\ldots\ldots$

17

$$\left|\frac{d\mathbf{r}}{du}\right| = 2\sqrt{6}; \quad \mathbf{T} = \frac{1}{\sqrt{6}}\{\mathbf{i} + \mathbf{j} + 2\mathbf{k}\}$$

Because

$$\left|\frac{d\mathbf{r}}{du}\right| = \{4 + 4 + 16\}^{1/2} = 24^{1/2} = 2\sqrt{6}$$

$$\mathbf{T} = \frac{\dfrac{d\mathbf{r}}{du}}{\left|\dfrac{d\mathbf{r}}{du}\right|} = \frac{2\mathbf{i} + 2\mathbf{j} + 4\mathbf{k}}{2\sqrt{6}} = \frac{1}{\sqrt{6}}\{\mathbf{i} + \mathbf{j} + 2\mathbf{k}\}$$

Let us do another.

Example 2

Find the unit tangent vector at the point $(2, 0, \pi)$ for the curve with parametric equations $x = 2\sin\theta$; $y = 3\cos\theta$; $z = 2\theta$.

We see that the point $(2, 0, \pi)$ corresponds to $\theta = \pi/2$.

Writing the curve in vector form $\mathbf{r} = \ldots\ldots\ldots\ldots$

18

$$\mathbf{r} = 2\sin\theta\,\mathbf{i} + 3\cos\theta\,\mathbf{j} + 2\theta\,\mathbf{k}$$

Then, at $\theta = \pi/2$, $\dfrac{d\mathbf{r}}{d\theta} = \ldots\ldots\ldots\ldots$

$\left|\dfrac{d\mathbf{r}}{d\theta}\right| = \ldots\ldots\ldots\ldots$

$\mathbf{T} = \ldots\ldots\ldots\ldots$

Finish it off

19

$$\frac{d\mathbf{r}}{d\theta} = -3\mathbf{j} + 2\mathbf{k}; \quad \left|\frac{d\mathbf{r}}{d\theta}\right| = \sqrt{13}$$

$$\mathbf{T} = \frac{1}{\sqrt{13}}(-3\mathbf{j} + 2\mathbf{k})$$

▶

Vector Analysis

And now

Example 3

Determine the unit tangent vector for the curve
$$x = 3t; \quad y = 2t^2; \quad z = t^2 + t$$
at the point (6, 8, 6).

On your own. **T** = ...

20

$$\boxed{\mathbf{T} = \frac{1}{\sqrt{98}}(3\mathbf{i} + 8\mathbf{j} + 5\mathbf{k})}$$

The point (6, 8, 6) corresponds to $t = 2$
$$\mathbf{r} = 3t\mathbf{i} + 2t^2\mathbf{j} + (t^2 + t)\mathbf{k}$$
$$\therefore \frac{d\mathbf{r}}{dt} = 3\mathbf{i} + 4t\mathbf{j} + (2t + 1)\mathbf{k}$$

At $t = 2$, $\mathbf{r} = 6\mathbf{i} + 8\mathbf{j} + 6\mathbf{k}$ and $\dfrac{d\mathbf{r}}{dt} = 3\mathbf{i} + 8\mathbf{j} + 5\mathbf{k}$

$$\therefore \left|\frac{d\mathbf{r}}{dt}\right| = (9 + 64 + 25)^{1/2} = \sqrt{98}$$

$$\therefore \mathbf{T} = \frac{d\mathbf{r}/dt}{|d\mathbf{r}/dt|} = \frac{1}{\sqrt{98}}(3\mathbf{i} + 8\mathbf{j} + 5\mathbf{k})$$

Partial differentiation of vectors

21

If a vector **F** is a function of two independent variables u and v, then the rules of differentiation follow the usual pattern.

If $\mathbf{F} = x\mathbf{i} + y\mathbf{j} + z\mathbf{k}$ then x, y, z will also be functions of u and v.

Then
$$\frac{\partial \mathbf{F}}{\partial u} = \frac{\partial x}{\partial u}\mathbf{i} + \frac{\partial y}{\partial u}\mathbf{j} + \frac{\partial z}{\partial u}\mathbf{k}$$

$$\frac{\partial \mathbf{F}}{\partial v} = \frac{\partial x}{\partial v}\mathbf{i} + \frac{\partial y}{\partial v}\mathbf{j} + \frac{\partial z}{\partial v}\mathbf{k}$$

$$\frac{\partial^2 \mathbf{F}}{\partial u^2} = \frac{\partial^2 x}{\partial u^2}\mathbf{i} + \frac{\partial^2 y}{\partial u^2}\mathbf{j} + \frac{\partial^2 z}{\partial u^2}\mathbf{k}$$

$$\frac{\partial^2 \mathbf{F}}{\partial v^2} = \frac{\partial^2 x}{\partial v^2}\mathbf{i} + \frac{\partial^2 y}{\partial v^2}\mathbf{j} + \frac{\partial^2 z}{\partial v^2}\mathbf{k}$$

$$\frac{\partial^2 \mathbf{F}}{\partial u \partial v} = \frac{\partial^2 x}{\partial u \partial v}\mathbf{i} + \frac{\partial^2 y}{\partial u \partial v}\mathbf{j} + \frac{\partial^2 z}{\partial u \partial v}\mathbf{k}$$

and for small finite changes du and dv in u and v, we have

$$d\mathbf{F} = \frac{\partial \mathbf{F}}{\partial u}du + \frac{\partial \mathbf{F}}{\partial v}dv$$

▶

Vector differentiation

Example

If $\mathbf{F} = 2uv\mathbf{i} + (u^2 - 2v)\mathbf{j} + (u + v^2)\mathbf{k}$

$\dfrac{\partial \mathbf{F}}{\partial u} = \ldots\ldots\ldots$; $\dfrac{\partial \mathbf{F}}{\partial v} = \ldots\ldots\ldots$

$\dfrac{\partial^2 \mathbf{F}}{\partial u^2} = \ldots\ldots\ldots$; $\dfrac{\partial^2 \mathbf{F}}{\partial u \partial v} = \ldots\ldots\ldots$

22

$\dfrac{\partial \mathbf{F}}{\partial u} = 2v\mathbf{i} + 2u\mathbf{j} + \mathbf{k}$; $\dfrac{\partial \mathbf{F}}{\partial v} = 2u\mathbf{i} - 2\mathbf{j} + 2v\mathbf{k}$

$\dfrac{\partial^2 \mathbf{F}}{\partial u^2} = 2\mathbf{j}$; $\dfrac{\partial^2 \mathbf{F}}{\partial u \partial v} = 2\mathbf{i}$

This is straightforward enough.

Integration of vector functions

The process is the reverse of that for differentiation. If a vector $\mathbf{F} = x\mathbf{i} + y\mathbf{j} + z\mathbf{k}$ where $\mathbf{F}$, x, y, z are expressed as functions of u, then

$$\int_a^b \mathbf{F}\,du = \mathbf{i} \int_a^b x\,du + \mathbf{j} \int_a^b y\,du + \mathbf{k} \int_a^b z\,du.$$

Example 1

If $\mathbf{F} = (3t^2 + 4t)\mathbf{i} + (2t - 5)\mathbf{j} + 4t^3\mathbf{k}$, then

$$\int_1^3 \mathbf{F}\,dt = \mathbf{i} \int_1^3 (3t^2 + 4t)\,dt + \mathbf{j} \int_1^3 (2t - 5)\,dt + \mathbf{k} \int_1^3 4t^3\,dt = \ldots\ldots\ldots$$

23

$$42\mathbf{i} - 2\mathbf{j} + 80\mathbf{k}$$

Because

$$\int_1^3 \mathbf{F}\,dt = \left[\mathbf{i}(t^3 + 2t^2) + \mathbf{j}(t^2 - 5t) + \mathbf{k}t^4\right]_1^3$$
$$= (45\mathbf{i} - 6\mathbf{j} + 81\mathbf{k}) - (3\mathbf{i} - 4\mathbf{j} + \mathbf{k}) = 42\mathbf{i} - 2\mathbf{j} + 80\mathbf{k}$$

Here is a slightly different one.

Example 2

If $\quad \mathbf{F} = 3u\mathbf{i} + u^2\mathbf{j} + (u + 2)\mathbf{k}$
and $\quad \mathbf{V} = 2u\mathbf{i} - 3u\mathbf{j} + (u - 2)\mathbf{k}$

evaluate $\displaystyle\int_0^2 (\mathbf{F} \times \mathbf{V})\,du.$

First we must determine $\mathbf{F} \times \mathbf{V}$ in terms of u.

$$\mathbf{F} \times \mathbf{V} = \ldots\ldots\ldots$$

240 Vector Analysis

24

$$\mathbf{F} \times \mathbf{V} = (u^3 + u^2 + 6u)\mathbf{i} - (u^2 - 10u)\mathbf{j} - (2u^3 + 9u^2)\mathbf{k}$$

Because

$$\mathbf{F} \times \mathbf{V} = \begin{vmatrix} \mathbf{i} & \mathbf{j} & \mathbf{k} \\ 3u & u^2 & (u+2) \\ 2u & -3u & (u-2) \end{vmatrix}$$

which gives the result above.

Then $\int_0^2 (\mathbf{F} \times \mathbf{V})\, du = \ldots\ldots\ldots$

25

$$\tfrac{4}{3}\{14\mathbf{i} + 13\mathbf{j} - 24\mathbf{k}\}$$

Because

$$\int (\mathbf{F} \times \mathbf{V})\,du = \left(\frac{u^4}{4} + \frac{u^3}{3} + 3u^2\right)\mathbf{i} - \left(\frac{u^3}{3} - 5u^2\right)\mathbf{j} - \left(\frac{u^4}{2} + 3u^3\right)\mathbf{k}$$

$$\therefore \int_0^2 (\mathbf{F} \times \mathbf{V})\,du = (4 + \tfrac{8}{3} + 12)\mathbf{i} - (\tfrac{8}{3} - 20)\mathbf{j} - (8 + 24)\mathbf{k}$$

$$= \tfrac{4}{3}\{14\mathbf{i} + 13\mathbf{j} - 24\mathbf{k}\}$$

Example 3

If $\mathbf{F} = \mathbf{A} \times (\mathbf{B} \times \mathbf{C})$ where

$\mathbf{A} = 3t^2\mathbf{i} + (2t - 3)\mathbf{j} + 4t\mathbf{k}$
$\mathbf{B} = 2\mathbf{i} + 4t\mathbf{j} + 3(1 - t)\mathbf{k}$
$\mathbf{C} = 2t\mathbf{i} - 3t^2\mathbf{j} - 2t\mathbf{k}$

determine $\int_0^1 \mathbf{F}\, dt$.

First we need to find $\mathbf{A} \times (\mathbf{B} \times \mathbf{C})$. The simplest way to do this is to use the relationship

$$\mathbf{A} \times (\mathbf{B} \times \mathbf{C}) = \ldots\ldots\ldots$$

26

$$\mathbf{A} \times (\mathbf{B} \times \mathbf{C}) = (\mathbf{A} \cdot \mathbf{C})\mathbf{B} - (\mathbf{A} \cdot \mathbf{B})\mathbf{C}$$

So $\quad \mathbf{A} \cdot \mathbf{C} = \ldots\ldots\ldots$
and $\quad \mathbf{A} \cdot \mathbf{B} = \ldots\ldots\ldots$

Vector differentiation

$$\boxed{\begin{aligned}\mathbf{A}\cdot\mathbf{C} &= 6t^3 - 6t^3 + 9t^2 - 8t^2 = t^2 \\ \mathbf{A}\cdot\mathbf{B} &= 6t^2 + 8t^2 - 12t + 12t - 12t^2 = 2t^2\end{aligned}}$$

Then $\mathbf{F} = \mathbf{A} \times (\mathbf{B} \times \mathbf{C})$
$$= t^2\{2\mathbf{i} + 4t\mathbf{j} + 3(1-t)\mathbf{k}\} - 2t^2\{2t\mathbf{i} - 3t^2\mathbf{j} - 2t\mathbf{k}\}$$

$$\therefore \int_0^1 \mathbf{F}\,dt = \ldots\ldots\ldots\ldots$$

Finish off the simplification and complete the integration.

$$\boxed{\tfrac{1}{60}\{-20\mathbf{i} + 132\mathbf{j} + 75\mathbf{k}\}}$$

Because

$$\mathbf{F} = \mathbf{A} \times (\mathbf{B} \times \mathbf{C}) = (2t^2 - 4t^3)\mathbf{i} + (4t^3 + 6t^4)\mathbf{j} + (3t^2 + t^3)\mathbf{k}$$

Integration with respect to t then gives the result stated above.

Now let us move on to the next stage of our development

Scalar and vector fields

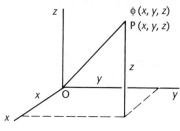

If every point P (x, y, z) of a region R of space has associated with it a scalar quantity $\phi(x, y, z)$, then $\phi(x, y, z)$ is a *scalar function* and a *scalar field* is said to exist in the region R.

Examples of scalar fields are temperature, potential, etc.

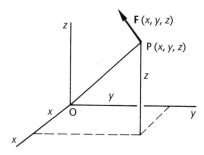

Similarly, if every point P (x, y, z) of a region R has associated with it a vector quantity $\mathbf{F}(x, y, z)$, then $\mathbf{F}(x, y, z)$ is a *vector function* and a *vector field* is said to exist in the region R.

Examples of vector fields are force, velocity, acceleration, etc. $\mathbf{F}(x, y, z)$ can be defined in terms of its components parallel to the coordinate axes, OX, OY, OZ.

That is, $\mathbf{F}(x, y, z) = F_x \mathbf{i} + F_y \mathbf{j} + F_z \mathbf{k}$.

Note these important definitions:
we shall be making good use of them as we proceed

---

30 Grad (gradient of a scalar function)

If a scalar function $\phi(x, y, z)$ is continuously differentiable with respect to its variables x, y, z, throughout the region, then the *gradient* of ϕ, written *grad* ϕ, is defined as the vector

$$\text{grad } \phi = \frac{\partial \phi}{\partial x}\mathbf{i} + \frac{\partial \phi}{\partial y}\mathbf{j} + \frac{\partial \phi}{\partial z}\mathbf{k} \tag{12}$$

Note that, while ϕ is a scalar function, grad ϕ is a vector function. For example, if ϕ depends upon the position of P and is defined by $\phi = 2x^2yz^3$, then

$$\text{grad } \phi = 4xyz^3\mathbf{i} + 2x^2z^3\mathbf{j} + 6x^2yz^2\mathbf{k}$$

Notation

The expression (12) above can be written

$$\text{grad } \phi = \left\{ \mathbf{i}\frac{\partial}{\partial x} + \mathbf{j}\frac{\partial}{\partial y} + \mathbf{k}\frac{\partial}{\partial z} \right\} \phi$$

where $\left(\mathbf{i}\frac{\partial}{\partial x} + \mathbf{j}\frac{\partial}{\partial y} + \mathbf{k}\frac{\partial}{\partial z} \right)$ is called a *vector differential operator* and is denoted by the symbol ∇ (pronounced 'del' or sometimes 'nabla')

i.e. $\quad \nabla \equiv \left(\mathbf{i}\frac{\partial}{\partial x} + \mathbf{j}\frac{\partial}{\partial y} + \mathbf{k}\frac{\partial}{\partial z} \right)$

Beware! ∇ cannot exist alone: it is an operator and must operate on a stated scalar function $\phi(x, y, z)$.

If $\mathbf{F}$ is a vector function, $\nabla \mathbf{F}$ has no meaning.

So we have:

$$\nabla \phi = \text{grad } \phi = \left(\mathbf{i}\frac{\partial}{\partial x} + \mathbf{j}\frac{\partial}{\partial y} + \mathbf{k}\frac{\partial}{\partial z} \right) \phi$$

$$= \mathbf{i}\frac{\partial \phi}{\partial x} + \mathbf{j}\frac{\partial \phi}{\partial y} + \mathbf{k}\frac{\partial \phi}{\partial z} \tag{13}$$

Make a note of this definition and then let us see how to use it

Example 1

If $\phi = x^2yz^3 + xy^2z^2$, determine grad ϕ at the point P (1, 3, 2).

By the definition, $\quad$ grad $\phi = \nabla\phi = \dfrac{\partial \phi}{\partial x}\mathbf{i} + \dfrac{\partial \phi}{\partial y}\mathbf{j} + \dfrac{\partial \phi}{\partial z}\mathbf{k}$.

All we have to do then is to find the partial derivatives at $x = 1$, $y = 3$, $z = 2$ and insert their values.

$$\therefore \nabla\phi = \ldots\ldots\ldots\ldots$$

$$\boxed{4(21\mathbf{i} + 8\mathbf{j} + 18\mathbf{k})}$$

Because

$\phi = x^2yz^3 + xy^2z^2 \quad \therefore \dfrac{\partial \phi}{\partial x} = 2xyz^3 + y^2z^2$

$\dfrac{\partial \phi}{\partial y} = x^2z^3 + 2xyz^2 \quad \dfrac{\partial \phi}{\partial z} = 3x^2yz^2 + 2xy^2z$

Then, at (1, 3, 2) $\quad \dfrac{\partial \phi}{\partial x} = 48 + 36 \quad \therefore \dfrac{\partial \phi}{\partial x} = 84$

$\dfrac{\partial \phi}{\partial y} = 8 + 24 \quad \therefore \dfrac{\partial \phi}{\partial y} = 32$

$\dfrac{\partial \phi}{\partial z} = 36 + 36 \quad \therefore \dfrac{\partial \phi}{\partial z} = 72$

$\therefore$ grad $\phi = \nabla\phi = 84\mathbf{i} + 32\mathbf{j} + 72\mathbf{k} = 4(21\mathbf{i} + 8\mathbf{j} + 18\mathbf{k})$

Example 2

If $\quad \mathbf{A} = x^2z\mathbf{i} + xy\mathbf{j} + y^2z\mathbf{k}$

and $\quad \mathbf{B} = yz^2\mathbf{i} + xz\mathbf{j} + x^2z\mathbf{k}$

determine an expression for grad $(\mathbf{A} \cdot \mathbf{B})$.

This we can soon do since we know that $\mathbf{A} \cdot \mathbf{B}$ is a scalar function of x, y and z.

First then, $\mathbf{A} \cdot \mathbf{B} = \ldots\ldots\ldots\ldots$

$$\boxed{\mathbf{A} \cdot \mathbf{B} = x^2yz^3 + x^2yz + x^2y^2z^2}$$

Then $\quad \nabla(\mathbf{A} \cdot \mathbf{B}) = \ldots\ldots\ldots\ldots$

34

$$2xyz(z^2 + 1 + yz)\mathbf{i} + x^2z(z^2 + 1 + 2yz)\mathbf{j} + x^2y(3z^2 + 1 + 2yz)\mathbf{k}$$

Because

if $\phi = \mathbf{A} \cdot \mathbf{B} = (x^2z\mathbf{i} + xy\mathbf{j} + y^2z\mathbf{k}) \cdot (yz^2\mathbf{i} + xz\mathbf{j} + x^2z\mathbf{k})$

$= x^2yz^3 + x^2yz + x^2y^2z^2$

$\dfrac{\partial \phi}{\partial x} = 2xyz^3 + 2xyz + 2xy^2z^2 = 2xyz(z^2 + 1 + yz)$

$\dfrac{\partial \phi}{\partial y} = x^2z^3 + x^2z + 2x^2yz^2 = x^2z(z^2 + 1 + 2yz)$

$\dfrac{\partial \phi}{\partial z} = 3x^2yz^2 + x^2y + 2x^2y^2z = x^2y(3z^2 + 1 + 2yz)$

$\therefore \nabla(\mathbf{A} \cdot \mathbf{B}) = 2xyz(z^2 + 1 + yz)\mathbf{i} + x^2z(z^2 + 1 + 2yz)\mathbf{j}$
$\qquad\qquad\qquad + x^2y(3z^2 + 1 + 2yz)\mathbf{k}$

Now let us obtain another useful relationship.

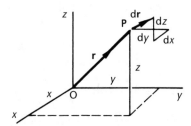

If $\overline{OP}$ is a position vector $\mathbf{r}$ where $\mathbf{r} = x\mathbf{i} + y\mathbf{j} + z\mathbf{k}$ and $d\mathbf{r}$ is a small displacement corresponding to changes dx, dy, dz in x, y, z respectively, then

$$d\mathbf{r} = dx\,\mathbf{i} + dy\,\mathbf{j} + dz\,\mathbf{k}$$

If $\phi(x, y, z)$ is a scalar function at P, we know that

$$\text{grad } \phi = \nabla \phi = \dfrac{\partial \phi}{\partial x}\mathbf{i} + \dfrac{\partial \phi}{\partial y}\mathbf{j} + \dfrac{\partial \phi}{\partial z}\mathbf{k}$$

Then grad $\phi \cdot d\mathbf{r} = \ldots\ldots\ldots\ldots$

35

$$\text{grad } \phi \cdot d\mathbf{r} = \dfrac{\partial \phi}{\partial x}dx + \dfrac{\partial \phi}{\partial y}dy + \dfrac{\partial \phi}{\partial z}dz$$

Because

$\text{grad } \phi \cdot d\mathbf{r} = \left(\dfrac{\partial \phi}{\partial x}\mathbf{i} + \dfrac{\partial \phi}{\partial y}\mathbf{j} + \dfrac{\partial \phi}{\partial z}\mathbf{k}\right) \cdot (dx\,\mathbf{i} + dy\,\mathbf{j} + dz\,\mathbf{k})$

$= \dfrac{\partial \phi}{\partial x}dx + \dfrac{\partial \phi}{\partial y}dy + \dfrac{\partial \phi}{\partial z}dz$

= the total differential $d\phi$ of ϕ

That is
$$d\phi = d\mathbf{r} \cdot \text{grad } \phi \qquad\qquad (14)$$

This will certainly be useful, so make a note of it

Directional derivatives

We have just established that

$$d\phi = \mathbf{dr} \cdot \text{grad}\, \phi$$

If ds is the small element of arc between P ($\mathbf{r}$) and Q ($\mathbf{r} + \mathbf{dr}$) then $ds = |\mathbf{dr}|$

$$\frac{\mathbf{dr}}{ds} = \frac{\mathbf{dr}}{|\mathbf{dr}|}$$

and $\dfrac{\mathbf{dr}}{ds}$ is thus a unit vector in the direction of $\mathbf{dr}$.

$$\therefore \frac{d\phi}{ds} = \frac{\mathbf{dr}}{ds} \cdot \text{grad}\, \phi$$

If we denote the unit vector $\dfrac{\mathbf{dr}}{ds}$ by $\hat{\mathbf{a}}$ then the result becomes

$$\frac{d\phi}{ds} = \hat{\mathbf{a}} \cdot \text{grad}\, \phi$$

$\dfrac{d\phi}{ds}$ is thus the projection of grad ϕ on the unit vector $\hat{\mathbf{a}}$ and is called the *directional derivative* of ϕ in the direction of $\hat{\mathbf{a}}$. It gives the rate of change of ϕ with distance measured in the direction of $\hat{\mathbf{a}}$ and $\dfrac{d\phi}{ds} = \hat{\mathbf{a}} \cdot \text{grad}\, \phi$ will be a maximum when $\hat{\mathbf{a}}$ and grad ϕ have the same direction, since then

$$\hat{\mathbf{a}} \cdot \text{grad}\, \phi = |\hat{\mathbf{a}}|\,|\text{grad}\, \phi|\cos\theta \text{ and } \theta \text{ will be zero.}$$

Thus the direction of grad ϕ gives the direction in which the maximum rate of change of ϕ occurs.

Example 1

Find the directional derivative of the function $\phi = x^2z + 2xy^2 + yz^2$ at the point (1, 2, −1) in the direction of the vector $\mathbf{A} = 2\mathbf{i} + 3\mathbf{j} - 4\mathbf{k}$.

We start off with $\phi = x^2z + 2xy^2 + yz^2$

$$\therefore \nabla\phi = \ldots\ldots\ldots\ldots$$

37

$$\nabla\phi = (2xz + 2y^2)\mathbf{i} + (4xy + z^2)\mathbf{j} + (x^2 + 2yz)\mathbf{k}$$

Because

$$\frac{\partial\phi}{\partial x} = 2xz + 2y^2; \quad \frac{\partial\phi}{\partial y} = 4xy + z^2; \quad \frac{\partial\phi}{\partial z} = x^2 + 2yz$$

Then, at $(1, 2, -1)$

$$\nabla\phi = (-2 + 8)\mathbf{i} + (8 + 1)\mathbf{j} + (1 - 4)\mathbf{k} = 6\mathbf{i} + 9\mathbf{j} - 3\mathbf{k}$$

Next we have to find the unit vector $\hat{\mathbf{a}}$ where $\mathbf{A} = 2\mathbf{i} + 3\mathbf{j} - 4\mathbf{k}$

$$\hat{\mathbf{a}} = \ldots\ldots\ldots\ldots$$

38

$$\hat{\mathbf{a}} = \frac{1}{\sqrt{29}}(2\mathbf{i} + 3\mathbf{j} - 4\mathbf{k})$$

Because

$$\mathbf{A} = 2\mathbf{i} + 3\mathbf{j} - 4\mathbf{k} \quad \therefore \quad |\mathbf{A}| = \sqrt{4 + 9 + 16} = \sqrt{29}$$

$$\hat{\mathbf{a}} = \frac{\mathbf{A}}{|\mathbf{A}|} = \frac{1}{\sqrt{29}}(2\mathbf{i} + 3\mathbf{j} - 4\mathbf{k})$$

So we have $\nabla\phi = 6\mathbf{i} + 9\mathbf{j} - 3\mathbf{k}$ and $\hat{\mathbf{a}} = \dfrac{1}{\sqrt{29}}(2\mathbf{i} + 3\mathbf{j} - 4\mathbf{k})$

$$\therefore \frac{d\phi}{ds} = \hat{\mathbf{a}} \cdot \nabla\phi$$

$$= \ldots\ldots\ldots\ldots$$

39

$$\frac{d\phi}{ds} = \frac{51}{\sqrt{29}} = 9.47$$

Because

$$\frac{d\phi}{ds} = \hat{\mathbf{a}} \cdot \nabla\phi = \frac{1}{\sqrt{29}}(2\mathbf{i} + 3\mathbf{j} - 4\mathbf{k}) \cdot (6\mathbf{i} + 9\mathbf{j} - 3\mathbf{k})$$

$$= \frac{1}{\sqrt{29}}(12 + 27 + 12) = \frac{51}{\sqrt{29}} = 9.47$$

That is all there is to it.

(a) From the given scalar function ϕ, determine $\nabla\phi$.

(b) Find the unit vector $\hat{\mathbf{a}}$ in the direction of the given vector $\mathbf{A}$.

(c) Then $\dfrac{d\phi}{ds} = \hat{\mathbf{a}} \cdot \nabla\phi$.

▶

Vector differentiation

Example 2

Find the directional derivative of $\phi = x^2y + y^2z + z^2x$ at the point $(1, -1, 2)$ in the direction of the vector $\mathbf{A} = 4\mathbf{i} + 2\mathbf{j} - 5\mathbf{k}$.

Same as before. *Work through it and check the result with the next frame*

$$\frac{d\phi}{ds} = \frac{-23}{3\sqrt{5}} = -3.43$$

40

Because

$\phi = x^2y + y^2z + z^2x$

$\therefore \nabla\phi = (2xy + z^2)\mathbf{i} + (x^2 + 2yz)\mathbf{j} + (y^2 + 2zx)\mathbf{k}$

$\therefore$ At $(1, -1, 2)$, $\nabla\phi = 2\mathbf{i} - 3\mathbf{j} + 5\mathbf{k}$

$\mathbf{A} = 4\mathbf{i} + 2\mathbf{j} - 5\mathbf{k}$ $\therefore$ $|\mathbf{A}| = \sqrt{16 + 4 + 25} = \sqrt{45} = 3\sqrt{5}$

$$\therefore \hat{\mathbf{a}} = \frac{1}{3\sqrt{5}}(4\mathbf{i} + 2\mathbf{j} - 5\mathbf{k})$$

$$\therefore \frac{d\phi}{ds} = \hat{\mathbf{a}} \cdot \nabla\phi = \frac{1}{3\sqrt{5}}(4\mathbf{i} + 2\mathbf{j} - 5\mathbf{k}) \cdot (2\mathbf{i} - 3\mathbf{j} + 5\mathbf{k})$$

$$= \frac{1}{3\sqrt{5}}(8 - 6 - 25) = \frac{-23}{3\sqrt{5}} = -3.43$$

Example 3

Find the direction from the point $(1, 1, 0)$ which gives the greatest rate of increase of the function $\phi = (x + 3y)^2 + (2y - z)^2$.

This appears to be different, but it rests on the fact that the greatest rate of increase of ϕ with respect to distance is in

....

---

the direction of $\nabla\phi$

41

All we need then is to find the vector $\nabla\phi$, which is

....

42

$$\nabla\phi = 4(2\mathbf{i} + 8\mathbf{j} - \mathbf{k})$$

Because

$$\phi = (x+3y)^2 + (2y-z)^2$$

$$\therefore \frac{\partial\phi}{\partial x} = 2(x+3y); \quad \frac{\partial\phi}{\partial y} = 6(x+3y) + 4(2y-z); \quad \frac{\partial\phi}{\partial z} = -2(2y-z)$$

$$\therefore \text{At } (1,\,1,\,0), \quad \frac{\partial\phi}{\partial x} = 8; \quad \frac{\partial\phi}{\partial y} = 32; \quad \frac{\partial\phi}{\partial z} = -4$$

$$\therefore \nabla\phi = 8\mathbf{i} + 32\mathbf{j} - 4\mathbf{k} = 4(2\mathbf{i} + 8\mathbf{j} - \mathbf{k})$$

∴ greatest rate of increase occurs in direction $2\mathbf{i} + 8\mathbf{j} - \mathbf{k}$

So on we go

43

Unit normal vectors

The equation of $\phi(x,\,y,\,z) = $ constant represents a surface in space. For example, $3x - 4y + 2z = 1$ is the equation of a plane and $x^2 + y^2 + z^2 = 4$ represents a sphere centred on the origin and of radius 2.

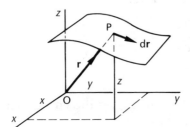

If $d\mathbf{r}$ is a displacement in this surface, then $d\phi = 0$ since ϕ is constant over the surface.

Therefore our previous relationship $d\mathbf{r} \cdot \text{grad } \phi = d\phi$ becomes

$$d\mathbf{r} \cdot \text{grad } \phi = 0$$

for all such small displacements $d\mathbf{r}$ in the surface.

But $d\mathbf{r} \cdot \text{grad } \phi = |\,d\mathbf{r}\,|\,|\,\text{grad }\phi\,|\cos\theta = 0$.

$$\therefore \theta = \frac{\pi}{2} \quad \therefore \text{grad }\phi \text{ is perpendicular to } d\mathbf{r}, \text{ i.e. grad }\phi \text{ is a vector perpendicular}$$

to the surface at P, in the direction of maximum rate of change of ϕ. The magnitude of that maximum rate of change is given by $|\,\text{grad }\phi\,|$.

The unit vector $\mathbf{N}$ in the direction of grad ϕ is called the *unit normal vector* at P.

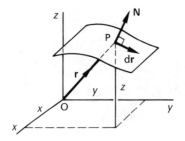

∴ Unit normal vector

$$\mathbf{N} = \frac{\nabla\phi}{|\nabla\phi|} \tag{15}$$

Vector differentiation

Example 1

Find the unit normal vector to the surface $x^3y + 4xz^2 + xy^2z + 2 = 0$ at the point $(1, 3, -1)$.

$$\text{Vector normal} = \nabla \phi = \ldots\ldots\ldots$$

$$\nabla \phi = (3x^2y + 4z^2 + y^2z)\mathbf{i} + (x^3 + 2xyz)\mathbf{j} + (8xz + xy^2)\mathbf{k}$$

44

Then, at $(1, 3, -1)$, $\quad \nabla \phi = 4\mathbf{i} - 5\mathbf{j} + \mathbf{k}$

and the unit normal at $(1, 3, -1)$ is ....

$$\frac{1}{\sqrt{42}}(4\mathbf{i} - 5\mathbf{j} + \mathbf{k})$$

45

Because

$$|\nabla \phi| = \sqrt{16 + 25 + 1} = \sqrt{42}$$

and $\quad \mathbf{N} = \dfrac{\nabla \phi}{|\nabla \phi|} = \dfrac{1}{\sqrt{42}}(4\mathbf{i} - 5\mathbf{j} + \mathbf{k})$

One more.

Example 2

Determine the unit normal to the surface

$xyz + x^2y - 5yz - 5 = 0$ at the point $(3, 1, 2)$.

All very straightforward. Complete it.

$$\text{Unit normal} = \mathbf{N} = \frac{1}{\sqrt{93}}(8\mathbf{i} + 5\mathbf{j} - 2\mathbf{k})$$

46

Because

$$\phi = xyz + x^2y - 5yz - 5$$

$$\therefore \nabla \phi = (yz + 2xy)\mathbf{i} + (xz + x^2 - 5z)\mathbf{j} + (xy - 5y)\mathbf{k}$$

At $(3, 1, 2)$, $\quad \nabla \phi = 8\mathbf{i} + 5\mathbf{j} - 2\mathbf{k}; \quad |\nabla \phi| = \sqrt{64 + 25 + 4} = \sqrt{93}$

$$\therefore \text{Unit normal} = \mathbf{N} = \frac{\nabla \phi}{|\nabla \phi|} = \frac{1}{\sqrt{93}}(8\mathbf{i} + 5\mathbf{j} - 2\mathbf{k})$$

Collecting our results so far, we have, for $\phi(x, y, z)$ a scalar function

(a) $\operatorname{grad} \phi = \nabla \phi = \dfrac{\partial \phi}{\partial x}\mathbf{i} + \dfrac{\partial \phi}{\partial y}\mathbf{j} + \dfrac{\partial \phi}{\partial z}\mathbf{k}$

(b) $d\phi = d\mathbf{r} \cdot \operatorname{grad} \phi$ where $d\phi = \dfrac{\partial \phi}{\partial x} dx + \dfrac{\partial \phi}{\partial y} dy + \dfrac{\partial \phi}{\partial z} dz$

(c) directional derivative $\dfrac{d\phi}{ds} = \hat{\mathbf{a}} \cdot \operatorname{grad} \phi$

(d) unit normal vector $\mathbf{N} = \dfrac{\nabla \phi}{|\nabla \phi|}$.

Copy out this brief summary for future reference. It will help

47 Grad of sums and products of scalars

(a) $\nabla(A + B) = \mathbf{i}\left\{\dfrac{\partial}{\partial x}(A + B)\right\} + \mathbf{j}\left\{\dfrac{\partial}{\partial y}(A + B)\right\} + \mathbf{k}\left\{\dfrac{\partial}{\partial z}(A + B)\right\}$

$= \left\{\dfrac{\partial A}{\partial x}\mathbf{i} + \dfrac{\partial A}{\partial y}\mathbf{j} + \dfrac{\partial A}{\partial z}\mathbf{k}\right\} + \left\{\dfrac{\partial B}{\partial x}\mathbf{i} + \dfrac{\partial B}{\partial y}\mathbf{j} + \dfrac{\partial B}{\partial z}\mathbf{k}\right\}$

$\therefore \nabla(A + B) = \nabla A + \nabla B$

(b) $\nabla(AB) = \mathbf{i}\left\{\dfrac{\partial}{\partial x}(AB)\right\} + \mathbf{j}\left\{\dfrac{\partial}{\partial y}(AB)\right\} + \mathbf{k}\left\{\dfrac{\partial}{\partial z}(AB)\right\}$

$= \mathbf{i}\left\{A\dfrac{\partial B}{\partial x} + B\dfrac{\partial A}{\partial x}\right\} + \mathbf{j}\left\{A\dfrac{\partial B}{\partial y} + B\dfrac{\partial A}{\partial y}\right\} + \mathbf{k}\left\{A\dfrac{\partial B}{\partial z} + B\dfrac{\partial A}{\partial z}\right\}$

$= \left\{A\dfrac{\partial B}{\partial x}\mathbf{i} + A\dfrac{\partial B}{\partial y}\mathbf{j} + A\dfrac{\partial B}{\partial z}\mathbf{k}\right\} + \left\{B\dfrac{\partial A}{\partial x}\mathbf{i} + B\dfrac{\partial A}{\partial y}\mathbf{j} + B\dfrac{\partial A}{\partial z}\mathbf{k}\right\}$

$= A\left\{\dfrac{\partial B}{\partial x}\mathbf{i} + \dfrac{\partial B}{\partial y}\mathbf{j} + \dfrac{\partial B}{\partial z}\mathbf{k}\right\} + B\left\{\dfrac{\partial A}{\partial x}\mathbf{i} + \dfrac{\partial A}{\partial y}\mathbf{j} + \dfrac{\partial A}{\partial z}\mathbf{k}\right\}$

$\therefore \nabla(AB) = A(\nabla B) + B(\nabla A)$

Remember that in these results A and B are scalars. The operator ∇ acting on a vector ....

48

has no meaning

Example

If $A = x^2 yz + xz^2$ and $B = xy^2 z - z^3$, evaluate $\nabla(AB)$ at the point $(2, 1, 3)$.

We know that $\nabla(AB) = A(\nabla B) + B(\nabla A)$

At $(2, 1, 3)$,

$\nabla B = \ldots\ldots\ldots\ldots$; $\nabla A = \ldots\ldots\ldots\ldots$

Vector differentiation

$$\boxed{\nabla B = 3\mathbf{i} + 12\mathbf{j} - 25\mathbf{k}; \quad \nabla A = 21\mathbf{i} + 12\mathbf{j} + 16\mathbf{k}}$$

$$\nabla B = \frac{\partial B}{\partial x}\mathbf{i} + \frac{\partial B}{\partial y}\mathbf{j} + \frac{\partial B}{\partial z}\mathbf{k} = y^2 z\mathbf{i} + 2xyz\mathbf{j} + (xy^2 - 3z^2)\mathbf{k}$$

$$= 3\mathbf{i} + 12\mathbf{j} - 25\mathbf{k} \quad \text{at } (2, 1, 3)$$

$$\nabla A = \frac{\partial A}{\partial x}\mathbf{i} + \frac{\partial A}{\partial y}\mathbf{j} + \frac{\partial A}{\partial z}\mathbf{k} = (2xyz + z^2)\mathbf{i} + x^2 z\mathbf{j} + (x^2 y + 2xz)\mathbf{k}$$

$$= 21\mathbf{i} + 12\mathbf{j} + 16\mathbf{k} \quad \text{at } (2, 1, 3)$$

Now $\nabla(AB) = A(\nabla B) + B(\nabla A) = \ldots\ldots\ldots\ldots$

Finish it

$$\boxed{\nabla(AB) = 3(-117\mathbf{i} + 36\mathbf{j} - 362\mathbf{k})}$$

Because

$\nabla(AB) = A(\nabla B) + B(\nabla A)$
$A = x^2 yz + xz^2 \quad \therefore \text{ at } (2, 1, 3), \quad A = 12 + 18 = 30$
$B = xy^2 z - z^3 \quad \therefore \text{ at } (2, 1, 3), \quad B = 6 - 27 = -21$
$\therefore \nabla(AB) = 30(3\mathbf{i} + 12\mathbf{j} - 25\mathbf{k}) - 21(21\mathbf{i} + 12\mathbf{j} + 16\mathbf{k})$

$$= -351\mathbf{i} + 108\mathbf{j} - 1086\mathbf{k}$$
$$= 3(-117\mathbf{i} + 36\mathbf{j} - 362\mathbf{k})$$

So add these to the list of results.

$\nabla(A + B) = \nabla A + \nabla B$
$\nabla(AB) = A(\nabla B) + B(\nabla A)$

where A and B are scalars.

Now on to the next page

Div (divergence of a vector function)

The operator $\nabla \cdot$ (notice the 'dot'; it makes all the difference) can be applied to a vector function $\mathbf{A}(x, y, z)$ to give the *divergence* of $\mathbf{A}$, written in short as *div* $\mathbf{A}$.

If $\mathbf{A} = a_x \mathbf{i} + a_y \mathbf{j} + a_z \mathbf{k}$

$$\text{div } \mathbf{A} = \nabla \cdot \mathbf{A} = \left(\mathbf{i}\frac{\partial}{\partial x} + \mathbf{j}\frac{\partial}{\partial y} + \mathbf{k}\frac{\partial}{\partial z} \right) \cdot \left(a_x \mathbf{i} + a_y \mathbf{j} + a_z \mathbf{k} \right)$$

$$\therefore \text{ div } \mathbf{A} = \nabla \cdot \mathbf{A} = \frac{\partial a_x}{\partial x} + \frac{\partial a_y}{\partial y} + \frac{\partial a_z}{\partial z}$$

Note that
(a) the grad operator ∇ acts on a scalar and gives a vector
(b) the div operation $\nabla \cdot$ acts on a vector and gives a scalar.

▶

Example 1

If $\mathbf{A} = x^2 y \mathbf{i} - xyz \mathbf{j} + yz^2 \mathbf{k}$ then
$$\text{div } \mathbf{A} = \nabla \cdot \mathbf{A} = \ldots\ldots\ldots$$

52

$$\boxed{\text{div } \mathbf{A} = \nabla \cdot \mathbf{A} = 2xy - xz + 2yz}$$

We simply take the appropriate partial derivatives of the coefficients of $\mathbf{i}$, $\mathbf{j}$ and $\mathbf{k}$. It could hardly be easier.

Example 2

If $\mathbf{A} = 2x^2 y \mathbf{i} - 2(xy^2 + y^3 z)\mathbf{j} + 3y^2 z^2 \mathbf{k}$, determine $\nabla \cdot \mathbf{A}$, i.e. div $\mathbf{A}$.

Complete it. $\nabla \cdot \mathbf{A} = \ldots\ldots\ldots$

53

$$\boxed{\nabla \cdot \mathbf{A} = 0}$$

Because
$$\mathbf{A} = 2x^2 y \mathbf{i} - 2(xy^2 + y^3 z)\mathbf{j} + 3y^2 z^2 \mathbf{k}$$
$$\nabla \cdot \mathbf{A} = \frac{\partial a_x}{\partial x} + \frac{\partial a_y}{\partial y} + \frac{\partial a_z}{\partial z}$$
$$= 4xy - 2(2xy + 3y^2 z) + 6y^2 z$$
$$= 4xy - 4xy - 6y^2 z + 6y^2 z = 0$$

Such a vector $\mathbf{A}$ for which $\nabla \cdot \mathbf{A} = 0$ at all points, i.e. for all values of x, y, z, is called a *solenoidal vector*. It is rather a special case.

Curl (curl of a vector function)

The *curl operator* denoted by $\nabla \times$, acts on a vector and gives another vector as a result.

If $\mathbf{A} = a_x \mathbf{i} + a_y \mathbf{j} + a_z \mathbf{k}$, then curl $\mathbf{A} = \nabla \times \mathbf{A}$.

i.e. curl $\mathbf{A} = \nabla \times \mathbf{A} = \left(\mathbf{i} \dfrac{\partial}{\partial x} + \mathbf{j} \dfrac{\partial}{\partial y} + \mathbf{k} \dfrac{\partial}{\partial z} \right) \times (a_x \mathbf{i} + a_y \mathbf{j} + a_z \mathbf{k})$

$$= \begin{vmatrix} \mathbf{i} & \mathbf{j} & \mathbf{k} \\ \dfrac{\partial}{\partial x} & \dfrac{\partial}{\partial y} & \dfrac{\partial}{\partial z} \\ a_x & a_y & a_z \end{vmatrix}$$

$\therefore \; \nabla \times \mathbf{A} = \mathbf{i} \left(\dfrac{\partial a_z}{\partial y} - \dfrac{\partial a_y}{\partial z} \right) + \mathbf{j} \left(\dfrac{\partial a_x}{\partial z} - \dfrac{\partial a_z}{\partial x} \right) + \mathbf{k} \left(\dfrac{\partial a_y}{\partial x} - \dfrac{\partial a_x}{\partial y} \right)$

Curl $\mathbf{A}$ is thus a vector function. *It is best remembered in its determinant form, so make a note of it.*

If $\nabla \times \mathbf{A} = \mathbf{0}$ then $\mathbf{A}$ is said to be *irrotational*.

Then on for an example

Vector differentiation

Example 1

54

If $\mathbf{A} = (y^4 - x^2 z^2)\mathbf{i} + (x^2 + y^2)\mathbf{j} - x^2 yz\mathbf{k}$, determine curl $\mathbf{A}$ at the point $(1, 3, -2)$.

$$\text{curl } \mathbf{A} = \nabla \times \mathbf{A} = \begin{vmatrix} \mathbf{i} & \mathbf{j} & \mathbf{k} \\ \dfrac{\partial}{\partial x} & \dfrac{\partial}{\partial y} & \dfrac{\partial}{\partial z} \\ y^4 - x^2 z^2 & x^2 + y^2 & -x^2 yz \end{vmatrix}$$

Now we expand the determinant

$$\nabla \times \mathbf{A} = \mathbf{i}\left\{\frac{\partial}{\partial y}(-x^2 yz) - \frac{\partial}{\partial z}(x^2 + y^2)\right\} - \mathbf{j}\left\{\frac{\partial}{\partial x}(-x^2 yz) - \frac{\partial}{\partial z}(y^4 - x^2 z^2)\right\}$$
$$+ \mathbf{k}\left\{\frac{\partial}{\partial x}(x^2 + y^2) - \frac{\partial}{\partial y}(y^4 - x^2 z^2)\right\}$$

All that now remains is to obtain the partial derivatives and substitute the values of x, y, z.

$$\therefore \nabla \times \mathbf{A} = \ldots\ldots\ldots\ldots$$

55

$$\boxed{2\mathbf{i} - 8\mathbf{j} - 106\mathbf{k}}$$

$\nabla \times \mathbf{A} = \mathbf{i}\{-x^2 z\} - \mathbf{j}\{-2xyz + 2x^2 z\} + \mathbf{k}\{2x - 4y^3\}$.

$\therefore$ At $(1, 3, -2)$, $\nabla \times \mathbf{A} = \mathbf{i}(2) - \mathbf{j}(12 - 4) + \mathbf{k}(2 - 108)$
$$= 2\mathbf{i} - 8\mathbf{j} - 106\mathbf{k}$$

Example 2

Determine curl $\mathbf{F}$ at the point $(2, 0, 3)$ given that

$$\mathbf{F} = ze^{2xy}\mathbf{i} + 2xz \cos y\mathbf{j} + (x + 2y)\mathbf{k}.$$

In determinant form, curl $\mathbf{F} = \nabla \times \mathbf{F} = \ldots\ldots\ldots\ldots$

56

$$\begin{vmatrix} \mathbf{i} & \mathbf{j} & \mathbf{k} \\ \dfrac{\partial}{\partial x} & \dfrac{\partial}{\partial y} & \dfrac{\partial}{\partial z} \\ ze^{2xy} & 2xz \cos y & x + 2y \end{vmatrix}$$

Now expand the determinant and substitute the values of x, y and z, finally obtaining curl $\mathbf{F} = \ldots\ldots\ldots\ldots$

57

$$\boxed{\text{curl } \mathbf{F} = \nabla \times \mathbf{F} = -2(\mathbf{i} + 3\mathbf{k})}$$

Because

$$\nabla \times \mathbf{F} = \mathbf{i}\{2 - 2x \cos y\} - \mathbf{j}\{1 - e^{2xy}\} + \mathbf{k}\{2z \cos y - 2xze^{2xy}\}$$

∴ At (2, 0, 3) $\nabla \times \mathbf{F} = \mathbf{i}(2 - 4) - \mathbf{j}(1 - 1) + \mathbf{k}(6 - 12)$

$$= -2\mathbf{i} - 6\mathbf{k} = -2(\mathbf{i} + 3\mathbf{k})$$

Every one is done in the same way.

Summary of grad, div and curl

(a) *Grad* operator ∇ acts on a *scalar* field to give a *vector* field.
(b) *Div* operator $\nabla \cdot$ acts on a *vector* field to give a *scalar* field.
(c) *Curl* operator $\nabla \times$ acts on a *vector* field to give a *vector* field.
(d) With a *scalar function* $\phi(x, y, z)$

$$\text{grad } \phi = \nabla \phi = \frac{\partial \phi}{\partial x}\mathbf{i} + \frac{\partial \phi}{\partial y}\mathbf{j} + \frac{\partial \phi}{\partial z}\mathbf{k}$$

(e) With a *vector function* $\mathbf{A} = a_x \mathbf{i} + a_y \mathbf{j} + a_z \mathbf{k}$

(1) $\text{div } \mathbf{A} = \nabla \cdot \mathbf{A} = \dfrac{\partial a_x}{\partial x} + \dfrac{\partial a_y}{\partial y} + \dfrac{\partial a_z}{\partial z}$

(2) $\text{curl } \mathbf{A} = \nabla \times \mathbf{A} = \begin{vmatrix} \mathbf{i} & \mathbf{j} & \mathbf{k} \\ \dfrac{\partial}{\partial x} & \dfrac{\partial}{\partial y} & \dfrac{\partial}{\partial z} \\ a_x & a_y & a_z \end{vmatrix}$

Check through that list, just to make sure. We shall need them all

58

By way of revision, here is one further example.

Example 3

If $\phi = x^2 y^2 + x^3 yz - yz^2$
and $\mathbf{F} = xy^2 \mathbf{i} - 2yz \mathbf{j} + xyz \mathbf{k}$
determine for the point P (1, −1, 2),

(a) $\nabla \phi$,   (b) unit normal,   (c) $\nabla \cdot \mathbf{F}$,   (d) $\nabla \times \mathbf{F}$.

Complete all four parts and then check the results with the next frame

Vector differentiation

Here is the working in full. $\phi = x^2y^2 + x^3yz - yz^2$

(a) $\nabla\phi = \dfrac{\partial\phi}{\partial x}\mathbf{i} + \dfrac{\partial\phi}{\partial y}\mathbf{j} + \dfrac{\partial\phi}{\partial z}\mathbf{k}$

$= (2xy^2 + 3x^2yz)\mathbf{i} + (2x^2y + x^3z - z^2)\mathbf{j} + (x^3y - 2yz)\mathbf{k}$

$\therefore$ At $(1, -1, 2)$ $\quad \nabla\phi = -4\mathbf{i} - 4\mathbf{j} + 3\mathbf{k}$

(b) $\mathbf{N} = \dfrac{\nabla\phi}{|\nabla\phi|}$ $\quad |\nabla\phi| = \sqrt{16 + 16 + 9} = \sqrt{41}$

$\therefore \mathbf{N} = \dfrac{-1}{\sqrt{41}}(4\mathbf{i} + 4\mathbf{j} - 3\mathbf{k})$

(c) $\mathbf{F} = xy^2\mathbf{i} - 2yz\mathbf{j} + xyz\mathbf{k} \quad \nabla\cdot\mathbf{F} = \dfrac{\partial a_x}{\partial x} + \dfrac{\partial a_y}{\partial y} + \dfrac{\partial a_z}{\partial z}$

$\therefore \nabla\cdot\mathbf{F} = y^2 - 2z + xy$

$\therefore$ At $(1, -1, 2)$ $\quad \nabla\cdot\mathbf{F} = 1 - 4 - 1 = -4 \quad \therefore \nabla\cdot\mathbf{F} = -4$

(d) $\nabla\times\mathbf{F} = \begin{vmatrix} \mathbf{i} & \mathbf{j} & \mathbf{k} \\ \dfrac{\partial}{\partial x} & \dfrac{\partial}{\partial y} & \dfrac{\partial}{\partial z} \\ xy^2 & -2yz & xyz \end{vmatrix}$

$\therefore \nabla\times\mathbf{F} = \mathbf{i}(xz + 2y) - \mathbf{j}(yz - 0) + \mathbf{k}(0 - 2xy)$

$= (xz + 2y)\mathbf{i} - yz\mathbf{j} - 2xy\mathbf{k}$

$\therefore$ At $(1, -1, 2)$ $\quad \nabla\times\mathbf{F} = 2\mathbf{j} + 2\mathbf{k} \quad \therefore \nabla\times\mathbf{F} = 2(\mathbf{j} + \mathbf{k})$

Now let us combine some of these operations.

Multiple operations

We can combine the operators grad, div and curl in multiple operations, as in the examples that follow.

Example 1

If $\mathbf{A} = x^2y\mathbf{i} + yz^3\mathbf{j} - zx^3\mathbf{k}$

then $\quad$ div $\mathbf{A} = \nabla\cdot\mathbf{A} = \left(\dfrac{\partial}{\partial x}\mathbf{i} + \dfrac{\partial}{\partial y}\mathbf{j} + \dfrac{\partial}{\partial z}\mathbf{k}\right)\cdot(x^2y\mathbf{i} + yz^3\mathbf{j} - zx^3\mathbf{k})$

$= 2xy + z^3 + x^3 = \phi \quad$ say

Then $\quad$ grad (div $\mathbf{A}$) $= \nabla(\nabla\cdot\mathbf{A}) = \dfrac{\partial\phi}{\partial x}\mathbf{i} + \dfrac{\partial\phi}{\partial y}\mathbf{j} + \dfrac{\partial\phi}{\partial z}\mathbf{k}$

$= (2y + 3x^2)\mathbf{i} + (2x)\mathbf{j} + (3z^2)\mathbf{k}$

i.e. $\quad$ grad div $\mathbf{A} = \nabla(\nabla\cdot\mathbf{A}) = (2y + 3x^2)\mathbf{i} + 2x\mathbf{j} + 3z^2\mathbf{k}$

Move on for the next example

256 Vector Analysis

61

Example 2

If $\phi = xyz - 2y^2z + x^2z^2$, determine div grad ϕ at the point (2, 4, 1).

First find grad ϕ and then the div of the result.

At (2, 4, 1), div grad $\phi = \nabla \cdot (\nabla \phi) = \ldots\ldots\ldots\ldots$

62

$$\boxed{\text{div grad } \phi = 6}$$

Because we have $\phi = xyz - 2y^2z + x^2z^2$

$$\text{grad } \phi = \nabla \phi = \frac{\partial \phi}{\partial x}\mathbf{i} + \frac{\partial \phi}{\partial y}\mathbf{j} + \frac{\partial \phi}{\partial z}\mathbf{k}$$

$$= (yz + 2xz^2)\mathbf{i} + (xz - 4yz)\mathbf{j} + (xy - 2y^2 + 2x^2z)\mathbf{k}$$

$\therefore$ div grad $\phi = \nabla \cdot (\nabla \phi) = 2z^2 - 4z + 2x^2$

$\therefore$ At (2, 4, 1), div grad $\phi = \nabla \cdot (\nabla \phi) = 2 - 4 + 8 = 6$

Example 3

If $\mathbf{F} = x^2yz\mathbf{i} + xyz^2\mathbf{j} + y^2z\mathbf{k}$ determine curl curl $\mathbf{F}$ at the point (2, 1, 1).

Determine an expression for curl $\mathbf{F}$ in the usual way, which will be a vector, and then the curl of the result. Finally substitute values.

$$\text{curl curl } \mathbf{F} = \ldots\ldots\ldots\ldots$$

63

$$\boxed{\text{curl curl } \mathbf{F} = \nabla \times (\nabla \times \mathbf{F}) = \mathbf{i} + 2\mathbf{j} + 6\mathbf{k}}$$

Because

$$\text{curl } \mathbf{F} = \begin{vmatrix} \mathbf{i} & \mathbf{j} & \mathbf{k} \\ \dfrac{\partial}{\partial x} & \dfrac{\partial}{\partial y} & \dfrac{\partial}{\partial z} \\ x^2yz & xyz^2 & y^2z \end{vmatrix}$$

$$= (2yz - 2xyz)\mathbf{i} + x^2y\mathbf{j} + (yz^2 - x^2z)\mathbf{k}$$

Then $\text{curl curl } \mathbf{F} = \begin{vmatrix} \mathbf{i} & \mathbf{j} & \mathbf{k} \\ \dfrac{\partial}{\partial x} & \dfrac{\partial}{\partial y} & \dfrac{\partial}{\partial z} \\ 2yz - 2xyz & x^2y & yz^2 - x^2z \end{vmatrix}$

$$= z^2\mathbf{i} - (-2xz - 2y + 2xy)\mathbf{j} + (2xy - 2z + 2xz)\mathbf{k}$$

$\therefore$ At (2, 1, 1), curl curl $\mathbf{F} = \nabla \times (\nabla \times \mathbf{F}) = \mathbf{i} + 2\mathbf{j} + 6\mathbf{k}$

Vector differentiation

Remember that grad, div and curl are operators and that they must act on a scalar or vector as appropriate. They cannot exist alone and must be followed by a function.

One or two interesting general results appear.

(a) *Curl grad ϕ* where ϕ is a scalar

$$\operatorname{grad} \phi = \frac{\partial \phi}{\partial x}\mathbf{i} + \frac{\partial \phi}{\partial y}\mathbf{j} + \frac{\partial \phi}{\partial z}\mathbf{k}$$

$$\therefore \operatorname{curl\ grad} \phi = \begin{vmatrix} \mathbf{i} & \mathbf{j} & \mathbf{k} \\ \dfrac{\partial}{\partial x} & \dfrac{\partial}{\partial y} & \dfrac{\partial}{\partial z} \\ \dfrac{\partial \phi}{\partial x} & \dfrac{\partial \phi}{\partial y} & \dfrac{\partial \phi}{\partial z} \end{vmatrix}$$

$$= \mathbf{i}\left\{\frac{\partial^2 \phi}{\partial y \partial z} - \frac{\partial^2 \phi}{\partial z \partial y}\right\} - \mathbf{j}\left\{\frac{\partial^2 \phi}{\partial z \partial x} - \frac{\partial^2 \phi}{\partial x \partial z}\right\} + \mathbf{k}\left\{\frac{\partial^2 \phi}{\partial x \partial y} - \frac{\partial^2 \phi}{\partial y \partial x}\right\}$$

$$= 0$$

$$\therefore \operatorname{curl\ grad} \phi = \nabla \times (\nabla \phi) = 0$$

(b) *Div curl $\mathbf{A}$* where $\mathbf{A}$ is a vector. $\mathbf{A} = a_x \mathbf{i} + a_y \mathbf{j} + a_z \mathbf{k}$

$$\operatorname{curl} \mathbf{A} = \nabla \times \mathbf{A} = \begin{vmatrix} \mathbf{i} & \mathbf{j} & \mathbf{k} \\ \dfrac{\partial}{\partial x} & \dfrac{\partial}{\partial y} & \dfrac{\partial}{\partial z} \\ a_x & a_y & a_z \end{vmatrix}$$

$$= \mathbf{i}\left(\frac{\partial a_z}{\partial y} - \frac{\partial a_y}{\partial z}\right) - \mathbf{j}\left(\frac{\partial a_z}{\partial x} - \frac{\partial a_x}{\partial z}\right) + \mathbf{k}\left(\frac{\partial a_y}{\partial x} - \frac{\partial a_x}{\partial y}\right)$$

Then $\operatorname{div\ curl} \mathbf{A} = \nabla \cdot (\nabla \times \mathbf{A}) = \left(\mathbf{i}\dfrac{\partial}{\partial x} + \mathbf{j}\dfrac{\partial}{\partial y} + \mathbf{k}\dfrac{\partial}{\partial z}\right) \cdot (\nabla \times \mathbf{A})$

$$= \frac{\partial^2 a_z}{\partial x \partial y} - \frac{\partial^2 a_y}{\partial z \partial x} - \frac{\partial^2 a_z}{\partial x \partial y} + \frac{\partial^2 a_x}{\partial y \partial z} + \frac{\partial^2 a_y}{\partial z \partial x} - \frac{\partial^2 a_x}{\partial y \partial z}$$

$$= 0$$

$$\therefore \operatorname{div\ curl} \mathbf{A} = \nabla \cdot (\nabla \times \mathbf{A}) = \mathbf{0}$$

(c) *Div grad ϕ* where ϕ is a scalar

$$\operatorname{grad} \phi = \frac{\partial \phi}{\partial x}\mathbf{i} + \frac{\partial \phi}{\partial y}\mathbf{j} + \frac{\partial \phi}{\partial z}\mathbf{k}$$

Then $\operatorname{div\ grad} \phi = \nabla \cdot (\nabla \phi)$

$$= \left(\mathbf{i}\frac{\partial}{\partial x} + \mathbf{j}\frac{\partial}{\partial y} + \mathbf{k}\frac{\partial}{\partial z}\right) \cdot \left(\frac{\partial \phi}{\partial x}\mathbf{i} + \frac{\partial \phi}{\partial y}\mathbf{j} + \frac{\partial \phi}{\partial z}\mathbf{k}\right)$$

$$= \frac{\partial^2 \phi}{\partial x^2} + \frac{\partial^2 \phi}{\partial y^2} + \frac{\partial^2 \phi}{\partial z^2}$$

$$\therefore \operatorname{div\ grad} \phi = \nabla \cdot (\nabla \phi) = \frac{\partial^2 \phi}{\partial x^2} + \frac{\partial^2 \phi}{\partial y^2} + \frac{\partial^2 \phi}{\partial z^2}$$

$$= \nabla^2 \phi, \text{ the Laplacian of } \phi$$

The operator ∇^2 is called the Laplacian.

So these general results are
(a) curl grad $\phi = \nabla \times (\nabla \phi) = 0$
(b) div curl $\mathbf{A} = \nabla \cdot (\nabla \times \mathbf{A}) = 0$
(c) div grad $\phi = \nabla \cdot (\nabla \phi) = \dfrac{\partial^2 \phi}{\partial x^2} + \dfrac{\partial^2 \phi}{\partial y^2} + \dfrac{\partial^2 \phi}{\partial z^2}$.

That brings us to the end of this particular Program. We have covered quite a lot of new material, so check carefully through the **Review summary** and **Can You?** checklist that follow: then you can deal with the **Test exercise**. The **Further problems** provide an opportunity for additional practice.

65 Review summary

1 Differentiation of vectors

If $\mathbf{A}$, a_x, a_y, a_z are functions of u

$$\dfrac{d\mathbf{A}}{du} = \dfrac{da_x}{du}\mathbf{i} + \dfrac{da_y}{du}\mathbf{j} + \dfrac{da_z}{du}\mathbf{k}$$

2 Unit tangent vector T

$$\mathbf{T} = \dfrac{\dfrac{d\mathbf{A}}{du}}{\left|\dfrac{d\mathbf{A}}{du}\right|}$$

3 Integration of vectors

$$\int_a^b \mathbf{A}\, du = \mathbf{i} \int_a^b a_x\, du + \mathbf{j} \int_a^b a_y\, du + \mathbf{k} \int_a^b a_z\, du$$

4 Grad (gradient of a scalar function ϕ)

$$\operatorname{grad} \phi = \nabla \phi = \dfrac{\partial \phi}{\partial x}\mathbf{i} + \dfrac{\partial \phi}{\partial y}\mathbf{j} + \dfrac{\partial \phi}{\partial z}\mathbf{k}$$

$$\text{'del'} = \text{operator } \nabla = \left(\mathbf{i}\dfrac{\partial}{\partial x} + \mathbf{j}\dfrac{\partial}{\partial y} + \mathbf{k}\dfrac{\partial}{\partial z}\right)$$

(a) *Directional derivative* $\dfrac{d\phi}{ds} = \hat{\mathbf{a}} \cdot \operatorname{grad} \phi = \hat{\mathbf{a}} \cdot \nabla \phi$ where $\hat{\mathbf{a}}$ is a unit vector in a stated direction. Grad ϕ gives the direction for maximum rate of change of ϕ.

(b) *Unit normal vector* $\mathbf{N}$ to surface $\phi(x, y, z) = $ constant.

$$\mathbf{N} = \dfrac{\nabla \phi}{|\nabla \phi|}$$

Vector differentiation

5 *Div* (divergence of a vector function **A**)

$$\text{div } \mathbf{A} = \nabla \cdot \mathbf{A} = \frac{\partial a_x}{\partial x} + \frac{\partial a_y}{\partial y} + \frac{\partial a_z}{\partial z}$$

If $\nabla \cdot \mathbf{A} = 0$ for all points, **A** is a solenoidal vector.

6 *Curl* (curl of a vector function **A**)

$$\text{curl } \mathbf{A} = \nabla \times \mathbf{A} = \begin{vmatrix} \mathbf{i} & \mathbf{j} & \mathbf{k} \\ \dfrac{\partial}{\partial x} & \dfrac{\partial}{\partial y} & \dfrac{\partial}{\partial z} \\ a_x & a_y & a_z \end{vmatrix}$$

If $\nabla \times \mathbf{A} = 0$ then **A** is an irrotational vector.

7 *Operators*

grad (∇) acts on a *scalar* and gives a *vector*
div ($\nabla \cdot$) acts on a *vector* and gives a *scalar*
curl ($\nabla \times$) acts on a *vector* and gives a *vector*

8 *Multiple operations*

(a) curl grad $\phi = \nabla \times (\nabla \phi) = 0$
(b) div curl $\mathbf{A} = \nabla \cdot (\nabla \times \mathbf{A}) = 0$
(c) div grad $\phi = \nabla \cdot (\nabla \phi) = \dfrac{\partial^2 \phi}{\partial x^2} + \dfrac{\partial^2 \phi}{\partial y^2} + \dfrac{\partial^2 \phi}{\partial z^2}$

$= \nabla^2 \phi$, the Laplacian of ϕ

✅ Can You?

Checklist 8

66

Check this list before and after you try the end of Program test.

On a scale of 1 to 5 how confident are you that you can: Frames

- Differentiate a vector field and derive a unit vector tangential to the vector field at a point?   1 to 21
  Yes ☐ ☐ ☐ ☐ ☐ No

- Integrate a vector field?   22 to 28
  Yes ☐ ☐ ☐ ☐ ☐ No

- Obtain the gradient of a scalar field, the directional derivative and a unit normal to a surface?   29 to 50
  Yes ☐ ☐ ☐ ☐ ☐ No

▶

- Obtain the divergence of a vector field and recognise a solenoidal vector field? [51 to 53]
  Yes ☐ ☐ ☐ ☐ ☐ No

- Obtain the curl of a vector field? [53 to 59]
  Yes ☐ ☐ ☐ ☐ ☐ No

- Obtain combinations of div, grad and curl acting on scalar and vector fields as appropriate? [60 to 64]
  Yes ☐ ☐ ☐ ☐ ☐ No

 Test exercise 8

67

1. If $\mathbf{A} = (u^2 + 5)\mathbf{i} - (u^2 + 3)\mathbf{j} + 2u^3\mathbf{k}$, determine
   (a) $\dfrac{d\mathbf{A}}{du}$; (b) $\dfrac{d^2\mathbf{A}}{du^2}$; (c) $\left|\dfrac{d\mathbf{A}}{du}\right|$; all at $u = 2$.

2. Determine the unit tangent vector at the point $(2, 4, 3)$ for the curve with parametric equations
   $x = 2u^2$; $y = u + 3$; $z = 4u^2 - u$.

3. If $\mathbf{F} = 2\mathbf{i} + 4u\mathbf{j} + u^2\mathbf{k}$ and $\mathbf{G} = u^2\mathbf{i} - 2u\mathbf{j} + 4\mathbf{k}$, determine
   $\displaystyle\int_0^2 (\mathbf{F} \times \mathbf{G})\,du$.

4. Find the directional derivative of the function $\phi = x^2y - 2xz^2 + y^2z$ at the point $(1, 3, 2)$ in the direction of the vector $\mathbf{A} = 3\mathbf{i} + 2\mathbf{j} - \mathbf{k}$.

5. Find the unit normal to the surface $\phi = 2x^3z + x^2y^2 + xyz - 4 = 0$ at the point $(2, 1, 0)$.

6. If $\mathbf{A} = x^2y\mathbf{i} + (xy + yz)\mathbf{j} + xz^2\mathbf{k}$; $\mathbf{B} = yz\mathbf{i} - 3xz\mathbf{j} + 2xy\mathbf{k}$; and $\phi = 3x^2y + xyz - 4y^2z^2 - 3$; determine, at the point $(1, 2, 1)$
   (a) $\nabla\phi$; (b) $\nabla \cdot \mathbf{A}$; (c) $\nabla \times \mathbf{B}$; (d) grad div $\mathbf{A}$; (e) curl curl $\mathbf{A}$.

Further problems 8

1. If $\mathbf{F} = x^2\mathbf{i} + (3x+2)\mathbf{j} + \sin x\mathbf{k}$, find
   (a) $\dfrac{d\mathbf{F}}{dx}$; (b) $\dfrac{d^2\mathbf{F}}{dx^2}$; (c) $\left|\dfrac{d\mathbf{F}}{dx}\right|$; (d) $\dfrac{d}{dx}(\mathbf{F}\cdot\mathbf{F})$ at $x=1$.

2. If $\mathbf{F} = u\mathbf{i} + (1-u)\mathbf{j} + 3u\mathbf{k}$ and $\mathbf{G} = 2\mathbf{i} - (1+u)\mathbf{j} - u^2\mathbf{k}$, determine
   (a) $\dfrac{d}{du}(\mathbf{F}\cdot\mathbf{G})$; (b) $\dfrac{d}{du}(\mathbf{F}\times\mathbf{G})$; (c) $\dfrac{d}{du}(\mathbf{F}+\mathbf{G})$.

3. Find the unit normal to the surface $4x^2y^2 - 3xz^2 - 2y^2z + 4 = 0$ at the point $(2, -1, -2)$.

4. Find the unit normal to the surface $2xy^2 + y^2z + x^2z - 11 = 0$ at the point $(-2, 1, 3)$.

5. Determine the unit vector normal to the surface
   $xz^2 + 3xy - 2yz^2 + 1 = 0$ at the point $(1, -2, -1)$.

6. Find the unit normal to the surface $x^2y - 2yz^2 + y^2z = 3$ at the point $(2, -3, 1)$.

7. Determine the directional derivative of $\phi = xe^y + yz^2 + xyz$ at the point $(2, 0, 3)$ in the direction of $\mathbf{A} = 3\mathbf{i} - 2\mathbf{j} + \mathbf{k}$.

8. Find the directional derivative of $\phi = (x+2y+z)^2 - (x-y-z)^2$ at the point $(2, 1, -1)$ in the direction of $\mathbf{A} = \mathbf{i} - 4\mathbf{j} + 2\mathbf{k}$.

9. If $\mathbf{F} = 4t^3\mathbf{i} - 2t^2\mathbf{j} + 4t\mathbf{k}$, determine when $t=1$
   (a) $\dfrac{d\mathbf{F}}{dt}$; (b) $\dfrac{d^2\mathbf{F}}{dt^2}$; (c) $\dfrac{d}{dt}(\mathbf{F}\cdot\mathbf{F})$.

10. If $\phi = x^2\sin z + ze^y$ find, at the point $(1, 3, 2)$, the values of
    (a) grad ϕ and (b) $|\text{grad }\phi|$.

11. Given that $\phi = xy^2 + yz^2 - x^2$, find the derivative of ϕ with respect to distance at the point $(1, 2, -1)$, measured parallel to the vector $2\mathbf{i} - 3\mathbf{j} + 4\mathbf{k}$.

12. Find unit vectors normal to the surfaces $x^2 + y^2 - z^2 + 3 = 0$ and $xy - yz + zx - 10 = 0$ at the point $(3, 2, 4)$ and hence find the angle between the two surfaces at that point.

13. If $\mathbf{r} = (t^2 + 3t)\mathbf{i} - 2\sin 3t\mathbf{j} + 3e^{2t}\mathbf{k}$, determine
    (a) $\dfrac{d\mathbf{r}}{dt}$; (b) $\dfrac{d^2\mathbf{r}}{dt^2}$; (c) the value of $\left|\dfrac{d^2\mathbf{r}}{dt^2}\right|$ at $t=0$.

14. (a) Show that curl $(-y\mathbf{i} + x\mathbf{j})$ is a constant vector.
    (b) Show that the vector field $(yz\mathbf{i} + zx\mathbf{j} + xy\mathbf{k})$ has zero divergence and zero curl.

15. If $\mathbf{A} = 2xz^2\mathbf{i} - xz\mathbf{j} + (y+z)\mathbf{k}$, find curl curl $\mathbf{A}$.

16. Determine grad ϕ where $\phi = x^2\cos(2yz - 0.5)$ and obtain its value at the point $(1, 3, 1)$.

Program 9

Vector integration

Frames

Learning outcomes

When you have completed this Program you will be able to:
- Evaluate the line integral of a scalar and a vector field in Cartesian coordinates
- Evaluate the volume integral of a vector field
- Evaluate the surface integral of a scalar and a vector field
- Determine whether or not a vector field is a conservative vector field
- Apply Gauss' divergence theorem
- Apply Stokes' theorem
- Determine the direction of unit normal vectors to a surface
- Apply Green's theorem in the plane

264 Vector Analysis

1

We dealt in some detail with line, surface and volume integrals in an earlier Program, when we approached the subject analytically. In many practical problems, it is more convenient to express these integrals in vector form and the methods often lead to more concise working.

Line integrals

Let a point P on the curve c joining A and B be denoted by the position vector **r** with respect to a fixed origin O.

If Q is a neighboring point on the curve with position vector $\mathbf{r} + d\mathbf{r}$, then $\overline{PQ} = d\mathbf{r}$.

The curve c can be divided up into many (n) such small arcs, approximating to $d\mathbf{r}_1, d\mathbf{r}_2, d\mathbf{r}_3 \ldots d\mathbf{r}_p \ldots$ so that

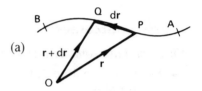

$$\overline{AB} = \sum_{p=1}^{n} d\mathbf{r}_p$$

where $d\mathbf{r}_p$ is a vector representing the element of arc in both magnitude and direction.

Scalar field

If a scalar field V exists for all points on the curve, then $\sum_{p=1}^{n} V \, d\mathbf{r}_p$ with $d\mathbf{r} \to 0$, defines the *line integral* of V along the curve c from A to B,

i.e. line integral $= \int_c V \, d\mathbf{r}$

We can illustrate this integral by erecting a continuous ordinate proportional to V at each point of the curve. $\int_c V \, d\mathbf{r}$ is then represented by the area of the curved surface between the ends A and B of the curve c.

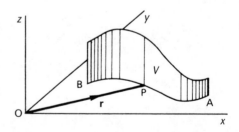

To evaluate a line integral, the integrand is expressed in terms of x, y, z, with $d\mathbf{r} = \ldots\ldots\ldots\ldots$

Vector integration

$$\mathbf{dr} = \mathbf{i}\,dx + \mathbf{j}\,dy + \mathbf{k}\,dz$$

2

In practice, x, y and z are often expressed in terms of parametric equations of a fourth variable (say u), i.e. $x = x(u)$; $y = y(u)$; $z = z(u)$. From these, dx, dy and dz can be written in terms of u and the integral evaluated in terms of this parameter u.

The following examples will show the method.

Example 1

If $V = xy^2z$, evaluate $\int_c V\,\mathbf{dr}$ along the curve c having parametric equations $x = 3u$; $y = 2u^2$; $z = u^3$ between A (0, 0, 0) and B (3, 2, 1).

$V = xy^2z = (3u)(4u^4)(u^3) = 12u^8$

$\mathbf{dr} = \mathbf{i}\,dx + \mathbf{j}\,dy + \mathbf{k}\,dz = \ldots\ldots\ldots$

$$\mathbf{dr} = \mathbf{i}\,3\,du + \mathbf{j}\,4u\,du + \mathbf{k}\,3u^2\,du$$

3

Because

$x = 3u, \quad \therefore\ dx = 3\,du$
$y = 2u^2, \quad \therefore\ dy = 4u\,du$
$z = u^3, \quad \therefore\ dz = 3u^2\,du$

Limits: A (0, 0, 0) corresponds to $u = \ldots\ldots\ldots$
B (3, 2, 1) corresponds to $u = \ldots\ldots\ldots$

$$A\,(0,0,0) \equiv u = 0 \quad B\,(3,2,1) \equiv u = 1$$

4

$$\therefore \int_c V\,\mathbf{dr} = \int_0^1 12u^8\,(\mathbf{i}\,3\,du + \mathbf{j}\,4u\,du + \mathbf{k}\,3u^2\,du)$$

$= \ldots\ldots\ldots$

Finish it off

$$4\mathbf{i} + \frac{24}{5}\mathbf{j} + \frac{36}{11}\mathbf{k}$$

5

Because

$$\int_c V\,\mathbf{dr} = 12\int_0^1 (\mathbf{i}\,3u^8\,du + \mathbf{j}\,4u^9\,du + \mathbf{k}\,3u^{10}\,du)$$

which integrates directly to give the result quoted above.

Now for another example.

6

Example 2

If $V = xy + y^2z$, evaluate $\int_c V \, d\mathbf{r}$ along the curve c defined by
$x = t^2$; $y = 2t$; $z = t + 5$ between A (0, 0, 5) and B (4, 4, 7).
As before, expressing V and $d\mathbf{r}$ in terms of the parameter t we have

$$V = \ldots\ldots\ldots\ldots \qquad d\mathbf{r} = \ldots\ldots\ldots\ldots$$

7

$$\boxed{V = 6t^3 + 20t^2; \quad d\mathbf{r} = \mathbf{i}\,2t\,dt + \mathbf{j}\,2\,dt + \mathbf{k}\,dt}$$

Because

$$V = xy + y^2z = (t^2)(2t) + (4t^2)(t+5) = 6t^3 + 20t^2.$$

Also $x = t^2 \quad dx = 2t\,dt$
$y = 2t \quad dy = 2\,dt \quad \therefore d\mathbf{r} = \mathbf{i}\,dx + \mathbf{j}\,dy + \mathbf{k}\,dz$
$z = t+5 \quad dz = dt \qquad\qquad\quad = \mathbf{i}\,2t\,dt + \mathbf{j}\,2\,dt + \mathbf{k}\,dt$

$$\therefore \int_c V\,d\mathbf{r} = \int_c (6t^3 + 20t^2)(\mathbf{i}\,2t + \mathbf{j}\,2 + \mathbf{k})\,dt$$

Limits: A (0, 0, 5) $\equiv t = \ldots\ldots\ldots\ldots$
B (4, 4, 7) $\equiv t = \ldots\ldots\ldots\ldots$

8

$$\boxed{A\,(0,0,5) \equiv t = 0; \quad B\,(4,4,7) \equiv t = 2}$$

$$\therefore \int_c V\,d\mathbf{r} = \int_0^2 (6t^3 + 20t^2)(\mathbf{i}\,2t + \mathbf{j}\,2 + \mathbf{k})\,dt$$
$$= \ldots\ldots\ldots\ldots \text{ Complete the integration.}$$

9

$$\boxed{\frac{8}{15}(444\,\mathbf{i} + 290\,\mathbf{j} + 145\,\mathbf{k})}$$

$$\int_c V\,d\mathbf{r} = 2\int_0^2 \{(6t^4 + 20t^3)\mathbf{i} + (6t^3 + 20t^2)\mathbf{j} + (3t^3 + 10t^2)\mathbf{k}\}\,dt$$

The actual integration is simple enough and gives the result shown. All line integrals in scalar fields are done in the same way.

Vector field

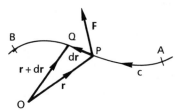

If a vector field **F** exists for all points of the curve c, then for each element of arc we can form the scalar product **F** · **dr**. Summing these products for all elements of arc, we have $\sum_{p=1}^{n} \mathbf{F} \cdot \mathbf{dr}_p$

Then, if $\mathbf{dr}_p \to 0$, the sum becomes the integral $\int_c \mathbf{F} \cdot \mathbf{dr}$,

i.e. the line integral of **F** from A to B along the stated curve

$$= \int_c \mathbf{F} \cdot \mathbf{dr}$$

In this case, since **F** · **dr** is a scalar product, then the line integral is a scalar.

To evaluate the line integral, **F** and **dr** are expressed in terms of x, y, z and the curve in parametric form. We have

$$\mathbf{F} = F_x \mathbf{i} + F_y \mathbf{j} + F_z \mathbf{k}$$

and $\quad \mathbf{dr} = \mathbf{i}\, dx + \mathbf{j}\, dy + \mathbf{k}\, dz$

Then $\mathbf{F} \cdot \mathbf{dr} = (F_x \mathbf{i} + F_y \mathbf{j} + F_z \mathbf{k}) \cdot (\mathbf{i}\, dx + \mathbf{j}\, dy + \mathbf{k}\, dz)$
$\qquad = F_x\, dx + F_y\, dy + F_z\, dz$

$$\therefore \int_c \mathbf{F} \cdot \mathbf{dr} = \int_c F_x\, dx + \int_c F_y\, dy + \int_c F_z\, dz$$

Now for an example to show it in operation.

Example 1

If $\mathbf{F} = x^2 y \mathbf{i} + xz \mathbf{j} - 2yz \mathbf{k}$, evaluate $\int_c \mathbf{F} \cdot \mathbf{dr}$ between A (0, 0, 0) and B (4, 2, 1) along the curve having parametric equations $x = 4t$; $y = 2t^2$; $z = t^3$.

Expressing everything in terms of the parameter t, we have

$$\mathbf{F} = \ldots\ldots\ldots\ldots$$

$dx = \ldots\ldots\ldots\ldots$; $\quad dy = \ldots\ldots\ldots\ldots$; $\quad dz = \ldots\ldots\ldots\ldots$

11

$$\mathbf{F} = 32t^4\,\mathbf{i} + 4t^4\,\mathbf{j} - 4t^5\,\mathbf{k}$$
$$dx = 4\,dt; \quad dy = 4t\,dt; \quad dz = 3t^2\,dt$$

Because

$$x^2y = (16t^2)(2t^2) = 32t^4 \qquad x = 4t \qquad \therefore\ dx = 4\,dt$$
$$xz = (4t)(t^3) = 4t^4 \qquad y = 2t^2 \qquad \therefore\ dy = 4t\,dt$$
$$2yz = (4t^2)(t^3) = 4t^5 \qquad z = t^3 \qquad \therefore\ dz = 3t^2\,dt$$

Then $\displaystyle \int \mathbf{F}\cdot d\mathbf{r} = \int (32t^4\mathbf{i} + 4t^4\mathbf{j} - 4t^5\mathbf{k})\cdot(\mathbf{i}\,4\,dt + \mathbf{j}\,4t\,dt + \mathbf{k}\,3t^2\,dt)$

$$= \int (128t^4 + 16t^5 - 12t^7)\,dt$$

Limits: $A\,(0, 0, 0) \equiv t = \ldots\ldots\ldots\ldots;\quad B\,(4, 2, 1) \equiv t = \ldots\ldots\ldots\ldots$

12

$$A \equiv t = 0; \quad B \equiv t = 1$$

$$\therefore \int_c \mathbf{F}\cdot d\mathbf{r} = \int_0^1 (128t^4 + 16t^5 - 12t^7)\,dt = \ldots\ldots\ldots\ldots$$

13

$$\frac{128}{5} + \frac{8}{3} - \frac{3}{2} = \frac{803}{30} = 26.77$$

If the vector field $\mathbf{F}$ is a *force field*, then the line integral $\displaystyle \int_c \mathbf{F}\cdot d\mathbf{r}$ represents the work done in moving a unit particle along the prescribed curve c from A to B.

Now for another example.

Example 2

If $\mathbf{F} = x^2y\mathbf{i} + 2yz\mathbf{j} + 3z^2x\mathbf{k}$, evaluate $\displaystyle \int_c \mathbf{F}\cdot d\mathbf{r}$ between $A\,(0, 0, 0)$ and $B\,(1, 2, 3)$
(a) along the straight lines c_1 from $(0, 0, 0)$ to $(1, 0, 0)$
then c_2 from $(1, 0, 0)$ to $(1, 2, 0)$
and c_3 from $(1, 2, 0)$ to $(1, 2, 3)$
(b) along the straight line c_4 joining $(0, 0, 0)$ to $(1, 2, 3)$.

As before, we first obtain an expression for $\mathbf{F}\cdot d\mathbf{r}$ which is

$\ldots\ldots\ldots\ldots$

Vector integration

14

$$\mathbf{F} \cdot \mathbf{dr} = x^2 y\, dx + 2yz\, dy + 3z^2 x\, dz$$

Because

$$\mathbf{F} \cdot \mathbf{dr} = (x^2 y\, \mathbf{i} + 2yz\, \mathbf{j} + 3z^2 x\, \mathbf{k}) \cdot (\mathbf{i}\, dx + \mathbf{j}\, dy + \mathbf{k}\, dz)$$

$$\therefore \int \mathbf{F} \cdot \mathbf{dr} = \int x^2 y\, dx + \int 2yz\, dy + \int 3z^2 x\, dz$$

(a) Here the integration is made in three sections, along c_1, c_2 and c_3.

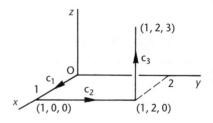

(1) c_1: $y = 0,\ z = 0,\ dy = 0,\ dz = 0$

$$\therefore \int_{c_1} \mathbf{F} \cdot \mathbf{dr} = 0 + 0 + 0 = 0$$

(2) c_2: The conditions along c_2 are

....

15

$$c_2:\quad x = 1,\quad z = 0,\quad dx = 0,\quad dz = 0$$

$$\therefore \int_{c_2} \mathbf{F} \cdot \mathbf{dr} = 0 + 0 + 0 = 0$$

(3) c_3: $x = 1,\ y = 2,\ dx = 0,\ dy = 0$

$$\therefore \int_{c_3} \mathbf{F} \cdot \mathbf{dr} = \ldots\ldots\ldots$$

16

27

Because

$$\int_{c_3} \mathbf{F} \cdot \mathbf{dr} = 0 + 0 + \int_0^3 3z^2\, dz = 27$$

Summing the three partial results

$$\int_{(0,0,0)}^{(1,2,3)} \mathbf{F} \cdot \mathbf{dr} = 0 + 0 + 27 = 27 \quad \therefore \int_{c_1 + c_2 + c_3} \mathbf{F} \cdot \mathbf{dr} = 27$$

▶

270 Vector Analysis

(b) If t is taken as the parameter, the parametric equations of c are

$x = \ldots\ldots\ldots$
$y = \ldots\ldots\ldots$
$z = \ldots\ldots\ldots$

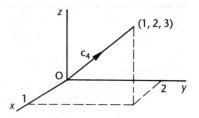

17

$$x = t; \quad y = 2t: \quad z = 3t$$

and the limits of t are $\ldots\ldots\ldots$

18

$$t = 0 \quad \text{and} \quad t = 1$$

As in Example 1, we now express everything in terms of t and complete the integral, finally getting

$$\int_{C_4} \mathbf{F} \cdot d\mathbf{r} = \ldots\ldots\ldots$$

19

$$\int_{C_4} \mathbf{F} \cdot d\mathbf{r} = \frac{115}{4} = 28.75$$

Because

$\mathbf{F} = 2t^3\mathbf{i} + 12t^2\mathbf{j} + 27t^3\mathbf{k}$
$d\mathbf{r} = \mathbf{i}\,dx + \mathbf{j}\,dy + \mathbf{k}\,dz = \mathbf{i}\,dt + \mathbf{j}\,2\,dt + \mathbf{k}\,3\,dt$

$$\therefore \int_{C_4} \mathbf{F} \cdot d\mathbf{r} = \int_0^1 (2t^3\mathbf{i} + 12t^2\mathbf{j} + 27t^3\mathbf{k}) \cdot (\mathbf{i} + 2\mathbf{j} + 3\mathbf{k})\,dt$$

$$= \int_0^1 (2t^3 + 24t^2 + 81t^3)\,dt = \int_0^1 (83t^3 + 24t^2)\,dt$$

$$= \left[83\frac{t^4}{4} + 8t^3\right]_0^1 = \frac{115}{4} = 28.75$$

So the value of the line integral depends on the path taken between the two end points A and B

(a) $\int \mathbf{F} \cdot d\mathbf{r}$ via c_1, c_2 and $c_3 = 27$

(b) $\int \mathbf{F} \cdot d\mathbf{r}$ via c_4 $\quad = 28.75$

We shall refer to this topic later.
One further example on your own. The working is just the same as before.

▶

Vector integration

Example 3

If $\mathbf{F} = x^2y^2\mathbf{i} + y^3z\mathbf{j} + z^2\mathbf{k}$, evaluate $\int_C \mathbf{F} \cdot d\mathbf{r}$ along the curve $x = 2u^2$, $y = 3u$, $z = u^3$ between A $(2, -3, -1)$ and B $(2, 3, 1)$. Proceed as before. You will have no difficulty.

$$\int_C \mathbf{F} \cdot d\mathbf{r} = \ldots\ldots\ldots\ldots$$

20

$$\boxed{\int_C \mathbf{F} \cdot d\mathbf{r} = \frac{500}{21} = 23.8}$$

Here is the working for you to check.

$x = 2u^2 \quad y = 3u \quad z = u^3$

$x^2y^2 = (4u^4)(9u^2) = 36u^6 \qquad dx = 4u\, du$

$y^3z = (27u^3)(u^3) = 27u^6 \qquad dy = 3\, du$

$z^2 = u^6 \qquad\qquad\qquad\qquad dz = 3u^2\, du$

Limits: A $(2, -3, -1)$ corresponds to $u = -1$

$\qquad\quad$ B $(2, 3, 1)$ $\quad$ corresponds to $u = 1$

$\therefore \int_C \mathbf{F} \cdot d\mathbf{r} = \int_{-1}^{1} (x^2y^2\mathbf{i} + y^3z\mathbf{j} + z^2\mathbf{k}) \cdot (\mathbf{i}\, dx + \mathbf{j}\, dy + \mathbf{k}\, dz)$

$\qquad\qquad\quad = \int_{-1}^{1} (36u^6\mathbf{i} + 27u^6\mathbf{j} + u^6\mathbf{k}) \cdot (\mathbf{i}\, 4u\, du + \mathbf{j}\, 3\, du + \mathbf{k}\, 3u^2\, du)$

$\qquad\qquad\quad = \int_{-1}^{1} (144u^7 + 81u^6 + 3u^8)\, du$

$\qquad\qquad\quad = \left[18u^8 + \frac{81u^7}{7} + \frac{u^9}{3}\right]_{-1}^{1} = \frac{500}{21} = 23.8$

Now on to the next section

Volume integrals

21

If V is a closed region bounded by a surface $\mathbf{S}$ and $\mathbf{F}$ is a vector field at each point of V and on its boundary surface $\mathbf{S}$, then $\int_V \mathbf{F}\, dV$ is the *volume integral* of $\mathbf{F}$ throughout the region.

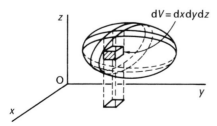

$$\int_V \mathbf{F}\, dV = \int_{x_1}^{x_2} \int_{y_1}^{y_2} \int_{z_1}^{z_2} \mathbf{F}\, dz\, dy\, dx$$

Example 1

Evaluate $\int_V \mathbf{F}\,dV$ where V is the region bounded by the planes $x = 0$, $x = 2$, $y = 0$, $y = 3$, $z = 0$, $z = 4$, and $\mathbf{F} = xy\mathbf{i} + z\mathbf{j} - x^2\mathbf{k}$.

We start, as in most cases, by sketching the diagram, which is

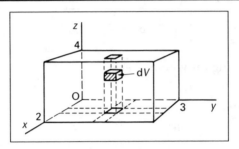

Then $\mathbf{F} = xy\mathbf{i} + z\mathbf{j} - x^2\mathbf{k}$ and $dV = dx\,dy\,dz$

$$\therefore \int_V \mathbf{F}\,dV = \int_0^4 \int_0^3 \int_0^2 (xy\mathbf{i} + z\mathbf{j} - x^2\mathbf{k})\,dx\,dy\,dz$$

$$= \int_0^4 \int_0^3 \left[\frac{x^2 y}{2}\mathbf{i} + xz\mathbf{j} - \frac{x^3}{3}\mathbf{k}\right]_{x=0}^{x=2} dy\,dz$$

$$= \int_0^4 \int_0^3 \left(2y\mathbf{i} + 2z\mathbf{j} - \frac{8}{3}\mathbf{k}\right) dy\,dz$$

$= \ldots\ldots\ldots\ldots$ Complete the integral.

$$\boxed{\int_V \mathbf{F}\,dV = 4(9\mathbf{i} + 12\mathbf{j} - 8\mathbf{k})}$$

Because

$$\int_V \mathbf{F}\,dV = \int_0^4 \left[y^2\mathbf{i} + 2yz\mathbf{j} - \frac{8}{3}y\mathbf{k}\right]_{y=0}^{y=3} dz$$

$$= \int_0^4 (9\mathbf{i} + 6z\mathbf{j} - 8\mathbf{k})\,dz$$

$$= \left[9z\mathbf{i} + 3z^2\mathbf{j} - 8z\mathbf{k}\right]_0^4$$

$$= 36\mathbf{i} + 48\mathbf{j} - 32\mathbf{k}$$

$$= 4(9\mathbf{i} + 12\mathbf{j} - 8\mathbf{k})$$

Now another.

Vector integration

Example 2

Evaluate $\int_V \mathbf{F}\,dV$ where V is the region bounded by the planes $x = 0$, $y = 0$, $z = 0$ and $2x + y + z = 2$, and $\mathbf{F} = 2z\mathbf{i} + y\mathbf{k}$.

To sketch the surface $2x + y + z = 2$, note that

when $z = 0$, $2x + y = 2$ i.e. $y = 2 - 2x$
when $y = 0$, $2x + z = 2$ i.e. $z = 2 - 2x$
when $x = 0$, $y + z = 2$ i.e. $z = 2 - y$

Inserting these in the planes $x = 0$, $y = 0$, $z = 0$ will help.
The diagram is therefore

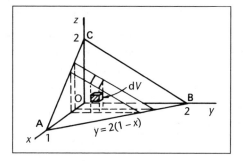

So $2x + y + z = 2$ cuts the axes at A $(1, 0, 0)$; B $(0, 2, 0)$; C $(0, 0, 2)$.

Also $\mathbf{F} = 2z\mathbf{i} + y\mathbf{k}$; $z = 2 - 2x - y = 2(1 - x) - y$

$$\therefore \int_V \mathbf{F}\,dV = \int_0^1 \int_0^{2(1-x)} \int_0^{2(1-x)-y} (2z\mathbf{i} + y\mathbf{k})\,dz\,dy\,dx$$

$$= \int_0^1 \int_0^{2(1-x)} \left[z^2\mathbf{i} + yz\mathbf{k}\right]_{z=0}^{z=2(1-x)-y} dy\,dx$$

$$= \int_0^1 \int_0^{2(1-x)} \{[4(1-x)^2 - 4(1-x)y + y^2]\mathbf{i}$$
$$+ [2(1-x)y - y^2]\mathbf{k}\}\,dy\,dx$$

$$= \int_0^1 \left[\left\{4(1-x)^2 y - 2(1-x)y^2 + \frac{y^3}{3}\right\}\mathbf{i}\right.$$
$$\left. + \left\{(1-x)y^2 - \frac{y^3}{3}\right\}\mathbf{k}\right]_{y=0}^{2(1-x)} dx$$

$= \ldots\ldots\ldots\ldots$

Finish the last stage

25

$$\int_V \mathbf{F}\,dV = \frac{1}{3}(2\mathbf{i} + \mathbf{k})$$

Because

$$\int_V \mathbf{F}\,dV = \int_0^1 \left\{ \frac{8}{3}(1-x)^3\mathbf{i} + \frac{4}{3}(1-x)^3\mathbf{k} \right\} dx$$

$$= \left[-\frac{2}{3}(1-x)^4\mathbf{i} - \frac{1}{3}(1-x)^4\mathbf{k} \right]_0^1 = \frac{1}{3}(2\mathbf{i} + \mathbf{k})$$

And now one more, slightly different.

Example 3

Evaluate $\int_V \mathbf{F}\,dV$ where $\mathbf{F} = 2\mathbf{i} + 2z\mathbf{j} + y\mathbf{k}$ and V is the region bounded by the planes $z = 0$, $z = 4$ and the surface $x^2 + y^2 = 9$.

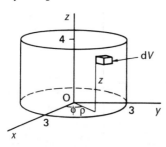

It will be convenient to use cylindrical polar coordinates (ρ, ϕ, z) so the relevant transformations are

$x = \ldots\ldots\ldots\ldots$; $y = \ldots\ldots\ldots\ldots$
$z = \ldots\ldots\ldots\ldots$; $dV = \ldots\ldots\ldots\ldots$

26

$$x = \rho\cos\phi; \qquad y = \rho\sin\phi$$
$$z = z; \qquad dV = \rho\,d\rho\,d\phi\,dz$$

Then $\int_V \mathbf{F}\,dV = \iiint_V (2\mathbf{i} + 2z\mathbf{j} + y\mathbf{k})\,dx\,dy\,dz$.

Changing into cylindrical polar coordinates with appropriate change of limits this becomes

$$\int_V \mathbf{F}\,dV = \int_{\phi=0}^{2\pi}\int_{\rho=0}^{3}\int_{z=0}^{4} (2\mathbf{i} + 2z\mathbf{j} + \rho\sin\phi\,\mathbf{k})\,dz\,\rho\,d\rho\,d\phi$$

$$= \int_{\phi=0}^{2\pi}\int_{\rho=0}^{3} \left[2z\mathbf{i} + z^2\mathbf{j} + \rho\sin\phi\,z\mathbf{k} \right]_{z=0}^{4} \rho\,d\rho\,d\phi$$

$$= \int_0^{2\pi}\int_0^3 (8\mathbf{i} + 16\mathbf{j} + 4\rho\sin\phi\,\mathbf{k})\,\rho\,d\rho\,d\phi$$

$$= 4\int_0^{2\pi}\int_0^3 (2\rho\mathbf{i} + 4\rho\mathbf{j} + \rho^2\sin\phi\,\mathbf{k})\,d\rho\,d\phi$$

Completing the working, we finally get

$$\int_V \mathbf{F}\,dV = \ldots\ldots\ldots\ldots$$

$$\boxed{72\pi(\mathbf{i}+2\mathbf{j})}$$

Because

$$\int_V \mathbf{F}\, dV = 4\int_0^{2\pi}\left[\rho^2\mathbf{i}+2\rho^2\mathbf{j}+\frac{\rho^3}{3}\sin\phi\,\mathbf{k}\right]_0^3 d\phi$$

$$= 4\int_0^{2\pi}(9\mathbf{i}+18\mathbf{j}+9\sin\phi\,\mathbf{k})\,d\phi$$

$$= 36\int_0^{2\pi}(\mathbf{i}+2\mathbf{j}+\sin\phi\,\mathbf{k})\,d\phi$$

$$= 36\left[\phi\mathbf{i}+2\phi\mathbf{j}-\cos\phi\,\mathbf{k}\right]_0^{2\pi}$$

$$= 36\{(2\pi\mathbf{i}+4\pi\mathbf{j}-\mathbf{k})-(-\mathbf{k})\}$$

$$= 72\pi(\mathbf{i}+2\mathbf{j})$$

You will, of course, remember that in appropriate cases, the use of cylindrical polar coordinates or spherical polar coordinates often simplifies the subsequent calculations. So keep them in mind.

Now let us turn to surface integrals – in the next frame

Surface integrals

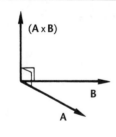

The vector product of two vectors **A** and **B** has magnitude $|\mathbf{A}\times\mathbf{B}|=AB\sin\theta$ at right angles to the plane of **A** and **B** to form a right-handed set.

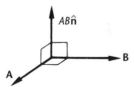

If $\theta=\dfrac{\pi}{2}$, then $|\mathbf{A}\times\mathbf{B}|=AB$ in the direction of the normal. Therefore, if $\hat{\mathbf{n}}$ is a unit normal then

$$\mathbf{A}\times\mathbf{B}=|\mathbf{A}|\,|\mathbf{B}|\hat{\mathbf{n}}=AB\,\hat{\mathbf{n}}$$

Vector Analysis

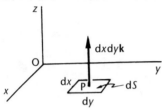

If P (x, y) is a point in the x–y plane, the element of area $dx\,dy$ has a vector area $d\mathbf{S} = (\mathbf{i}\,dx) \times (\mathbf{j}\,dy)$.

i.e. $d\mathbf{S} = dx\,dy(\mathbf{i} \times \mathbf{j}) = dx\,dy\,\mathbf{k}$

i.e. a vector of magnitude $dx\,dy$ acting in the direction of $\mathbf{k}$ and referred to as the *vector area*.

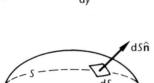

For a general surface S in space, each element of surface dS has a *vector area* $d\mathbf{S}$ such that $d\mathbf{S} = dS\,\hat{\mathbf{n}}$.

You will remember we established previously that for a surface S given by the equation $\phi(x, y, z) = $ constant, the unit normal $\hat{\mathbf{n}}$ is given by

$$\hat{\mathbf{n}} = \frac{\text{grad}\,\phi}{|\text{grad}\,\phi|} = \frac{\nabla \phi}{|\nabla \phi|}$$

Let us see how we can apply these results to the following examples.

29 Scalar fields

Example 1

A scalar field $V = xyz$ exists over the curved surface S defined by $x^2 + y^2 = 4$ between the planes $z = 0$ and $z = 3$ in the first octant. Evaluate $\int_S V\,d\mathbf{S}$ over this surface.

We have $V = xyz$ S: $x^2 + y^2 - 4 = 0$, $z = 0$ to $z = 3$

$d\mathbf{S} = \hat{\mathbf{n}}\,dS$ where $\hat{\mathbf{n}} = \dfrac{\nabla \phi}{|\nabla \phi|}$

Now $\nabla \phi = \dfrac{\partial \phi}{\partial x}\mathbf{i} + \dfrac{\partial \phi}{\partial y}\mathbf{j} + \dfrac{\partial \phi}{\partial z}\mathbf{k} = 2x\mathbf{i} + 2y\mathbf{j}$ and

$|\nabla \phi| = \sqrt{4x^2 + 4y^2} = 2\sqrt{x^2 + y^2} = 2\sqrt{4} = 4$

Therefore

$\hat{\mathbf{n}} = \dfrac{\nabla \phi}{|\nabla \phi|} = \dfrac{x\mathbf{i} + y\mathbf{j}}{2}$ so that $d\mathbf{S} = \hat{\mathbf{n}}\,dS = \dfrac{x\mathbf{i} + y\mathbf{j}}{2}\,dS$

$$\therefore \int_S V\,d\mathbf{S} = \int_S V\hat{\mathbf{n}}\,dS$$

$$= \frac{1}{2}\int_S xyz(x\mathbf{i} + y\mathbf{j})\,dS$$

$$= \frac{1}{2}\int_S (x^2 yz\mathbf{i} + xy^2 z\mathbf{j})\,dS \qquad (1)$$

Vector integration

We have to evaluate this integral over the prescribed surface.

Changing to cylindrical coordinates with $\rho = 2$

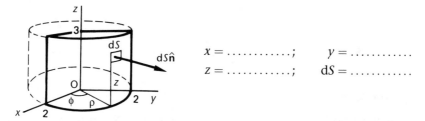

$x = \ldots\ldots\ldots ;\quad y = \ldots\ldots\ldots$
$z = \ldots\ldots\ldots ;\quad dS = \ldots\ldots\ldots$

$$x = 2\cos\phi; \quad y = 2\sin\phi$$
$$z = z; \quad dS = 2\,d\phi\,dz$$

$\therefore\ x^2 yz = (4\cos^2 \phi)(2\sin\phi)(z)$
$\quad\quad\quad = 8\cos^2 \phi \sin\phi\, z$
$\quad xy^2 z = (2\cos\phi)(4\sin^2\phi)(z)$
$\quad\quad\quad = 8\cos\phi \sin^2\phi\, z$

Then result (1) above becomes

$$\int_S \mathbf{V}\,d\mathbf{S} = \frac{1}{2}\int_0^{\pi/2}\int_0^3 (8\cos^2\phi\sin\phi\, z\mathbf{i} + 8\cos\phi\sin^2\phi\, z\mathbf{j})\,2\,dz\,d\phi$$

$$= 4\int_0^{\pi/2}\int_0^3 (\cos^2\phi\sin\phi\,\mathbf{i} + \cos\phi\sin^2\phi\,\mathbf{j})\,2z\,dz\,d\phi$$

$$= 4\int_0^{\pi/2} (\cos^2\phi\sin\phi\,\mathbf{i} + \cos\phi\sin^2\phi\,\mathbf{j})\,9\,d\phi$$

and this eventually gives

$$\int_S \mathbf{V}\,d\mathbf{S} = \ldots\ldots\ldots$$

278 Vector Analysis

31

$$\boxed{\int_S V \, dS = 12(\mathbf{i}+\mathbf{j})}$$

Because

$$\int_S V \, dS = 36\left[-\frac{\cos^3\phi}{3}\mathbf{i} + \frac{\sin^3\phi}{3}\mathbf{j}\right]_0^{\pi/2} = 12(\mathbf{i}+\mathbf{j})$$

Example 2

A scalar field $V = x + y + z$ exists over the surface S defined by $2x + 2y + z = 2$ bounded by $x = 0$, $y = 0$, $z = 0$ in the first octant.

Evaluate $\int_S V \, dS$ over this surface.

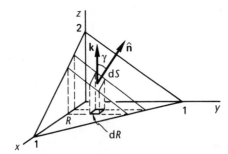

S: $2x + 2y + z = 2$
$x = 0 \quad z = 2 - 2y$
$y = 0 \quad z = 2 - 2x$
$z = 0 \quad y = 1 - x$

$d\mathbf{S} = \hat{\mathbf{n}} \, dS$ where $\hat{\mathbf{n}} = \dfrac{\nabla\phi}{|\nabla\phi|}$

Now $\nabla\phi = \dfrac{\partial\phi}{\partial x}\mathbf{i} + \dfrac{\partial\phi}{\partial y}\mathbf{j} + \dfrac{\partial\phi}{\partial z}\mathbf{k} = 2\mathbf{i} + 2\mathbf{j} + \mathbf{k}$ and

$|\nabla\phi| = \sqrt{4+4+1} = \sqrt{9} = 3$

Therefore

$\hat{\mathbf{n}} = \dfrac{\nabla\phi}{|\nabla\phi|} = \dfrac{2\mathbf{i} + 2\mathbf{j} + \mathbf{k}}{3}$ so that $d\mathbf{S} = \hat{\mathbf{n}} \, dS = \dfrac{1}{3}(2\mathbf{i} + 2\mathbf{j} + \mathbf{k}) \, dS$

If we now project dS onto the x–y plane, $dR = dS \cos\gamma$

$$\cos\gamma = \hat{\mathbf{n}} \cdot \mathbf{k} = \frac{1}{3}(2\mathbf{i} + 2\mathbf{j} + \mathbf{k}) \cdot (\mathbf{k}) = \frac{1}{3}$$

$\therefore \; dR = \dfrac{1}{3} dS \quad \therefore \; dS = 3\, dR = 3\, dx\, dy$

$\therefore \; \int_S V \, dS = \int_S V\hat{\mathbf{n}} \, dS = \int_S \int (x+y+z)\dfrac{1}{3}(2\mathbf{i}+2\mathbf{j}+\mathbf{k})3 \, dx\, dy$

But $z = 2 - 2x - 2y$

$\therefore \; \int_S V \, dS = \int_{x=0}^1 \int_{y=0}^{1-x} (2 - x - y)(2\mathbf{i} + 2\mathbf{j} + \mathbf{k}) \, dy\, dx$

$= \ldots\ldots\ldots$

$$\boxed{\frac{2}{3}(2\mathbf{i}+2\mathbf{j}+\mathbf{k})}$$

Because

$$\int_S V\,d\mathbf{S} = \int_0^1 \left[2y - xy - \frac{y^2}{2}\right]_0^{1-x}(2\mathbf{i}+2\mathbf{j}+\mathbf{k})\,dx$$

$$= \left[\frac{3}{2}x - x^2 + \frac{x^3}{6}\right]_0^1 (2\mathbf{i}+2\mathbf{j}+\mathbf{k})$$

$$= \frac{2}{3}(2\mathbf{i}+2\mathbf{j}+\mathbf{k})$$

Vector fields

Example 1

A vector field $\mathbf{F} = y\mathbf{i} + 2\mathbf{j} + \mathbf{k}$ exists over a surface S defined by $x^2 + y^2 + z^2 = 9$ bounded by $x = 0$, $y = 0$, $z = 0$ in the first octant. Evaluate $\int_S \mathbf{F}\cdot d\mathbf{S}$ over the surface indicated.

$d\mathbf{S} = \hat{\mathbf{n}}\,dS$ where $\hat{\mathbf{n}} = \dfrac{\nabla\phi}{|\nabla\phi|}$ where $\phi = x^2 + y^2 + z^2 - 9 = 0$

Now $\nabla\phi = \dfrac{\partial\phi}{\partial x}\mathbf{i} + \dfrac{\partial\phi}{\partial y}\mathbf{j} + \dfrac{\partial\phi}{\partial z}\mathbf{k} = 2x\mathbf{i} + 2y\mathbf{j} + 2z\mathbf{k}$ and

$|\nabla\phi| = \sqrt{4x^2 + 4y^2 + 4z^2} = 2\sqrt{x^2 + y^2 + z^2} = 2\sqrt{9} = 6$

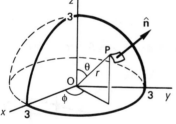

$\therefore \hat{\mathbf{n}} = \dfrac{1}{6}(2x\mathbf{i} + 2y\mathbf{j} + 2z\mathbf{k})$

$= \dfrac{1}{3}(x\mathbf{i} + y\mathbf{j} + z\mathbf{k})$

$$\int_S \mathbf{F}\cdot d\mathbf{S} = \int_S \mathbf{F}\cdot\hat{\mathbf{n}}\,dS = \int_S (y\mathbf{i} + 2\mathbf{j} + \mathbf{k})\cdot\frac{1}{3}(x\mathbf{i} + y\mathbf{j} + z\mathbf{k})\,dS$$

$$= \frac{1}{3}\int_S (xy + 2y + z)\,dS$$

Before integrating over the surface, we convert to spherical polar coordinates.

$x = \ldots\ldots\ldots\ldots;\qquad y = \ldots\ldots\ldots\ldots$

$z = \ldots\ldots\ldots\ldots;\qquad dS = \ldots\ldots\ldots\ldots$

34

$$x = 3\sin\theta\cos\phi; \quad y = 3\sin\theta\sin\phi$$
$$z = 3\cos\theta; \quad dS = 9\sin\theta\,d\theta\,d\phi$$

Limits of θ and ϕ are $\theta = 0$ to $\dfrac{\pi}{2}$; $\phi = 0$ to $\dfrac{\pi}{2}$.

$$\therefore \int_S \mathbf{F}\cdot d\mathbf{S} = \frac{1}{3}\int_0^{\pi/2}\int_0^{\pi/2}(9\sin^2\theta\sin\phi\cos\phi + 6\sin\theta\sin\phi$$
$$+ 3\cos\theta)\,9\sin\theta\,d\theta\,d\phi$$

$$= 9\int_0^{\pi/2}\int_0^{\pi/2}(3\sin^3\theta\sin\phi\cos\phi + 2\sin^2\theta\sin\phi$$
$$+ \sin\theta\cos\theta)\,d\theta\,d\phi$$

$$= \ldots\ldots\ldots\ldots$$

Complete the integral

35

$$\int_S \mathbf{F}\cdot d\mathbf{S} = 9\left(1 + \frac{3\pi}{4}\right)$$

Because

$$\int_S \mathbf{F}\cdot d\mathbf{S} = 9\int_0^{\pi/2}\left(2\sin\phi\cos\phi + \frac{\pi}{2}\sin\phi + \frac{1}{2}\right)d\phi$$

$$= 9\left[\sin^2\phi - \frac{\pi}{2}\cos\phi - \frac{\phi}{2}\right]_0^{\pi/2} = 9\left(1 + \frac{3\pi}{4}\right)$$

Example 2

Evaluate $\displaystyle\int_S \mathbf{F}\cdot d\mathbf{S}$ where $\mathbf{F} = 2y\mathbf{j} + z\mathbf{k}$ and S is the surface $x^2 + y^2 = 4$ in the first two octants bounded by the planes $z = 0$, $z = 5$ and $y = 0$.

$\phi: x^2 + y^2 - 4 = 0 \quad \hat{\mathbf{n}} = \dfrac{\nabla\phi}{|\nabla\phi|}$

$\nabla\phi = \dfrac{\partial\phi}{\partial x}\mathbf{i} + \dfrac{\partial\phi}{\partial y}\mathbf{j} + \dfrac{\partial\phi}{\partial z}\mathbf{k} = 2x\mathbf{i} + 2y\mathbf{j}$

$\therefore |\nabla\phi| = \sqrt{4x^2 + 4y^2} = 2\sqrt{x^2 + y^2}$

$= 2\sqrt{4} = 4$

$\therefore \hat{\mathbf{n}} = \dfrac{\nabla\phi}{|\nabla\phi|} = \dfrac{2x\mathbf{i} + 2y\mathbf{j}}{4} = \dfrac{1}{2}(x\mathbf{i} + y\mathbf{j})$

$\therefore \displaystyle\int_S \mathbf{F}\cdot d\mathbf{S} = \int_S \mathbf{F}\cdot\hat{\mathbf{n}}\,dS = \ldots\ldots\ldots\ldots$

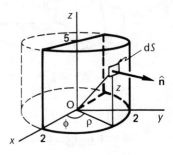

Vector integration

> $$\boxed{\int_S y^2 \, dS}$$

36

Because

$$\int_S \mathbf{F} \cdot \hat{\mathbf{n}} \, dS = \int_S (2y\mathbf{j} + z\mathbf{k}) \cdot \frac{1}{2}(x\mathbf{i} + y\mathbf{j}) \, dS$$

$$= \frac{1}{2} \int_S (2y^2) \, dS = \int_S y^2 \, dS$$

This is clearly a case for using cylindrical polar coordinates.

$$x = \ldots\ldots\ldots ; \qquad y = \ldots\ldots\ldots$$
$$z = \ldots\ldots\ldots ; \qquad dS = \ldots\ldots\ldots$$

> $$\boxed{\begin{array}{ll} x = 2\cos\phi; & y = 2\sin\phi \\ z = z; & dS = 2\, d\phi\, dz \end{array}}$$

37

$$\therefore \int_S \mathbf{F} \cdot d\mathbf{S} = \int_S y^2 \, dS = \int_S \int 4\sin^2\phi \cdot 2 \, d\phi \, dz = 8 \int_S \int \sin^2\phi \, d\phi \, dz$$

Limits: $\phi = 0$ to $\phi = \pi$; $z = 0$ to $z = 5$

$$\therefore \int_S \mathbf{F} \cdot d\mathbf{S} = \ldots\ldots\ldots$$

> $$\boxed{20\pi}$$

38

Because

$$\int_S \mathbf{F} \cdot d\mathbf{S} = 4 \int_{z=0}^{5} \int_{\phi=0}^{\pi} (1 - \cos 2\phi) \, d\phi \, dz$$

$$= 4 \int_0^5 \left[\phi - \frac{\sin 2\phi}{2}\right]_0^{\pi} dz$$

$$= 4 \int_0^5 \pi \, dz = 4\pi \left[z\right]_0^5 = 20\pi$$

Example 3

Evaluate $\int_S \mathbf{F} \cdot d\mathbf{S}$ where $\mathbf{F}$ is the field $x^2\mathbf{i} - y\mathbf{j} + 2z\mathbf{k}$ and S is the surface $2x + y + 2z = 2$ bounded by $x = 0$, $y = 0$, $z = 0$ in the first octant.

We can sketch the diagram by putting $x = 0$, $y = 0$, $z = 0$ in turn in the equation for S.

When $x = 0$ $y + 2z = 2$ $z = 1 - \dfrac{y}{2}$

           $y = 0$ $x + z = 1$ $z = 1 - x$

           $z = 0$ $2x + y = 2$ $y = 2 - 2x$

So the diagram is

$$\ldots\ldots\ldots$$

39

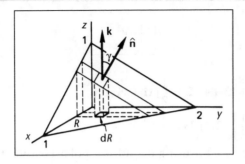

$$\mathbf{F} = x^2\mathbf{i} - y\mathbf{j} + 2z\mathbf{k}; \qquad \phi: \ 2x + y + 2z - 2 = 0$$

$$\nabla\phi = \frac{\partial \phi}{\partial x}\mathbf{i} + \frac{\partial \phi}{\partial y}\mathbf{j} + \frac{\partial \phi}{\partial z}\mathbf{k} = 2\mathbf{i} + \mathbf{j} + 2\mathbf{k} \qquad |\nabla\phi| = 3$$

$$\int_S \mathbf{F} \cdot d\mathbf{S} = \int_S \mathbf{F} \cdot \hat{\mathbf{n}}\, dS$$

$$= \ldots\ldots\ldots \text{ (next stage)}$$

40

$$\boxed{\frac{1}{3}\int_S (2x^2 - y + 4z)\, dS}$$

Because

$$\int_S \mathbf{F} \cdot \hat{\mathbf{n}}\, dS = \int_S (x^2\mathbf{i} - y\mathbf{j} + 2z\mathbf{k}) \cdot \frac{1}{3}(2\mathbf{i} + \mathbf{j} + 2\mathbf{k})\, dS$$

$$= \frac{1}{3}\int_S (2x^2 - y + 4z)\, dS$$

If we now project the element of surface dS onto the x–y plane

$$dR = dS \cos\gamma \quad \cos\gamma = \hat{\mathbf{n}} \cdot \mathbf{k} \quad \therefore\ dR = \hat{\mathbf{n}} \cdot \mathbf{k}\, dS \quad \therefore\ dS = \frac{dx\, dy}{\hat{\mathbf{n}} \cdot \mathbf{k}}$$

$$\therefore\ \hat{\mathbf{n}} \cdot \mathbf{k} = \frac{1}{3}(2\mathbf{i} + \mathbf{j} + 2\mathbf{k}) \cdot (\mathbf{k}) = \frac{2}{3} \quad \therefore\ dS = \frac{3}{2}\, dx\, dy$$

Using these new relationships, $\int_S \mathbf{F} \cdot d\mathbf{S} = \int_S \mathbf{F} \cdot \hat{\mathbf{n}}\, dS$

$$= \ldots\ldots\ldots$$

Vector integration

41

$$\int_R \int \frac{1}{2}(2x^2 - y + 4z)\,dx\,dy$$

Because

$$\int_S \mathbf{F} \cdot \hat{\mathbf{n}}\,dS = \frac{1}{3}\int_S (2x^2 - y + 4z)\,dS$$

$$= \frac{1}{3}\int_R \int (2x^2 - y + 4z)\frac{3}{2}\,dx\,dy$$

$$= \frac{1}{2}\int_R \int (2x^2 - y + 4z)\,dx\,dy$$

Limits: $y = 0$ to $y = 2 - 2x$; $x = 0$ to $x = 1$

$$\therefore \int_S \mathbf{F} \cdot \hat{\mathbf{n}}\,dS = \frac{1}{2}\int_0^1 \int_0^{2-2x} (2x^2 - y + 4z)\,dy\,dx$$

But $2x + y + 2z = 2$ $\quad \therefore z = \frac{1}{2}(2 - 2x - y)$

$$\therefore \int_S \mathbf{F} \cdot \hat{\mathbf{n}}\,dS = \ldots\ldots\ldots$$

Complete the integration

42

$$\boxed{\frac{1}{2}}$$

Here is the rest of the working.

$$\int_S \mathbf{F} \cdot d\mathbf{S} = \int_S \mathbf{F} \cdot \hat{\mathbf{n}}\,dS = \frac{1}{2}\int_0^1 \int_0^{2-2x} (2x^2 - y + 4 - 4x - 2y)\,dy\,dx$$

$$= \frac{1}{2}\int_0^1 \int_0^{2-2x} (2x^2 - 4x + 4 - 3y)\,dy\,dx$$

$$= \frac{1}{2}\int_0^1 \left[(2x^2 - 4x + 4)y - \frac{3y^2}{2}\right]_0^{2-2x}\,dx$$

$$= \frac{1}{2}\int_0^1 (4x^2 - 8x + 8 - 4x^3 + 8x^2 - 8x - 6 + 12x - 6x^2)\,dx$$

$$= \frac{1}{2}\int_0^1 (6x^2 - 4x^3 - 4x + 2)\,dx = \int_0^1 (3x^2 - 2x^3 - 2x + 1)\,dx$$

$$= \left[x^3 - \frac{x^4}{2} - x^2 + x\right]_0^1 = \frac{1}{2}$$

While we are concerned with vector fields, let us move on to a further point of interest.

Conservative vector fields

43

In general, the value of the line integral $\int_c \mathbf{F} \cdot d\mathbf{r}$ between two stated points A and B depends on the particular path of integration followed.

If, however, the line integral between A and B is independent of the path of integration between the two end points, then the vector field $\mathbf{F}$ is said to be *conservative*.

It follows that, for a closed path in a conservative field, $\oint_c \mathbf{F} \cdot d\mathbf{r} = 0$.

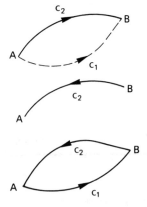

Because, if the field is conservative

$$\int_{c_1(AB)} \mathbf{F} \cdot d\mathbf{r} = \int_{c_2(AB)} \mathbf{F} \cdot d\mathbf{r}$$

But $\int_{c_2(BA)} \mathbf{F} \cdot d\mathbf{r} = -\int_{c_2(AB)} \mathbf{F} \cdot d\mathbf{r}$

Hence, for the closed path $\mathbf{AB}_{c_1} + \mathbf{BA}_{c_2}$

$$\oint \mathbf{F} \cdot d\mathbf{r} = \int_{c_1(AB)} \mathbf{F} \cdot d\mathbf{r} + \int_{c_2(BA)} \mathbf{F} \cdot d\mathbf{r}$$

$$= \int_{c_1(AB)} \mathbf{F} \cdot d\mathbf{r} - \int_{c_2(AB)} \mathbf{F} \cdot d\mathbf{r}$$

$$= \int_{c_1(AB)} \mathbf{F} \cdot d\mathbf{r} - \int_{c_1(AB)} \mathbf{F} \cdot d\mathbf{r} = 0$$

$$\therefore \oint \mathbf{F} \cdot d\mathbf{r} = 0$$

Note that this result holds good only for a closed curve and when the vector field is a conservative field.

Now for an example.

Example

If $\mathbf{F} = 2xyz\mathbf{i} + x^2z\mathbf{j} + x^2y\mathbf{k}$, evaluate the line integral $\int \mathbf{F} \cdot d\mathbf{r}$ between A (0, 0, 0) and B (2, 4, 6)

(a) along the curve c whose parametric equations are $x = u$, $y = u^2$, $z = 3u$

(b) along the three straight lines c_1: (0, 0, 0) to (2, 0, 0); c_2: (2, 0, 0) to (2, 4, 0); c_3: (2, 4, 0) to (2, 4, 6).

Hence determine whether or not $\mathbf{F}$ is a conservative field.

First draw the diagram

. . . .

Vector integration

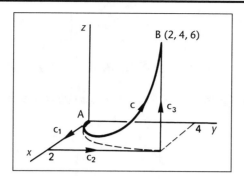

44

(a) $\mathbf{F} = 2xyz\mathbf{i} + x^2 z\mathbf{j} + x^2 y\mathbf{k}$

$x = u;$ $\quad y = u^2;$ $\quad z = 3u$

$\therefore \; dx = du;$ $\quad dy = 2u\,du;$ $\quad dz = 3\,du.$

$\mathbf{F} \cdot d\mathbf{r} = (2xyz\mathbf{i} + x^2 z\mathbf{j} + x^2 y\mathbf{k}) \cdot (\mathbf{i}\,dx + \mathbf{j}\,dy + \mathbf{k}\,dz)$

$\qquad = 2xyz\,dx + x^2 z\,dy + x^2 y\,dz$

Using the transformations shown above, we can now express $\mathbf{F} \cdot d\mathbf{r}$ in terms of u.

$$\mathbf{F} \cdot d\mathbf{r} = \ldots\ldots\ldots\ldots$$

45

$$\boxed{15u^4\,du}$$

Because

$2xyz\,dx = (2u)(u^2)(3u)\,du = 6u^4\,du$

$x^2 z\,dy = (u^2)(3u)(2u)\,du = 6u^4\,du$

$x^2 y\,dz = (u^2)(u^2)3\,du \quad\;\; = 3u^4\,du$

$\therefore \; \mathbf{F} \cdot d\mathbf{r} = 6u^4\,du + 6u^4\,du + 3u^4\,du = 15u^4\,du$

The limits of integration in u are

$$\ldots\ldots\ldots\ldots$$

46

$$u = 0 \text{ to } u = 2$$

$$\therefore \int_c \mathbf{F} \cdot d\mathbf{r} = \int_0^2 15u^4 \, du = [3u^5]_0^2 = 96 \qquad \int_c \mathbf{F} \cdot d\mathbf{r} = 96$$

(b) The diagram for (b) is as shown. We consider each straight line section in turn.

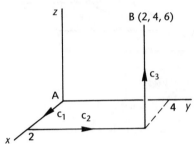

$$\int \mathbf{F} \cdot d\mathbf{r} = \int (2xyz \, dx + x^2 z \, dy + x^2 y \, dz)$$

c_1: $(0,0,0)$ to $(2,0,0)$; $y = 0, z = 0, dy = 0, dz = 0$

$$\therefore \int_{c_1} \mathbf{F} \cdot d\mathbf{r} = 0 + 0 + 0 = 0$$

In the same way, we evaluate the line integral along c_2 and c_3.

$$\int_{c_2} \mathbf{F} \cdot d\mathbf{r} = \ldots\ldots\ldots ; \qquad \int_{c_3} \mathbf{F} \cdot d\mathbf{r} = \ldots\ldots\ldots$$

47

$$\int_{c_2} \mathbf{F} \cdot d\mathbf{r} = 0; \qquad \int_{c_3} \mathbf{F} \cdot d\mathbf{r} = 96$$

Because we have $\int \mathbf{F} \cdot d\mathbf{r} = \int (2xyz \, dx + x^2 z \, dy + x^2 y \, dz)$

c_2: $(2,0,0)$ to $(2,4,0)$; $x = 2, \quad z = 0, \quad dx = 0, \quad dz = 0$

$$\therefore \int_{c_2} \mathbf{F} \cdot d\mathbf{r} = 0 + 0 + 0 = 0$$

$$\int_{c_2} \mathbf{F} \cdot d\mathbf{r} = 0$$

c_3: $(2,4,0)$ to $(2,4,6)$; $x = 2, \quad y = 4, \quad dx = 0, \quad dy = 0$

$$\therefore \int_{c_3} \mathbf{F} \cdot d\mathbf{r} = 0 + 0 + \int_0^6 16 \, dz = [16z]_0^6 = 96$$

$$\int_{c_3} \mathbf{F} \cdot d\mathbf{r} = 96$$

Collecting the three results together

$$\int_{c_1+c_2+c_3} \mathbf{F} \cdot d\mathbf{r} = 0 + 0 + 96 \qquad \therefore \int_{c_1+c_2+c_3} \mathbf{F} \cdot d\mathbf{r} = 96$$

Vector integration

In this particular example, the value of the line integral is independent of the two paths we have used joining the same two end points and indicates that **F** may be a conservative field. It follows that

$$\int_C \mathbf{F} \cdot d\mathbf{r} - \int_{c_1+c_2+c_3} \mathbf{F} \cdot d\mathbf{r} = 0 \quad \text{i.e.} \quad \oint \mathbf{F} \cdot d\mathbf{r} = 0$$

So, if **F** is a conservative field, $\oint \mathbf{F} \cdot d\mathbf{r} = 0$

Make a note of this for future use

Two tests can be applied to establish that a given vector field is conservative.

If **F** is a conservative field

(a) curl **F** = 0

(b) **F** can be expressed as grad V where V is a scalar field to be determined.

For example, in the work we have just completed, we showed that $\mathbf{F} = 2xyz\mathbf{i} + x^2 z\mathbf{j} + x^2 y\mathbf{k}$ is a conservative field.

(a) If we determine curl **F** in this case, we have

$$\text{curl } \mathbf{F} = \ldots\ldots\ldots\ldots$$

| 48 |

| curl **F** = 0 |

| 49 |

Because

$$\text{curl } \mathbf{F} = \begin{vmatrix} \mathbf{i} & \mathbf{j} & \mathbf{k} \\ \dfrac{\partial}{\partial x} & \dfrac{\partial}{\partial y} & \dfrac{\partial}{\partial z} \\ 2xyz & x^2 z & x^2 y \end{vmatrix}$$

$$= (x^2 - x^2)\mathbf{i} - (2xy - 2xy)\mathbf{j} + (2xz - 2xz)\mathbf{k} = \mathbf{0}$$

$\therefore$ curl **F** = **0**

(b) We can attempt to express **F** as grad V where V is a scalar in x, y, z.

If $V = f(x, y, z)$

$$\text{grad } V = \frac{\partial V}{\partial x}\mathbf{i} + \frac{\partial V}{\partial y}\mathbf{j} + \frac{\partial V}{\partial z}\mathbf{k}$$

and we have $\mathbf{F} = 2xyz\mathbf{i} + x^2 z\mathbf{j} + x^2 y\mathbf{k}$

$\therefore \dfrac{\partial V}{\partial x} = 2xyz \qquad \therefore V = x^2 yz + f(y, z)$

$\dfrac{\partial V}{\partial y} = x^2 z \qquad \therefore V = \ldots\ldots\ldots\ldots$

$\dfrac{\partial V}{\partial z} = x^2 y \qquad \therefore V = \ldots\ldots\ldots\ldots$

We therefore have to find a scalar function V that satisfies the three requirements. $V = \ldots\ldots\ldots\ldots$

50

$$V = x^2 yz$$

Because

$\dfrac{\partial V}{\partial x} = 2xyz \qquad \therefore\ V = x^2 yz + f(y, z)$

$\dfrac{\partial V}{\partial y} = x^2 z \qquad \therefore\ V = x^2 yz + g(x, z)$

$\dfrac{\partial V}{\partial z} = x^2 y \qquad \therefore\ V = x^2 yz + h(x, y)$

These three are satisfied if $f(y, z) = g(z, x) = h(x, y) = 0$

$\therefore\ \mathbf{F} = \text{grad } V$ where $V = x^2 yz$

So two tests can be applied to determine whether or not a vector field is conservative. They are

(a) ....

(b) ....

51

(a) curl $\mathbf{F} = 0$
(b) $\mathbf{F} = \text{grad } V$

Any one of these conditions can be applied as is convenient.
Now what about these?

Exercise

Determine which of the following vector fields are conservative.

(a) $\mathbf{F} = (x+y)\mathbf{i} + (y-z)\mathbf{j} + (x+y+z)\mathbf{k}$
(b) $\mathbf{F} = (2xz+y)\mathbf{i} + (z+x)\mathbf{j} + (x^2+y)\mathbf{k}$
(c) $\mathbf{F} = y\sin z\,\mathbf{i} + x\sin z\,\mathbf{j} + (xy\cos z + 2z)\mathbf{k}$
(d) $\mathbf{F} = 2xy\mathbf{i} + (x^2+4yz)\mathbf{j} + 2y^2 z\mathbf{k}$
(e) $\mathbf{F} = y\cos x\cos z\,\mathbf{i} + \sin x\cos z\,\mathbf{j} - y\sin x\sin z\,\mathbf{k}$.

Complete all five and check your findings with the next frame.

52

(a) No (b) Yes (c) Yes (d) No (e) Yes

Divergence theorem (Gauss' theorem)

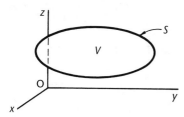

For a closed surface S, enclosing a region V in a vector field $\mathbf{F}$,

$$\int_V \operatorname{div} \mathbf{F}\, dV = \int_S \mathbf{F} \cdot d\mathbf{S}$$

In general, this means that the volume integral (triple integral) on the left-hand side can be expressed as a surface integral (double integral) on the right-hand side. Let us work through one or two examples.

Example 1

Verify the divergence theorem for the vector field $\mathbf{F} = x^2\mathbf{i} + z\mathbf{j} + y\mathbf{k}$ taken over the region bounded by the planes $z = 0$, $z = 2$, $x = 0$, $x = 1$, $y = 0$, $y = 3$.

Start off, as always, by sketching the relevant diagram, which is

....

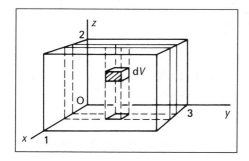

$dV = dx\, dy\, dz$

We have to show that

$$\int_V \operatorname{div} \mathbf{F}\, dV = \int_S \mathbf{F} \cdot d\mathbf{S}$$

(a) To find $\int_V \operatorname{div} \mathbf{F}\, dV$

$$\operatorname{div} \mathbf{F} = \nabla \cdot \mathbf{F} = \left(\frac{\partial}{\partial x}\mathbf{i} + \frac{\partial}{\partial y}\mathbf{j} + \frac{\partial}{\partial z}\mathbf{k}\right) \cdot (x^2\mathbf{i} + z\mathbf{j} + y\mathbf{k})$$

$$= \frac{\partial}{\partial x}(x^2) + \frac{\partial}{\partial y}(z) + \frac{\partial}{\partial z}(y) = 2x + 0 + 0 = 2x$$

$$\therefore \int_V \operatorname{div} \mathbf{F}\, dV = \int_V 2x\, dV = \iiint_V 2x\, dz\, dy\, dx$$

Inserting the limits and completing the integration

$$\int_V \operatorname{div} \mathbf{F}\, dV = \ldots\ldots\ldots\ldots$$

54

$$\int_V \text{div } \mathbf{F} \, dV = 6$$

Because

$$\int_V \text{div } \mathbf{F} \, dV = \int_0^1 \int_0^3 \int_0^2 2x \, dz \, dy \, dx = \int_0^1 \int_0^3 \left[2xz\right]_0^2 dy \, dx$$

$$= \int_0^1 \left[4xy\right]_0^3 dx = \int_0^1 12x \, dx = \left[6x^2\right]_0^1 = 6$$

Now we have to find $\int_S \mathbf{F} \cdot d\mathbf{S}$

(b) To find $\int_S \mathbf{F} \cdot d\mathbf{S}$ i.e. $\int_S \mathbf{F} \cdot \hat{\mathbf{n}} \, dS$

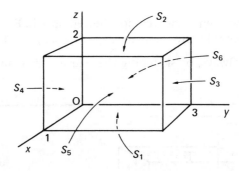

The enclosing surface S consists of six separate plane faces denoted as $S_1, S_2, \ldots, S_6$ as shown. We consider each face in turn.

$$\mathbf{F} = x^2 \mathbf{i} + z \mathbf{j} + y \mathbf{k}$$

(1) S_1 (base): $z = 0$; $\hat{\mathbf{n}} = -\mathbf{k}$ (outwards and downwards)

$$\therefore \mathbf{F} = x^2 \mathbf{i} + y \mathbf{k} \qquad dS_1 = dx \, dy$$

$$\therefore \int_{S_1} \mathbf{F} \cdot \hat{\mathbf{n}} \, dS = \iint_{S_1} (x^2 \mathbf{i} + y \mathbf{k}) \cdot (-\mathbf{k}) \, dy \, dx$$

$$= \int_0^1 \int_0^3 (-y) \, dy \, dx$$

$$= \int_0^1 \left[-\frac{y^2}{2}\right]_0^3 dx$$

$$= -\frac{9}{2}$$

(2) S_2 (top): $z = 2$; $\hat{\mathbf{n}} = \mathbf{k}$ $dS_2 = dx \, dy$

$$\therefore \int_{S_2} \mathbf{F} \cdot \hat{\mathbf{n}} \, dS = \ldots\ldots\ldots\ldots$$

Vector integration

55

$$\boxed{\dfrac{9}{2}}$$

Because

$$\int_{S_2} \mathbf{F} \cdot \hat{\mathbf{n}}\, dS = \iint_{S_2} (x^2\mathbf{i} + 2\mathbf{j} + y\mathbf{k}) \cdot (\mathbf{k})\, dy\, dx$$

$$= \int_0^1 \int_0^3 y\, dy\, dx = \dfrac{9}{2}$$

So we go on.

(3) S_3 (right-hand end): $y = 3;$ $\hat{\mathbf{n}} = \mathbf{j}$ $dS_3 = dx\, dz$

$\mathbf{F} = x^2\mathbf{i} + z\mathbf{j} + y\mathbf{k}$

$$\therefore \int_{S_3} \mathbf{F} \cdot \hat{\mathbf{n}}\, dS = \iint_{S_3} (x^2\mathbf{i} + z\mathbf{j} + 3\mathbf{k}) \cdot (\mathbf{j})\, dz\, dx$$

$$= \int_0^1 \int_0^2 z\, dz\, dx$$

$$= \int_0^1 \left[\dfrac{z^2}{2}\right]_0^2 dx = \int_0^1 2\, dx = 2$$

(4) S_4 (left-hand end): $y = 0;$ $\hat{\mathbf{n}} = -\mathbf{j}$ $dS_4 = dx\, dz$

$$\therefore \int_{S_4} \mathbf{F} \cdot \hat{\mathbf{n}}\, dS = \ldots\ldots\ldots\ldots$$

56

$$\boxed{-2}$$

Because

$$\int_{S_4} \mathbf{F} \cdot \hat{\mathbf{n}}\, dS = \iint_{S_4} (x^2\mathbf{i} + z\mathbf{j} + y\mathbf{k}) \cdot (-\mathbf{j})\, dz\, dx = \int_0^1 \int_0^2 (-z)\, dz\, dx$$

$$= \int_0^1 \left[-\dfrac{z^2}{2}\right]_0^2 dx = \int_0^1 (-2)\, dx = -2$$

Now for the remaining two sides S_5 and S_6.

Evaluate these in the same manner, obtaining

$$\int_{S_5} \mathbf{F} \cdot \hat{\mathbf{n}}\, dS = \ldots\ldots\ldots\ldots$$

$$\int_{S_6} \mathbf{F} \cdot \hat{\mathbf{n}}\, dS = \ldots\ldots\ldots\ldots$$

57

$$\int_{S_5} \mathbf{F} \cdot \hat{\mathbf{n}}\, dS = 6; \qquad \int_{S_6} \mathbf{F} \cdot \hat{\mathbf{n}}\, dS = 0$$

Check:

(5) S_5 (front): $\quad x = 1; \qquad \hat{\mathbf{n}} = \mathbf{i} \qquad dS_5 = dy\, dz$

$$\therefore \int_{S_5} \mathbf{F} \cdot \hat{\mathbf{n}}\, dS = \iint_{S_5} (\mathbf{i} + z\mathbf{j} + y\mathbf{k}) \cdot (\mathbf{i})\, dy\, dz = \iint_{S_5} 1\, dy\, dz = 6$$

(6) S_6 (back): $\quad x = 0; \qquad \hat{\mathbf{n}} = -\mathbf{i} \qquad dS_6 = dy\, dz$

$$\therefore \int_{S_6} \mathbf{F} \cdot \hat{\mathbf{n}}\, dS = \iint_{S_6} (z\mathbf{j} + y\mathbf{k}) \cdot (-\mathbf{i})\, dy\, dz = \iint_{S_6} 0\, dy\, dz = 0$$

Now on to the next frame where we will collect our results together

58

For the whole surface S we therefore have

$$\int_S \mathbf{F} \cdot dS = -\frac{9}{2} + \frac{9}{2} + 2 - 2 + 6 + 0 = 6$$

and from our previous work in section (a) $\int_V \operatorname{div} \mathbf{F}\, dV = 6$

We have therefore verified as required that, in this example

$$\int_V \operatorname{div} \mathbf{F}\, dV = \int_S \mathbf{F} \cdot dS$$

We have made rather a meal of this since we have set out the working in detail. In practice, the actual writing can often be considerably simplified. Let us move on to another example.

Example 2

Verify the Gauss divergence theorem for the vector field $\mathbf{F} = x\mathbf{i} + 2\mathbf{j} + z^2\mathbf{k}$ taken over the region bounded by the planes $z = 0$, $z = 4$, $x = 0$, $y = 0$ and the surface $x^2 + y^2 = 4$ in the first octant.

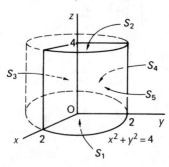

Divergence theorem

$$\int_V \operatorname{div} \mathbf{F}\, dV = \int_S \mathbf{F} \cdot dS$$

S consists of five surfaces $S_1, S_2, \ldots, S_5$ as shown.

(a) $\operatorname{div} \mathbf{F} = \nabla \cdot \mathbf{F} = \left(\dfrac{\partial}{\partial x}\mathbf{i} + \dfrac{\partial}{\partial y}\mathbf{j} + \dfrac{\partial}{\partial z}\mathbf{k}\right) \cdot (x\mathbf{i} + 2\mathbf{j} + z^2\mathbf{k})$

$= \ldots\ldots\ldots\ldots$

Vector integration

59

$$\boxed{1+2z}$$

$$\therefore \int_V \operatorname{div} \mathbf{F}\,dV = \int_V \nabla \cdot \mathbf{F}\,dV = \iiint_V (1+2z)\,dx\,dy\,dz$$

Changing to cylindrical polar coordinates (ρ, ϕ, z)

$$x = \rho\cos\phi \qquad y = \rho\sin\phi \qquad z = z \qquad dV = \rho\,d\rho\,d\phi\,dz$$

Transforming the variables and inserting the appropriate limits, we then have

$$\int_V \operatorname{div} \mathbf{F}\,dV = \ldots\ldots\ldots\ldots$$

Finish it

60

$$\boxed{20\pi}$$

Because

$$\int_V \operatorname{div} \mathbf{F}\,dV = \int_0^{\pi/2}\!\!\int_0^2\!\!\int_0^4 (1+2z)\,dz\,\rho\,d\rho\,d\phi$$

$$= \int_0^{\pi/2}\!\!\int_0^2 \left[z+z^2\right]_0^4 \rho\,d\rho\,d\phi = \int_0^{\pi/2}\!\!\int_0^2 20\rho\,d\rho\,d\phi$$

$$= \int_0^{\pi/2} \left[10\rho^2\right]_0^2 d\phi = \int_0^{\pi/2} 40\,d\phi = 20\pi \qquad (1)$$

(b) Now we evaluate $\displaystyle\int_S \mathbf{F}\cdot d\mathbf{S}$ over the closed surface.

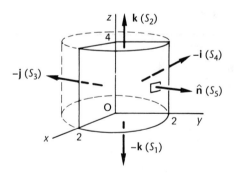

The unit normal vector for each surface is shown.

$$\mathbf{F} = x\mathbf{i} + 2\mathbf{j} + z^2\mathbf{k}$$

(1) S_1: $z=0$; $\quad \hat{\mathbf{n}} = -\mathbf{k} \quad \mathbf{F} = x\mathbf{i} + 2\mathbf{j}$

$$\therefore \int_{S_1} \mathbf{F}\cdot\hat{\mathbf{n}}\,dS = \int_{S_1}(x\mathbf{i}+2\mathbf{j})\cdot(-\mathbf{k})\,dS = 0$$

▶

(2) S_2: $z = 4$; $\hat{\mathbf{n}} = \mathbf{k}$ $\mathbf{F} = x\mathbf{i} + 2\mathbf{j} + 16\mathbf{k}$

$$\therefore \int_{S_2} \mathbf{F} \cdot \hat{\mathbf{n}} \, dS = \int_{S_2} (x\mathbf{i} + 2\mathbf{j} + 16\mathbf{k}) \cdot (\mathbf{k}) \, dS = \int_{S_2} 16 \, dS$$
$$= 16\left(\frac{\pi 4}{4}\right) = 16\pi$$

In the same way for S_3: $\int_{S_3} \mathbf{F} \cdot \hat{\mathbf{n}} \, dS = \ldots\ldots\ldots$

and for S_4: $\int_{S_4} \mathbf{F} \cdot \hat{\mathbf{n}} \, dS = \ldots\ldots\ldots$

61

$$\int_{S_3} \mathbf{F} \cdot \hat{\mathbf{n}} \, dS = -16; \quad \int_{S_4} \mathbf{F} \cdot \hat{\mathbf{n}} \, dS = 0$$

Because we have

(3) S_3: $y = 0$; $\hat{\mathbf{n}} = -\mathbf{j}$ $\mathbf{F} = x\mathbf{i} + 2\mathbf{j} + z^2\mathbf{k}$

$$\therefore \int_{S_3} \mathbf{F} \cdot \hat{\mathbf{n}} \, dS = \int_{S_3} (x\mathbf{i} + 2\mathbf{j} + z^2\mathbf{k}) \cdot (-\mathbf{j}) \, dS$$
$$= \int_{S_3} (-2) \, dS = -2(8) = -16$$

(4) S_4: $x = 0$; $\hat{\mathbf{n}} = -\mathbf{i}$ $\mathbf{F} = 2\mathbf{j} + z^2\mathbf{k}$

$$\therefore \int_{S_4} \mathbf{F} \cdot \hat{\mathbf{n}} \, dS = \int_{S_4} (2\mathbf{j} + z^2\mathbf{k}) \cdot (-\mathbf{i}) \, dS = 0$$

Finally we have

(5) S_5: $x^2 + y^2 - 4 = 0$ $\hat{\mathbf{n}} = \ldots\ldots\ldots$

62

$$\hat{\mathbf{n}} = \frac{1}{2}(x\mathbf{i} + y\mathbf{j})$$

Because

$$x^2 + y^2 - 4 = 0 \quad \hat{\mathbf{n}} = \frac{\nabla S}{|\nabla S|} = \frac{2x\mathbf{i} + 2y\mathbf{j}}{\sqrt{4x^2 + 4y^2}} = \frac{x\mathbf{i} + y\mathbf{j}}{2}$$

$$\therefore \int_{S_5} \mathbf{F} \cdot \hat{\mathbf{n}} \, dS = \int_{S_5} (x\mathbf{i} + 2\mathbf{j} + z^2\mathbf{k}) \cdot \left(\frac{x\mathbf{i} + y\mathbf{j}}{2}\right) dS = \frac{1}{2} \int_{S_5} (x^2 + 2y) \, dS$$

Converting to cylindrical polar coordinates, this gives

$$\int_{S_5} \mathbf{F} \cdot \hat{\mathbf{n}} \, dS = \ldots\ldots\ldots$$

$$\boxed{4\pi + 16}$$

Because we have

$$\int_{S_5} \mathbf{F} \cdot \hat{\mathbf{n}}\, dS = \frac{1}{2}\int_{S_5} (x^2 + 2y)\, dS$$

also $x = 2\cos\phi;$ $y = 2\sin\phi$
 $z = z;$ $dS = 2\, d\phi\, dz$

$$\therefore \int_{S_5} \mathbf{F} \cdot \hat{\mathbf{n}}\, dS = \frac{1}{2}\int_0^4 \int_0^{\pi/2} (4\cos^2\phi + 4\sin\phi)\, 2\, d\phi\, dz$$

$$= 2\int_0^4 \int_0^{\pi/2} \{(1 + \cos 2\phi) + 2\sin\phi\}\, d\phi\, dz$$

$$= 2\int_0^4 \left[\left(\phi - \frac{\sin 2\phi}{2}\right) - 2\cos\phi\right]_0^{\pi/2} dz$$

$$= 2\int_0^4 \left(\frac{\pi}{2} + 2\right) dz = 4\pi + 16$$

Therefore, for the total surface S

$$\int_S \mathbf{F} \cdot \hat{\mathbf{n}}\, dS = 0 + 16\pi - 16 + 0 + 4\pi + 16 = 20\pi \tag{2}$$

$$\therefore \int_V \operatorname{div} \mathbf{F}\, dV = \int_S \mathbf{F} \cdot d\mathbf{S} = 20\pi$$

Other examples are worked in much the same way. You will remember that, for a closed surface, the normal vectors at all points are drawn in an *outward* direction.

Now we move on to a further important theorem.

Stokes' theorem

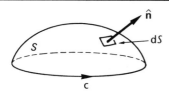

If $\mathbf{F}$ is a vector field existing over an open surface S and around its boundary, closed curve c, then

$$\int_S \operatorname{curl} \mathbf{F} \cdot d\mathbf{S} = \oint_c \mathbf{F} \cdot d\mathbf{r}$$

This means that we can express a surface integral in terms of a line integral round the boundary curve.

The proof of this theorem is rather lengthy and is to be found in the Appendix of the authors' *Advanced Engineering Mathematics (4th edition)*. Let us demonstrate its application in the following examples.

Example 1

A hemisphere S is defined by $x^2 + y^2 + z^2 = 4$ $(z \geq 0)$. A vector field $\mathbf{F} = 2y\mathbf{i} - x\mathbf{j} + xz\mathbf{k}$ exists over the surface and around its boundary c. Verify Stokes' theorem, that $\int_S \text{curl } \mathbf{F} \cdot d\mathbf{S} = \oint_c \mathbf{F} \cdot d\mathbf{r}$.

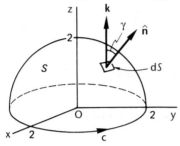

$S: \quad x^2 + y^2 + z^2 - 4 = 0$
$\mathbf{F} = 2y\mathbf{i} - x\mathbf{j} + xz\mathbf{k}$
c is the circle $x^2 + y^2 = 4$.

(a) $\oint_c \mathbf{F} \cdot d\mathbf{r} = \int_c (2y\mathbf{i} - x\mathbf{j} + xz\mathbf{k}) \cdot (\mathbf{i}\, dx + \mathbf{j}\, dy + \mathbf{k}\, dz)$

$= \int_c (2y\, dx - x\, dy + xz\, dz)$

Converting to polar coordinates

$x = 2 \cos \theta;$ $\quad y = 2 \sin \theta;$ $\quad z = 0$
$dx = -2 \sin \theta\, d\theta;$ $\quad dy = 2 \cos \theta\, d\theta;$ $\quad$ Limits $\theta = 0$ to 2π

Making the substitutions and completing the integral

$$\oint_c \mathbf{F} \cdot d\mathbf{r} = \ldots\ldots\ldots\ldots$$

65

$$\oint_c \mathbf{F} \cdot d\mathbf{r} = -12\pi$$

Because

$\oint_c \mathbf{F} \cdot d\mathbf{r} = \int_0^{2\pi} (4 \sin \theta [-2 \sin \theta\, d\theta] - 2 \cos \theta\, 2 \cos \theta\, d\theta)$

$= -4 \int_0^{2\pi} (2 \sin^2 \theta + \cos^2 \theta)\, d\theta$

$= -4 \int_0^{2\pi} (1 + \sin^2 \theta)\, d\theta = -2 \int_0^{2\pi} (3 - \cos 2\theta)\, d\theta$

$= -2 \left[3\theta - \dfrac{\sin 2\theta}{2} \right]_0^{2\pi} = -12\pi \qquad (1)$

On to the next frame

Vector integration

66

(b) Now we determine $\int_S \text{curl } \mathbf{F} \cdot d\mathbf{S}$

$$\int \text{curl } \mathbf{F} \cdot d\mathbf{S} = \int \text{curl } \mathbf{F} \cdot \hat{\mathbf{n}} \, dS \qquad \mathbf{F} = 2y\mathbf{i} - x\mathbf{j} + xz\mathbf{k}$$

$$\therefore \text{curl } \mathbf{F} = \ldots\ldots\ldots\ldots$$

67

$$\boxed{\text{curl } \mathbf{F} = -z\mathbf{j} - 3\mathbf{k}}$$

Because

$$\text{curl } \mathbf{F} = \begin{vmatrix} \mathbf{i} & \mathbf{j} & \mathbf{k} \\ \dfrac{\partial}{\partial x} & \dfrac{\partial}{\partial y} & \dfrac{\partial}{\partial z} \\ 2y & -x & xz \end{vmatrix} = \mathbf{i}(0-0) - \mathbf{j}(z-0) + \mathbf{k}(-1-2) = -z\mathbf{j} - 3\mathbf{k}$$

Now $\hat{\mathbf{n}} = \dfrac{\nabla S}{|\nabla S|} = \dfrac{2x\mathbf{i} + 2y\mathbf{j} + 2z\mathbf{k}}{\sqrt{4x^2 + 4y^2 + 4z^2}} = \dfrac{x\mathbf{i} + y\mathbf{j} + z\mathbf{k}}{2}$

Then $\int_S \text{curl } \mathbf{F} \cdot \hat{\mathbf{n}} \, dS = \int_S (-z\mathbf{j} - 3\mathbf{k}) \cdot \left(\dfrac{x\mathbf{i} + y\mathbf{j} + z\mathbf{k}}{2}\right) dS$

$$= \dfrac{1}{2} \int_S (-yz - 3z) \, dS$$

Expressing this in spherical polar coordinates and integrating, we get

$$\int_S \text{curl } \mathbf{F} \cdot \hat{\mathbf{n}} \, dS = \ldots\ldots\ldots\ldots$$

68

$$\boxed{-12\pi}$$

Because

$x = 2\sin\theta \cos\phi; \quad y = 2\sin\theta \sin\phi; \quad z = 2\cos\theta; \quad dS = 4\sin\theta \, d\theta \, d\phi$

$$\therefore \int_S \text{curl } \mathbf{F} \cdot \hat{\mathbf{n}} \, dS = \dfrac{1}{2} \int_S \int (-2\sin\theta \sin\phi \, 2\cos\theta - 6\cos\theta) 4\sin\theta \, d\theta \, d\phi$$

$$= -4 \int_0^{2\pi} \int_0^{\pi/2} (2\sin^2\theta \cos\theta \sin\phi + 3\sin\theta \, \cos\theta) \, d\theta \, d\phi$$

$$= -4 \int_0^{2\pi} \left[\dfrac{2\sin^3\theta \sin\phi}{3} + \dfrac{3\sin^2\theta}{2}\right]_0^{\pi/2} d\phi$$

$$= -4 \int_0^{2\pi} \left(\dfrac{2}{3}\sin\phi + \dfrac{3}{2}\right) d\phi = -12\pi \qquad (2)$$

So we have from our two results (1) and (2)

$$\int_S \text{curl } \mathbf{F} \cdot d\mathbf{S} = \oint_c \mathbf{F} \cdot d\mathbf{r}$$

Before we proceed with another example, let us clarify a point relating to the direction of unit normal vectors now that we are dealing with surfaces.

So on to the next frame

Direction of unit normal vectors to a surface S

69

When we were dealing with the divergence theorem, the normal vectors were drawn in a direction outward from the enclosed region.

With an open surface as we now have, there is in fact no inward or outward direction. With any general surface, a normal vector can be drawn in either of two opposite directions. To avoid confusion, a convention must therefore be agreed upon and the established rule is as follows.

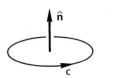 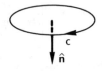

A unit normal $\hat{n}$ is drawn perpendicular to the surface S at any point in the direction indicated by applying a right-handed screw sense to the direction of integration round the boundary c.

Having noted that point, we can now deal with the next example.

Example 2

A surface consists of five sections formed by the planes $x = 0$, $x = 1$, $y = 0$, $y = 3$, $z = 2$ in the first octant. If the vector field $\mathbf{F} = y\mathbf{i} + z^2\mathbf{j} + xy\mathbf{k}$ exists over the surface and around its boundary, verify Stokes' theorem.

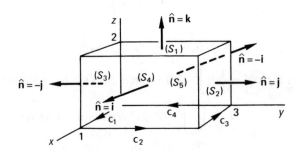

If we progress round the boundary along c_1, c_2, c_3, c_4 in an anti-clockwise manner, the normals to the surfaces will be as shown.

We have to verify that $\int_S \text{curl } \mathbf{F} \cdot d\mathbf{S} = \oint_c \mathbf{F} \cdot d\mathbf{r}$

(a) We will start off by finding $\oint_c \mathbf{F} \cdot d\mathbf{r}$

$$\int \mathbf{F} \cdot d\mathbf{r} = \ldots\ldots\ldots\ldots$$

Vector integration

70

$$\int \mathbf{F} \cdot d\mathbf{r} = \int (y\,dx + z^2\,dy + xy\,dz)$$

(1) Along c_1: $y = 0$; $z = 0$; $dy = 0$; $dz = 0$

$$\therefore \int_{c_1} \mathbf{F} \cdot d\mathbf{r} = \int (0 + 0 + 0) = 0$$

(2) Along c_2: $x = 1$; $z = 0$; $dx = 0$; $dz = 0$

$$\therefore \int_{c_2} \mathbf{F} \cdot d\mathbf{r} = \int (0 + 0 + 0) = 0$$

In the same way

$$\int_{c_3} \mathbf{F} \cdot d\mathbf{r} = \ldots\ldots\ldots \quad \text{and} \quad \int_{c_4} \mathbf{F} \cdot d\mathbf{r} = \ldots\ldots\ldots$$

71

$$\int_{c_3} \mathbf{F} \cdot d\mathbf{r} = -3; \quad \int_{c_4} \mathbf{F} \cdot d\mathbf{r} = 0$$

Because

(3) Along c_3: $y = 3$; $z = 0$; $dy = 0$; $dz = 0$

$$\therefore \int_{c_3} \mathbf{F} \cdot d\mathbf{r} = \int_1^0 (3\,dx + 0 + 0) = \left[3x\right]_1^0 = -3$$

(4) Along c_4: $x = 0$; $z = 0$; $dx = 0$; $dz = 0$

$$\therefore \int_{c_4} \mathbf{F} \cdot d\mathbf{r} = \int (0 + 0 + 0) = 0$$

$$\therefore \oint_c \mathbf{F} \cdot d\mathbf{r} = 0 + 0 - 3 + 0 = -3$$

$$\oint_c \mathbf{F} \cdot d\mathbf{r} = -3 \qquad (1)$$

(b) Now we have to find $\int_S \text{curl } \mathbf{F} \cdot d\mathbf{S}$.

First we need an expression for curl $\mathbf{F}$.

$$\mathbf{F} = y\mathbf{i} + z^2\mathbf{j} + xy\mathbf{k}$$

$\therefore$ curl $\mathbf{F} = \ldots\ldots\ldots$

72

$$\boxed{\operatorname{curl} \mathbf{F} = (x - 2z)\mathbf{i} - y\mathbf{j} - \mathbf{k}}$$

Because

$$\operatorname{curl} \mathbf{F} = \nabla \times \mathbf{F} = \begin{vmatrix} \mathbf{i} & \mathbf{j} & \mathbf{k} \\ \dfrac{\partial}{\partial x} & \dfrac{\partial}{\partial y} & \dfrac{\partial}{\partial z} \\ y & z^2 & xy \end{vmatrix}$$

$$= \mathbf{i}(x - 2z) - \mathbf{j}(y - 0) + \mathbf{k}(0 - 1) = (x - 2z)\mathbf{i} - y\mathbf{j} - \mathbf{k}$$

Then, for each section, we obtain $\displaystyle\int \operatorname{curl} \mathbf{F} \cdot d\mathbf{S} = \int \operatorname{curl} \mathbf{F} \cdot \hat{\mathbf{n}} \, dS$

(1) S_1 (top): $\hat{\mathbf{n}} = \mathbf{k}$

$$\therefore \int_{S_1} \operatorname{curl} \mathbf{F} \cdot \hat{\mathbf{n}} \, dS = \ldots\ldots\ldots$$

73

$$\boxed{-3}$$

Because

$$\int_{S_1} \operatorname{curl} \mathbf{F} \cdot \hat{\mathbf{n}} \, dS = \int_{S_1} \{(x - 2z)\mathbf{i} - y\mathbf{j} - \mathbf{k}\} \cdot (\mathbf{k}) \, dS$$

$$= \int_{S_1} (-1) \, dS = -(\text{area of } S_1) = -3$$

Then, likewise

(2) S_2 (right-hand end): $\hat{\mathbf{n}} = \mathbf{j}$

$$\therefore \int_{S_2} \operatorname{curl} \mathbf{F} \cdot \hat{\mathbf{n}} \, dS = \int_{S_2} \{(x - 2z)\mathbf{i} - y\mathbf{j} - \mathbf{k}\} \cdot (\mathbf{j}) \, dS$$

$$= \int_{S_2} (-y) \, dS$$

But $y = 3$ for this section

$$\therefore \int_{S_2} \operatorname{curl} \mathbf{F} \cdot \hat{\mathbf{n}} \, dS = \int_{S_2} (-3) \, dS = (-3)(2) = -6$$

(3) S_3 (left-hand end): $\hat{\mathbf{n}} = -\mathbf{j}$

$$\therefore \int_{S_3} \operatorname{curl} \mathbf{F} \cdot \hat{\mathbf{n}} \, dS = \ldots\ldots\ldots$$

Vector integration

74

$$\boxed{0}$$

Because

$$\int_{S_3} \text{curl } \mathbf{F} \cdot \hat{\mathbf{n}} \, dS = \int_{S_3} \{(x-2z)\mathbf{i} - y\mathbf{j} - \mathbf{k}\} \cdot (-\mathbf{j}) \, dS$$

$$= \int_{S_3} y \, dS$$

But $y = 0$ over S_3

$$\therefore \int_{S_3} \text{curl } \mathbf{F} \cdot \hat{\mathbf{n}} \, dS = 0$$

Working in the same way

$$\int_{S_4} \text{curl } \mathbf{F} \cdot \hat{\mathbf{n}} \, dS = \ldots\ldots\ldots ; \quad \int_{S_5} \text{curl } \mathbf{F} \cdot \hat{\mathbf{n}} \, dS = \ldots\ldots\ldots$$

75

$$\boxed{\int_{S_4} \text{curl } \mathbf{F} \cdot \hat{\mathbf{n}} \, dS = -6; \quad \int_{S_5} \text{curl } \mathbf{F} \cdot \hat{\mathbf{n}} \, dS = 12}$$

Because

(4) S_4 (front): $\hat{\mathbf{n}} = \mathbf{i}$

$$\therefore \int_{S_4} \text{curl } \mathbf{F} \cdot \hat{\mathbf{n}} \, dS = \int_{S_4} \{(x-2z)\mathbf{i} - y\mathbf{j} - \mathbf{k}\} \cdot (\mathbf{i}) \, dS$$

$$= \int_{S_4} (x - 2z) \, dS$$

But $x = 1$ over S_4

$$\therefore \int_{S_4} \text{curl } \mathbf{F} \cdot \hat{\mathbf{n}} \, dS = \int_0^3 \int_0^2 (1 - 2z) \, dz \, dy = \int_0^3 \left[z - z^2\right]_0^2 dy$$

$$= \int_0^3 (-2) \, dy = \left[-2y\right]_0^3 = -6$$

(5) S_5 (back): $\hat{\mathbf{n}} = -\mathbf{i}$ with $x = 0$ over S_5

Similar working to that above gives $\int_{S_5} \text{curl } \mathbf{F} \cdot \hat{\mathbf{n}} \, dS = 12$

Finally, collecting the five results together gives

$$\int_{S} \text{curl } \mathbf{F} \cdot \hat{\mathbf{n}} \, dS = \ldots\ldots\ldots$$

76

$$\int_S \text{curl } \mathbf{F} \cdot \hat{\mathbf{n}} \, dS = -3 - 6 + 0 - 6 + 12 = -3 \qquad (2)$$

So, referring back to our result for section (a) we see that

$$\int_S \text{curl } \mathbf{F} \cdot d\mathbf{S} = \oint_c \mathbf{F} \cdot d\mathbf{r}$$

Of course we can, on occasions, make use of Stokes' theorem to lighten the working – as in the next example.

Example 3

A surface S consists of that part of the cylinder $x^2 + y^2 = 9$ between $z = 0$ and $z = 4$ for $y \geq 0$ and the two semicircles of radius 3 in the planes $z = 0$ and $z = 4$. If $\mathbf{F} = z\mathbf{i} + xy\mathbf{j} + xz\mathbf{k}$, evaluate $\int_S \text{curl } \mathbf{F} \cdot d\mathbf{S}$ over the surface.

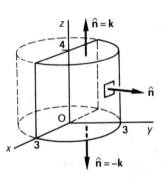

The surface S consists of three sections

(a) the curved surface of the cylinder
(b) the top and bottom semicircles.

We could therefore evaluate

$$\int_S \text{curl } \mathbf{F} \cdot d\mathbf{S}$$

over each of these separately.

However, we know by Stokes' theorem that

$$\int_S \text{curl } \mathbf{F} \cdot d\mathbf{S} = \ldots\ldots\ldots\ldots$$

77

$$\oint_c \mathbf{F} \cdot d\mathbf{r} \text{ where c is the boundary of } S$$

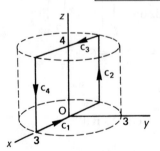

$$\mathbf{F} = z\mathbf{i} + xy\mathbf{j} + xz\mathbf{k}$$

$$\therefore \oint_c \mathbf{F} \cdot d\mathbf{r} = \oint_c (z\mathbf{i} + xy\mathbf{j} + xz\mathbf{k}) \cdot (\mathbf{i} \, dx + \mathbf{j} \, dy + \mathbf{k} \, dz)$$

$$= \oint_c (z \, dx + xy \, dy + xz \, dz)$$

Now we can work through this easily enough, taking c_1, c_2, c_3, c_4 in turn, and summing the results, which gives

$$\int_S \text{curl } \mathbf{F} \cdot d\mathbf{S} = \oint_c \mathbf{F} \cdot d\mathbf{r} = \ldots\ldots\ldots\ldots$$

Vector integration

$$\boxed{-24}$$

Here is the working in detail. $\oint_C \mathbf{F} \cdot d\mathbf{r} = \oint_C (z\,dx + xy\,dy + xz\,dz)$

(1) c_1: $y = 0$; $z = 0$; $dy = 0$; $dz = 0$

$$\int_{c_1} \mathbf{F} \cdot d\mathbf{r} = \int_{c_1} (0 + 0 + 0) = 0$$

(2) c_2: $x = -3$; $y = 0$; $dx = 0$; $dy = 0$

$$\int_{c_2} \mathbf{F} \cdot d\mathbf{r} = \int_{c_2} (0 + 0 - 3z\,dz) = \left[\frac{-3z^2}{2}\right]_0^4 = -24$$

(3) c_3: $y = 0$; $z = 4$; $dy = 0$; $dz = 0$

$$\int_{c_3} \mathbf{F} \cdot d\mathbf{r} = \int_{c_3} (4\,dx + 0 + 0) = \int_{-3}^{3} 4\,dx = 24$$

(4) c_4: $x = 3$; $y = 0$; $dx = 0$; $dy = 0$

$$\int_{c_4} \mathbf{F} \cdot d\mathbf{r} = \int_{c_4} (0 + 0 + 3z\,dz) = \left[\frac{3z^2}{2}\right]_4^0 = -24$$

Totalling up these four results, we have

$$\oint_C \mathbf{F} \cdot d\mathbf{r} = 0 - 24 + 24 - 24 = -24$$

But $\int_S \text{curl } \mathbf{F} \cdot d\mathbf{S} = \oint_C \mathbf{F} \cdot d\mathbf{r}$ $\therefore \int_S \text{curl } \mathbf{F} \cdot d\mathbf{S} = -24$

This working is a good deal easier than calculating $\int_S \text{curl } \mathbf{F} \cdot d\mathbf{S}$ over the three separate surfaces direct.

So, if you have not already done so, make a note of Stokes' theorem:

$$\int_S \text{curl } \mathbf{F} \cdot d\mathbf{S} = \oint_C \mathbf{F} \cdot d\mathbf{r}$$

Then on to the next section of the work

Green's theorem

79

Green's theorem enables an integral over a plane area to be expressed in terms of a line integral round its boundary curve.

We showed in Program 5 that, if P and Q are two single-valued functions of x and y, continuous over a plane surface S, and c is its boundary curve, then

$$\oint_c (P\,dx + Q\,dy) = \iint_S \left(\frac{\partial Q}{\partial x} - \frac{\partial P}{\partial y}\right) dx\,dy$$

where the line integral is taken round c in an anticlockwise manner.

In vector terms, this becomes:

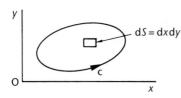

S is a two-dimensional space enclosed by a simple closed curve c.

$$dS = dx\,dy$$
$$d\mathbf{S} = \hat{\mathbf{n}}\,dS = \mathbf{k}\,dx\,dy$$

If $\mathbf{F} = P\mathbf{i} + Q\mathbf{j}$ where $P = P(x, y)$ and $Q = Q(x, y)$ then

$$\text{curl } \mathbf{F} = \ldots\ldots\ldots\ldots$$

80

$$\boxed{\mathbf{k}\left(\frac{\partial Q}{\partial x} - \frac{\partial P}{\partial y}\right)}$$

Because

$$\text{curl } \mathbf{F} = \begin{vmatrix} \mathbf{i} & \mathbf{j} & \mathbf{k} \\ \dfrac{\partial}{\partial x} & \dfrac{\partial}{\partial y} & \dfrac{\partial}{\partial z} \\ P & Q & 0 \end{vmatrix}$$

$$= \mathbf{i}\left(0 - \frac{\partial Q}{\partial z}\right) - \mathbf{j}\left(0 - \frac{\partial P}{\partial z}\right) + \mathbf{k}\left(\frac{\partial Q}{\partial x} - \frac{\partial P}{\partial y}\right)$$

But in the x–y plane, $\dfrac{\partial Q}{\partial z} = \dfrac{\partial P}{\partial z} = 0$. $\therefore$ curl $\mathbf{F} = \mathbf{k}\left(\dfrac{\partial Q}{\partial x} - \dfrac{\partial P}{\partial y}\right)$

So $\displaystyle\int \text{curl } \mathbf{F} \cdot d\mathbf{S} = \int \text{curl } \mathbf{F} \cdot \hat{\mathbf{n}}\,dS$ and in the x–y plane, $\hat{\mathbf{n}} = \mathbf{k}$

$$\therefore \int_S \text{curl } \mathbf{F} \cdot d\mathbf{S} = \int_S \mathbf{k}\left(\frac{\partial Q}{\partial x} - \frac{\partial P}{\partial y}\right) \cdot (\mathbf{k})\,dS = \iint_S \left(\frac{\partial Q}{\partial x} - \frac{\partial P}{\partial y}\right) dx\,dy$$

$$\therefore \int_S \text{curl } \mathbf{F} \cdot d\mathbf{S} = \iint_S \left(\frac{\partial Q}{\partial x} - \frac{\partial P}{\partial y}\right) dx\,dy \qquad (1)$$

Now by Stokes' theorem ....

Vector integration

$$\int_S \text{curl } \mathbf{F} \cdot d\mathbf{S} = \oint_c \mathbf{F} \cdot d\mathbf{r}$$

and, in this case,
$$\oint_c \mathbf{F} \cdot d\mathbf{r} = \oint_c (P\mathbf{i} + Q\mathbf{j}) \cdot (\mathbf{i}\, dx + \mathbf{j}\, dy + \mathbf{k}\, dz)$$
$$= \oint_c (P\, dx + Q\, dy)$$
$$\therefore \oint_c \mathbf{F} \cdot d\mathbf{r} = \oint_c (P\, dx + Q\, dy) \qquad (2)$$

Therefore from (1) and (2)

Stokes' theorem $\int_S \text{curl } \mathbf{F} \cdot d\mathbf{S} = \oint_c \mathbf{F} \cdot d\mathbf{r}$ in two dimensions becomes

Green's theorem $\iint_S \left(\dfrac{\partial Q}{\partial x} - \dfrac{\partial P}{\partial y} \right) dx\, dy = \oint_c (P\, dx + Q\, dy)$

Example

Verify Green's theorem for the integral $\oint_c \{(x^2 + y^2)\, dx + (x + 2y)\, dy\}$ taken round the boundary curve c defined by

$y = 0 \qquad 0 \le x \le 2$
$x^2 + y^2 = 4 \qquad 0 \le x \le 2$
$x = 0 \qquad 0 \le y \le 2$

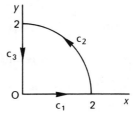

Green's theorem: $\iint_S \left(\dfrac{\partial Q}{\partial x} - \dfrac{\partial P}{\partial y} \right) dx\, dy = \oint_c (P\, dx + Q\, dy)$

In this case $(x^2 + y^2)\, dx + (x + 2y)\, dy = P\, dx + Q\, dy$
$\therefore P = x^2 + y^2 \quad \text{and} \quad Q = x + 2y$

We now take c_1, c_2, c_3 in turn.

(1) c_1: $y = 0$; $dy = 0$
$$\therefore \int_{c_1} (P\, dx + Q\, dy) = \int_0^2 x^2\, dx = \left[\dfrac{x^3}{3} \right]_0^2 = \dfrac{8}{3}$$

(2) c_2: $x^2 + y^2 = 4 \quad \therefore y^2 = 4 - x^2 \quad \therefore y = (4 - x^2)^{1/2}$
$x + 2y = x + 2(4 - x^2)^{1/2}$
$dy = \dfrac{1}{2}(4 - x^2)^{-1/2}(-2x)\, dx = \dfrac{-x}{\sqrt{4 - x^2}}\, dx$

$$\therefore \int_{c_2} (P\, dx + Q\, dy) = \ldots\ldots\ldots\ldots$$

Make any necessary substitutions and evaluate the line integral for c_2.

82

$$\boxed{\pi - 4}$$

Because we have

$$\int_{c_2}(P\,dx + Q\,dy) = \int_{c_2}\left\{4 + (x + 2\sqrt{4-x^2})\left(\frac{-x}{\sqrt{4-x^2}}\right)\right\}dx$$

$$= \int_{c_2}\left\{4 - 2x - \frac{x^2}{\sqrt{4-x^2}}\right\}dx$$

Putting $x = 2\sin\theta$, $\sqrt{4-x^2} = 2\cos\theta$, $dx = 2\cos\theta\,d\theta$

Limits: $x = 2$, $\theta = \frac{\pi}{2}$; $x = 0$, $\theta = 0$.

$$\therefore \int_{c_2}(P\,dx + Q\,dy) = \int_{\pi/2}^{0}\left\{4 - 4\sin\theta - \frac{4\sin^2\theta}{2\cos\theta}\right\}2\cos\theta\,d\theta$$

$$= 4\left[2\sin\theta - \sin^2\theta - \frac{1}{2}\left(\theta - \frac{\sin 2\theta}{2}\right)\right]_{\pi/2}^{0}$$

$$= 4\left[-\left(2 - 1 - \frac{\pi}{4}\right)\right] = \pi - 4$$

Finally

(3) c_3: $x = 0$; $dx = 0$

$$\therefore \int_{c_3}(P\,dx + Q\,dy) = \int_{2}^{0}2y\,dy = \left[y^2\right]_{2}^{0} = -4$$

$\therefore$ Collecting our three partial results

$$\oint_{c}(P\,dx + Q\,dy) = \frac{8}{3} + \pi - 4 - 4 = \pi - \frac{16}{3} \tag{1}$$

That is one part done. Now we have to evaluate $\iint_{S}\left(\frac{\partial Q}{\partial x} - \frac{\partial P}{\partial y}\right)dx\,dy$

$P = x^2 + y^2$ $\quad\therefore\quad \frac{\partial P}{\partial y} = 2y$

$Q = x + 2y$ $\quad\therefore\quad \frac{\partial Q}{\partial x} = 1$

$$\therefore \iint_{S}\left(\frac{\partial Q}{\partial x} - \frac{\partial P}{\partial y}\right)dx\,dy = \iint_{S}(1 - 2y)\,dy\,dx$$

It will be more convenient to work in polar coordinates, so we make the substitutions

$x = r\cos\theta$; $y = r\sin\theta$; $dS = dx\,dy = r\,dr\,d\theta$

$$\therefore \iint_{S}\left(\frac{\partial Q}{\partial x} - \frac{\partial P}{\partial y}\right)dx\,dy = \int_{0}^{\pi/2}\int_{0}^{2}(1 - 2r\sin\theta)r\,dr\,d\theta$$

$$= \ldots\ldots\ldots\ldots$$

Complete it

$$\boxed{\pi - \frac{16}{3}}$$

Here it is:

$$\iint_S \left(\frac{\partial Q}{\partial x} - \frac{\partial P}{\partial y}\right) dx\, dy = \int_0^{\pi/2} \int_0^2 (r - 2r^2 \sin\theta)\, dr\, d\theta$$

$$= \int_0^{\pi/2} \left[\frac{r^2}{2} - \frac{2r^3}{3}\sin\theta\right]_0^2 d\theta$$

$$= \int_0^{\pi/2} \left\{2 - \frac{16}{3}\sin\theta\right\} d\theta$$

$$= \left[2\theta + \frac{16}{3}\cos\theta\right]_0^{\pi/2} = \pi - \frac{16}{3} \qquad (2)$$

So we have established once again that

$$\oint_c (P\, dx + Q\, dy) = \iint_S \left(\frac{\partial Q}{\partial x} - \frac{\partial P}{\partial y}\right) dx\, dy$$

And that brings us to the end of this particular Program. We have covered a number of important sections, so check carefully down the **Review summary** and the **Can You?** checklist, and then work through the **Test exercise** that follows. The **Further problems** provide valuable additional practice.

Review summary

1 *Line integrals*

(a) Scalar field V: $\displaystyle\int_c V\, d\mathbf{r}$

The curve c is expressed in parametric form.

$d\mathbf{r} = \mathbf{i}\, dx + \mathbf{j}\, dy + \mathbf{k}\, dz$

(b) Vector field $\mathbf{F}$: $\displaystyle\int_c \mathbf{F} \cdot d\mathbf{r}$

$\mathbf{F} = F_x \mathbf{i} + F_y \mathbf{j} + F_z \mathbf{k}$
$d\mathbf{r} = \mathbf{i}\, dx + \mathbf{j}\, dy + \mathbf{k}\, dz$
$\mathbf{F} \cdot d\mathbf{r} = F_x\, dx + F_y\, dy + F_z\, dz$

2 *Volume integrals*

$\mathbf{F}$ is a vector field; V a closed region with boundary surface S.

$$\int_V \mathbf{F}\, dV = \int_{x_1}^{x_2} \int_{y_1}^{y_2} \int_{z_1}^{z_2} \mathbf{F}\, dz\, dy\, dx$$

3. *Surface integrals* (surface defined by $\phi(x, y, z) = $ constant)
   (a) Scalar field $V(x, y, z)$:
   $$\int_S V\, d\mathbf{S} = \int_S V\hat{\mathbf{n}}\, dS; \qquad \hat{\mathbf{n}} = \frac{\nabla \phi}{|\nabla \phi|} = \frac{\text{grad } \phi}{|\text{grad } \phi|}$$
   (b) Vector field $\mathbf{F} = F_x \mathbf{i} + F_y \mathbf{j} + F_z \mathbf{k}$:
   $$\int_S \mathbf{F} \cdot d\mathbf{S} = \int_S \mathbf{F} \cdot \hat{\mathbf{n}}\, dS; \qquad \hat{\mathbf{n}} = \frac{\nabla \phi}{|\nabla \phi|}$$

4. *Polar coordinates*
   (a) Plane polar coordinates (r, θ)

   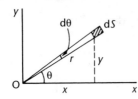

   $x = r\cos\theta; \quad y = r\sin\theta$
   $dS = r\, dr\, d\theta$

   (b) Cylindrical polar coordinates (ρ, ϕ, z)

   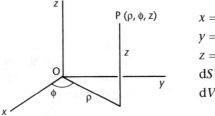

   $P(\rho, \phi, z)$
   $x = \rho\cos\phi$
   $y = \rho\sin\phi$
   $z = z$
   $dS = \rho\, d\phi\, dz$
   $dV = \rho\, d\rho\, d\phi\, dz$

   (c) Spherical polar coordinates (r, θ, ϕ)

   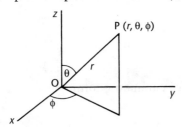

   $P(r, \theta, \phi)$
   $x = r\sin\theta\cos\phi$
   $y = r\sin\theta\sin\phi$
   $z = r\cos\theta$
   $dS = r^2 \sin\theta\, d\theta\, d\phi$
   $dV = r^2 \sin\theta\, dr\, d\theta\, d\phi$

5. *Conservative vector fields*
   A vector field $\mathbf{F}$ is conservative if
   (a) $\oint_c \mathbf{F} \cdot d\mathbf{r} = 0$ for all closed curves
   (b) curl $\mathbf{F} = 0$
   (c) $\mathbf{F} = \text{grad } V$ where V is a scalar

Vector integration

6 *Divergence theorem* (Gauss' theorem)

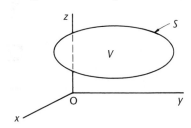

Closed surface S enclosing a region V in a vector field $\mathbf{F}$.

$$\int_V \text{div}\,\mathbf{F}\,dV = \int_S \mathbf{F}\cdot d\mathbf{S}$$

7 *Stokes' theorem*

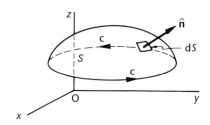

An open surface S bounded by a simple closed curve c, then

$$\int_S \text{curl}\,\mathbf{F}\cdot d\mathbf{S} = \oint_c \mathbf{F}\cdot d\mathbf{r}$$

8 *Green's theorem*

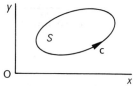

The curve c is a simple closed curve enclosing a plane space S in the x–y plane. P and Q are functions of both x and y.

Then $\displaystyle\iint_S \left(\frac{\partial Q}{\partial x} - \frac{\partial P}{\partial y}\right) dx\,dy = \oint_c (P\,dx + Q\,dy)$.

✅ Can You?

Checklist 9

85

Check this list before and after you try the end of Program test.

On a scale of 1 to 5 how confident are you that you can: Frames

- Evaluate the line integral of a scalar and a vector field in Cartesian coordinates?   1 to 20
  Yes ☐ ☐ ☐ ☐ ☐ No

- Evaluate the volume integral of a vector field?   21 to 27
  Yes ☐ ☐ ☐ ☐ ☐ No

- Evaluate the surface integral of a scalar and a vector field?   28 to 42
  Yes ☐ ☐ ☐ ☐ ☐ No

▶

- Determine whether or not a vector field is a conservative vector field?   43 to 52
  Yes ☐ ☐ ☐ ☐ ☐   No
- Apply Gauss' divergence theorem?   52 to 63
  Yes ☐ ☐ ☐ ☐ ☐   No
- Apply Stokes' theorem?   64 to 68
  Yes ☐ ☐ ☐ ☐ ☐   No
- Determine the direction of unit normal vectors to a surface?   69 to 78
  Yes ☐ ☐ ☐ ☐ ☐   No
- Apply Green's theorem in the plane?   79 to 83
  Yes ☐ ☐ ☐ ☐ ☐   No

Test exercise 9

1. If $V = x^3y + 2xy^2 + yz$, evaluate $\int_c V\,d\mathbf{r}$ between A $(0, 0, 0)$ and B $(2, 1, -3)$ along the curve with parametric equations $x = 2t$, $y = t^2$, $z = -3t^3$.

2. If $\mathbf{F} = x^2y^3\,\mathbf{i} + yz^2\,\mathbf{j} + zx^2\,\mathbf{k}$, evaluate $\int_c \mathbf{F}\cdot d\mathbf{r}$ along the curve $x = 3u^2$, $y = u$, $z = 2u^3$ between A $(3, -1, -2)$ and B $(3, 1, 2)$.

3. Evaluate $\int_V \mathbf{F}\,dV$ where $\mathbf{F} = 3\mathbf{i} + z\mathbf{j} + 2y\mathbf{k}$ and V is the region bounded by the planes $z = 0$, $z = 3$ and the surface $x^2 + y^2 = 4$.

4. If V is the scalar field $V = xyz^2$, evaluate $\int_S V\,d\mathbf{S}$ over the surface S defined by $x^2 + y^2 = 9$ between $z = 0$ and $z = 2$ in the first octant.

5. Evaluate $\int_S \mathbf{F}\cdot d\mathbf{S}$ over the surface S defined by $x^2 + y^2 + z^2 = 4$ for $z \geq 0$ and bounded by $x = 0$, $y = 0$, $z = 0$ in the first octant where $\mathbf{F} = x\mathbf{i} + 2z\mathbf{j} + y\mathbf{k}$.

6. Determine which of the following vector fields are conservative.
   (a) $\mathbf{F} = (2xy + z)\mathbf{i} + (x^2 + 2yz)\mathbf{j} + (x + y^2)\mathbf{k}$
   (b) $\mathbf{F} = (yz + 2y)\mathbf{i} + (xz + 2x)\mathbf{j} + (xy + 3)\mathbf{k}$
   (c) $\mathbf{F} = (yz^2 + 3)\mathbf{i} + (xz^2 + 2)\mathbf{j} + (2xyz + 4)\mathbf{k}$

Vector integration

7 By the use of the divergence theorem, determine $\int_S \mathbf{F} \cdot d\mathbf{S}$ where
$\mathbf{F} = x\mathbf{i} + xy\mathbf{j} + 2\mathbf{k}$, taken over the region bounded by the planes $z = 0$, $z = 4$, $x = 0$, $y = 0$ and the surface $x^2 + y^2 = 9$ in the first octant.

8 A surface consists of parts of the planes $x = 0$, $x = 2$, $y = 0$, $y = 2$ and $z = 3 - y$ in the region $z \geq 0$. Apply Stokes' theorem to evaluate $\int_S \text{curl } \mathbf{F} \cdot d\mathbf{S}$ over the surface where $\mathbf{F} = 2x\mathbf{i} + xz\mathbf{j} + yz\mathbf{k}$ where S lies in the $z = 0$ plane.

9 Verify Green's theorem in the plane for the integral
$$\oint_c \{(xy^2 - 2x)dx + (x + 2xy^2)dy\}$$
where c is the square with vertices at $(1, 1)$, $(-1, 1)$, $(-1, -1)$ and $(1, -1)$.

🚴 Further problems 9

1 If $V = x^2 yz$, evaluate $\int_c V \, d\mathbf{r}$ between A $(0, 0, 0)$ and B $(6, 2, 4)$

(a) along the straight lines c_1: $(0, 0, 0)$ to $(6, 0, 0)$
c_2: $(6, 0, 0)$ to $(6, 2, 0)$
c_3: $(6, 2, 0)$ to $(6, 2, 4)$

(b) along the path c_4 having parametric equations $x = 3t$, $y = t$, $z = 2t$.

2 If $V = xy^2 + yz$, evaluate to one decimal place $\int_c V \, d\mathbf{r}$ along the curve c having parametric equations $x = 2t^2$, $y = 4t$, $z = 3t + 5$ between A $(0, 0, 5)$ and B $(8, 8, 11)$.

3 Evaluate to one decimal place the integral $\int_c (xyz + 4x^2 y) \, d\mathbf{r}$ along the curve c with parametric equations $x = 2u$, $y = u^2$, $z = 3u^3$ between A $(2, 1, 3)$ and B $(4, 4, 24)$.

4 If $\mathbf{F} = xy\mathbf{i} + yz\mathbf{j} + 3xyz\mathbf{k}$, evaluate $\int_c \mathbf{F} \cdot d\mathbf{r}$ between A $(0, 2, 0)$ and B $(3, 6, 1)$ where c has the parametric equations $x = 3u$, $y = 4u + 2$, $z = u^2$.

5 $\mathbf{F} = x^2 \mathbf{i} - 2xy\mathbf{j} + yz\mathbf{k}$. Evaluate $\int_c \mathbf{F} \cdot d\mathbf{r}$ between A $(2, 1, 2)$ and B $(4, 4, 5)$ where c is the path with parametric equations $x = 2u$, $y = u^2$, $z = 3u - 1$.

6 A unit particle is moved in an anticlockwise manner round a circle with center $(0, 0, 4)$ and radius 2 in the plane $z = 4$ in a force field defined as $\mathbf{F} = (xy + z)\mathbf{i} + (2x + y)\mathbf{j} + (x + y + z)\mathbf{k}$. Find the work done.

7 Evaluate $\int_V \mathbf{F}\,dV$ where $\mathbf{F} = \mathbf{i} - y\mathbf{j} + \mathbf{k}$ and V is the region bounded by the plane $z = 0$ and the hemisphere $x^2 + y^2 + z^2 = 4$, for $z \geq 0$.

8 V is the region bounded by the planes $x = 0$, $y = 0$, $z = 0$ and the surfaces $y = 4 - x^2$ ($z \geq 0$) and $y = 4 - z^2$ ($y \geq 0$).

If $\mathbf{F} = 2\mathbf{i} + y^2\mathbf{j} - \mathbf{k}$, evaluate $\int_V \mathbf{F}\,dV$ throughout the region.

9 If $\mathbf{F} = 3\mathbf{i} + 2\mathbf{j} - 2x\mathbf{k}$, evaluate $\int_V \mathbf{F}\,dV$ where V is the region bounded by the planes $y = 0$, $z = 0$, $z = 4 - y$ ($z \geq 0$) and the surface $x^2 + y^2 = 16$.

10 A scalar field $V = x + y$ exists over a surface S defined by
$$x^2 + y^2 + z^2 = 9,$$
bounded by the planes $x = 0$, $y = 0$, $z = 0$ in the first octant. Evaluate $\int_S V\,d\mathbf{S}$ over the curved surface.

11 A surface S is defined by $y^2 + z = 4$ and is bounded by the planes $x = 0$, $x = 3$, $y = 0$, $z = 0$ in the first octant. Evaluate $\int_S V\,d\mathbf{S}$ over this curved surface where V denotes the scalar field $V = x^2yz$.

12 Evaluate $\int_S \text{curl } \mathbf{F} \cdot d\mathbf{S}$ over the surface S defined by $2x + 2y + z = 2$ and bounded by $x = 0$, $y = 0$, $z = 0$ in the first octant and where
$$\mathbf{F} = y^2\mathbf{i} + 2yz\mathbf{j} + xy\mathbf{k}.$$

13 Evaluate $\int_S \mathbf{F} \cdot d\mathbf{S}$ over the hemisphere defined by $x^2 + y^2 + z^2 = 25$ with $z \geq 0$, where $\mathbf{F} = (x + y)\mathbf{i} - 2z\mathbf{j} + y\mathbf{k}$.

14 A vector field $\mathbf{F} = 2x\mathbf{i} + z\mathbf{j} + y\mathbf{k}$ exists over a surface S defined by $x^2 + y^2 + z^2 = 16$, bounded by the planes $z = 0$, $z = 3$, $x = 0$, $y = 0$. Evaluate $\int_S \mathbf{F} \cdot d\mathbf{S}$ over the stated curved surface.

15 Evaluate $\int_S \mathbf{F} \cdot d\mathbf{S}$, where $\mathbf{F}$ is the vector field $x^2\mathbf{i} + 2z\mathbf{j} - y\mathbf{k}$, over the curved surface S defined by $x^2 + y^2 = 25$ and bounded by $z = 0$, $z = 6$, $y \geq 3$.

16 A region V is defined by the quartersphere $x^2 + y^2 + z^2 = 16$, $z \geq 0$, $y \geq 0$ and the planes $z = 0$, $y = 0$. A vector field $\mathbf{F} = xy\mathbf{i} + y^2\mathbf{j} + \mathbf{k}$ exists throughout and on the boundary of the region. Verify the Gauss divergence theorem for the region stated.

17 A surface consists of parts of the planes $x = 0$, $x = 1$, $y = 0$, $y = 2$, $z = 1$ in the first octant. If $\mathbf{F} = y\mathbf{i} + x^2z\mathbf{j} + xy\mathbf{k}$, verify Stokes' theorem.

18 S is the surface $z = x^2 + y^2$ bounded by the planes $z = 0$ and $z = 4$. Verify Stokes' theorem for a vector field $\mathbf{F} = xy\mathbf{i} + x^3\mathbf{j} + xz\mathbf{k}$.

19 A vector field $\mathbf{F} = xy\mathbf{i} + z^2\mathbf{j} + xyz\mathbf{k}$ exists over the surfaces
$x^2 + y^2 + z^2 = a^2$, $x = 0$ and $y = 0$
in the first octant. Verify Stokes' theorem that $\int_S \text{curl } \mathbf{F} \cdot d\mathbf{S} = \oint_c \mathbf{F} \cdot d\mathbf{r}$.

20 A surface is defined by $z^2 = 4(x^2 + y^2)$ where $0 \leq z \leq 6$. If a vector field $\mathbf{F} = z\mathbf{i} + xy^2\mathbf{j} + x^2 z\mathbf{k}$ exists over the surface and on the boundary circle c, show that $\oint_c \mathbf{F} \cdot d\mathbf{r} = \int_S \text{curl } \mathbf{F} \cdot d\mathbf{S}$.

21 Verify Green's theorem in the plane for the integral
$$\oint_c \{(x-y)\,dx - (y^2 + xy)\,dy\}$$
where c is the circle with unit radius, centered on the origin.

Program 10

Curvilinear coordinates

Frames
1 to 40

Learning outcomes

When you have completed this Program you will be able to:

- Derive the family of curves of constant coordinates for curvilinear coordinates
- Derive unit base vectors and scale factors in orthogonal curvilinear coordinates
- Obtain the element of arc ds and the element of volume dV in orthogonal curvilinear coordinates
- Obtain expressions for the operators grad, div and curl in orthogonal curvilinear coordinates

Curvilinear coordinates

This short Program is an extension of the two previous ones and may not be required for all courses. It can well be bypassed without adversely affecting the rest of the work.

Curvilinear coordinates

Let us consider two variables u and v, each of which is a function of x and y

i.e. $u = f(x, y)$
$v = g(x, y)$

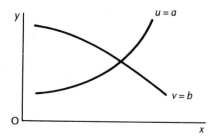

If u and v are each assigned a constant value a and b, the equations will, in general, define two intersecting curves.

If u and v are each given several such values, the equations define a network of curves covering the x–y plane.

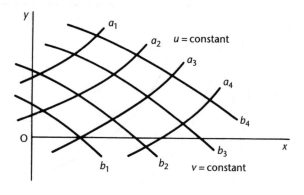

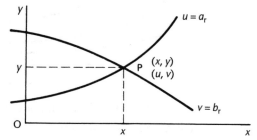

A pair of curves $u = a_r$ and $v = b_r$ pass through each point in the plane. Hence, any point in the plane can be expressed in *rectangular coordinates* (x, y) or in *curvilinear coordinates* (u, v).

Let us see how this works out in an example, so move on

Example 1

Let us consider the case where $u = xy$ and $v = x^2 - y$.

(a) With $u = xy$, if we put $u = 4$, then $y = \dfrac{4}{x}$ and we can plot y against x to obtain the relevant curve.

Similarly, putting $u = 8, 16, 32, \ldots$ we can build up a family of curves, all of the pattern $u = xy$.

x		0.5	1.0	2.0	3.0	4.0
y	$u = 4$	8	4	2	1.33	1.0
	$u = 8$	16	8	4	2.67	2
	$u = 16$	32	16	8	5.33	4
	$u = 32$	64	32	16	10.67	8

If we plot these on graph paper between $x = 0$ and $x = 4$ with a range of y from $y = 0$ to $y = 20$, we obtain

....

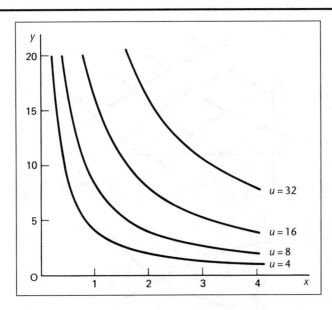

Note that each graph is labelled with its individual u-value.

(b) With $v = x^2 - y$, we proceed in just the same way. We rewrite the equation as $y = x^2 - v$; assign values such as $8, 4, 0, -4, -8, -12, -16, \ldots$ to v; and draw the relevant curve in each case. If we do that for $x = 0$ to $x = 4$ and limit the y-values to the range $y = 0$ to $y = 20$, we obtain the family of curves

....

Curvilinear coordinates

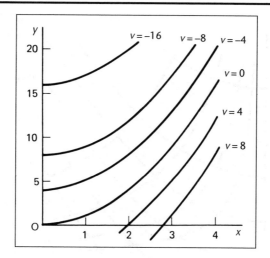

The table of function values is as follows.

x			0	1	2	3	4
y	v =	8	−8	−7	−4	1	8
	v =	4	−4	−3	0	5	12
	v =	0	0	1	4	9	16
	v = −4		4	5	8	13	20
	v = −8		8	9	12	17	24
	v = −12		12	13	16	21	28
	v = −16		16	17	20	25	32

Note again that we label each graph with its own v-value.

This again is a family of curves with the common pattern $v = x^2 - y$, the members being distinguished from each other by the value assigned to v in each case.

Now we draw both sets of curves on a common set of x–y axes, taking

the range of x from $x = 0$ to $x = 4$

and the range of y from $y = 0$ to $y = 20$.

It is worthwhile taking a little time over it – and good practice!

When you have the complete picture, move on to the next frame

5

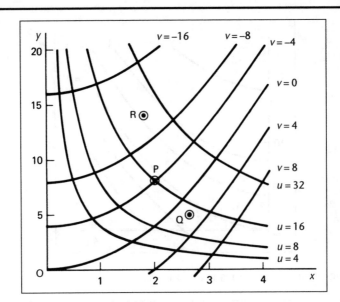

The position of any point in the plane can now be stated in two ways. For example, the point P has Cartesian rectangular coordinates $x = 2$, $y = 8$. It can also be stated in curvilinear coordinates $u = 16$, $v = -4$, for it is at the point of intersection of the two curves corresponding to $u = 16$ and $v = -4$.

Likewise, for the point Q, the position in rectangular coordinates is $x = 2.65$, $y = 5.0$ and for its position in curvilinear coordinates we must estimate it within the network. Approximate values are $u = 13$, $v = 2$.

Similarly, the curvilinear coordinates of R ($x = 1.8$, $y = 14$) are approximately

$$u = \ldots\ldots\ldots\ldots ; \quad v = \ldots\ldots\ldots\ldots$$

6

$$u = 26; \quad v = -11$$

Their actual values are in fact $u = 25.2$ and $v = -10.76$.

Now let us deal with another example.

7

Example 2

If $u = x^2 + 2y$ and $v = y - (x+1)^2$, these can be rewritten as $y = \frac{1}{2}(u - x^2)$ and $y = v + (x+1)^2$. We can now plot the family of curves, say between $x = 0$ and $x = 4$, with $u = 5(5)30$ and $v = -20(5)5$, i.e. values of u from 5 to 30 at intervals of 5 units and values of v from -20 to 5 at intervals of 5 units.

The resulting network is easily obtained and appears as

$$\ldots\ldots\ldots\ldots$$

Curvilinear coordinates

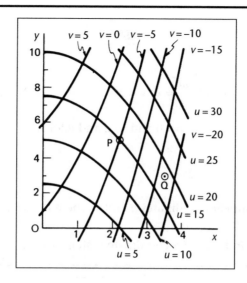

For P, the rectangular coordinates are $(x = 2.18, y = 5.1)$ and the curvilinear coordinates are $(u = 15, v = -5)$.
For Q, the rectangular coordinates are ....
and the curvilinear coordinates are ....

Q: $(x = 3.5, y = 3.0)$; $(u = 18.5, v = -17)$

Orthogonal curvilinear coordinates

If the coordinate curves for u and v forming the network cross at right angles, the system of coordinates is said to be *orthogonal*. The test for orthogonality is given by the dot product of the vectors formed from the partial derivatives. This is, if

$$\frac{\partial u}{\partial x}\frac{\partial v}{\partial x} + \frac{\partial u}{\partial y}\frac{\partial v}{\partial y} = 0 \text{ then } u \text{ and } v \text{ are orthogonal.}$$

Example 3

Given the curvilinear coordinates u and v where $u = xy$ and $v = x^2 - y^2$ then u and v form a coordinate system that is ....

10

> orthogonal

Because

$u = xy$ so $\dfrac{\partial u}{\partial x} = y$ and $\dfrac{\partial u}{\partial y} = x$, $v = x^2 - y^2$ so $\dfrac{\partial v}{\partial x} = 2x$ and $\dfrac{\partial v}{\partial y} = -2y$. Then

$\dfrac{\partial u}{\partial x}\dfrac{\partial v}{\partial x} + \dfrac{\partial u}{\partial y}\dfrac{\partial v}{\partial y} = 2xy - 2xy = 0$ and so u and v form a coordinate system that is orthogonal.

Example 4

Given the curvilinear coordinates u and v where $u = x^2 + 2y$ and $v = y - (x+1)^2$ then

u and v form a coordinate system that is ....

11

> not orthogonal

Because

$u = x^2 + 2y$ so $\dfrac{\partial u}{\partial x} = 2x$ and $\dfrac{\partial u}{\partial y} = 2$, $v = y - (x+1)^2$ so $\dfrac{\partial v}{\partial x} = -2(x+1)$ and $\dfrac{\partial v}{\partial y} = 1$.

Then

$\dfrac{\partial u}{\partial x}\dfrac{\partial v}{\partial x} + \dfrac{\partial u}{\partial y}\dfrac{\partial v}{\partial y} = -4x(x+1) + 2 \neq 0$ and so u and v form a coordinate system that is not orthogonal.

Let us extend these ideas to three dimensions. Move on

Orthogonal coordinate systems in space

12

Any vector **F** can be expressed in terms of its components in three mutually perpendicular directions, which have normally been the directions of the coordinate axes, i.e.

$\mathbf{F} = F_x\mathbf{i} + F_y\mathbf{j} + F_z\mathbf{k}$

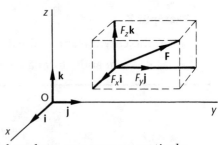

where **i**, **j**, **k** are the unit vectors parallel to the x, y, z axes respectively.

Curvilinear coordinates

Situations can arise, however, where the directions of the unit vectors do not remain fixed, but vary from point to point in space according to prescribed conditions. Examples of this occur in cylindrical and spherical polar coordinates, with which we are already familiar.

1 *Cylindrical polar coordinates* (ρ, ϕ, z)
Let P be a point with cylindrical coordinates (ρ, ϕ, z) as shown. The position of P is a function of the three variables ρ, ϕ, z

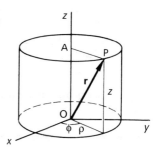

(a) If ϕ and z remain constant and ρ varies, then P will move out along AP by an amount $\dfrac{\partial \mathbf{r}}{\partial \rho}$ and the unit vector $\mathbf{I}$ in this direction will be given by

$$\mathbf{I} = \frac{\partial \mathbf{r}}{\partial \rho} \bigg/ \left|\frac{\partial \mathbf{r}}{\partial \rho}\right|$$

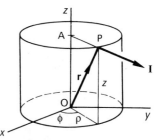

(b) If, instead, ρ and z remain constant and ϕ varies, P will move

....

round the circle with AP as radius

13

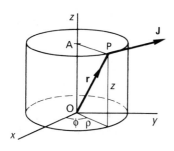

$\dfrac{\partial \mathbf{r}}{\partial \phi}$ is therefore a vector along the tangent to the circle at P and the unit vector $\mathbf{J}$ at P will be given by

$$\mathbf{J} = \frac{\partial \mathbf{r}}{\partial \phi} \bigg/ \left|\frac{\partial \mathbf{r}}{\partial \phi}\right|$$

▶

322 Vector Analysis

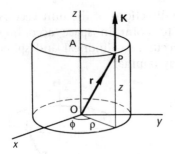

(c) Finally, if ρ and ϕ remain constant and z increases, the vector $\dfrac{\partial \mathbf{r}}{\partial z}$ will be parallel to the z-axis and the unit vector **K** in this direction will be given by

$$\mathbf{K} = \frac{\partial \mathbf{r}}{\partial z} \bigg/ \left|\frac{\partial \mathbf{r}}{\partial z}\right|$$

Putting our three unit vectors on to one diagram, we have

....

14

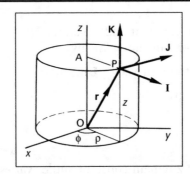

Note that **I**, **J**, **K** are mutually perpendicular and form a right-handed set. But note also that, unlike the unit vectors **i**, **j**, **k** in the Cartesian system, the unit vectors **I**, **J**, **K**, or *base vectors* as they are called, are not fixed in directions, but change as the position of P changes.

So we have, for cylindrical polar coordinates

$$\mathbf{I} = \frac{\partial \mathbf{r}}{\partial \rho} \bigg/ \left|\frac{\partial \mathbf{r}}{\partial \rho}\right|$$

$$\mathbf{J} = \frac{\partial \mathbf{r}}{\partial \phi} \bigg/ \left|\frac{\partial \mathbf{r}}{\partial \phi}\right|$$

$$\mathbf{K} = \frac{\partial \mathbf{r}}{\partial z} \bigg/ \left|\frac{\partial \mathbf{r}}{\partial z}\right|$$

If **F** is a vector associated with P, then $\mathbf{F(r)} = F_\rho \mathbf{I} + F_\phi \mathbf{J} + F_z \mathbf{K}$ where F_ρ, F_ϕ, F_z are the components of **F** in the directions of the unit base vectors **I**, **J**, **K**.

Now let us attend to spherical coordinates in the same way.

2 Spherical polar coordinates (r, θ, ϕ)

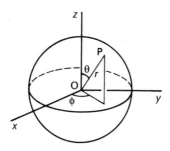

P is a function of the three variables r, θ, ϕ.

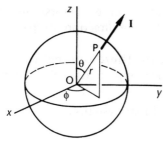

(a) If θ and ϕ remain constant and r increases, P moves outwards in the direction OP. $\dfrac{\partial \mathbf{r}}{\partial r}$ is thus a vector normal to the surface of the sphere at P and the unit vector $\mathbf{I}$ in that direction is therefore

$$\mathbf{I} = \frac{\partial \mathbf{r}}{\partial r} \bigg/ \left|\frac{\partial \mathbf{r}}{\partial r}\right|$$

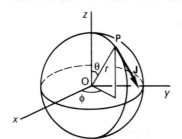

(b) If r and ϕ remain constant and θ increases, P will move along the 'meridian' through P, i.e. $\dfrac{\partial \mathbf{r}}{\partial \theta}$ is a tangent vector to this circle at P and the unit vector $\mathbf{J}$ is given by

$$\mathbf{J} = \frac{\partial \mathbf{r}}{\partial \theta} \bigg/ \left|\frac{\partial \mathbf{r}}{\partial \theta}\right|$$

(c) If r and θ remain constant and ϕ increases, P will move ....

---

along the circle through P perpendicular to the z-axis

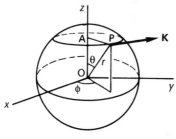

$\dfrac{\partial \mathbf{r}}{\partial \phi}$ is therefore a tangent vector at P and the unit vector $\mathbf{K}$ in this direction is given by

$$\mathbf{K} = \frac{\partial \mathbf{r}}{\partial \phi} \bigg/ \left|\frac{\partial \mathbf{r}}{\partial \phi}\right|$$

So, putting the three results on one diagram, we have ....

17

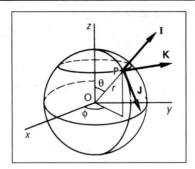

Once again, the three unit vectors at P (base vectors) are mutually perpendicular and form a right-handed set. Their directions in space, however, change as the position of P changes.

A vector **F** associated with P can therefore be expressed as $\mathbf{F} = F_r\mathbf{I} + F_\theta\mathbf{J} + F_\phi\mathbf{K}$ where F_r, F_θ, F_ϕ are the components of **F** in the directions of the base vectors **I**, **J**, **K**.

Both cylindrical and spherical polar coordinate systems are

....

18

orthogonal

Scale factors

Collecting the recent results together, we have:

1 For cylindrical polar coordinates, the unit base vectors are

$$\mathbf{I} = \frac{\partial \mathbf{r}}{\partial \rho} \bigg/ \left|\frac{\partial \mathbf{r}}{\partial \rho}\right| = \frac{1}{h_\rho}\frac{\partial \mathbf{r}}{\partial \rho} \qquad \text{where } h_\rho = \left|\frac{\partial \mathbf{r}}{\partial \rho}\right|$$

$$\mathbf{J} = \frac{\partial \mathbf{r}}{\partial \phi} \bigg/ \left|\frac{\partial \mathbf{r}}{\partial \phi}\right| = \frac{1}{h_\phi}\frac{\partial \mathbf{r}}{\partial \phi} \qquad \text{where } h_\phi = \left|\frac{\partial \mathbf{r}}{\partial \phi}\right|$$

$$\mathbf{K} = \frac{\partial \mathbf{r}}{\partial z} \bigg/ \left|\frac{\partial \mathbf{r}}{\partial z}\right| = \frac{1}{h_z}\frac{\partial \mathbf{r}}{\partial z} \qquad \text{where } h_z = \left|\frac{\partial \mathbf{r}}{\partial z}\right|$$

2 For spherical polar coordinates, the unit base vectors are

$$\mathbf{I} = \frac{\partial \mathbf{r}}{\partial r} \bigg/ \left|\frac{\partial \mathbf{r}}{\partial r}\right| = \frac{1}{h_r}\frac{\partial \mathbf{r}}{\partial r} \qquad \text{where } h_r = \left|\frac{\partial \mathbf{r}}{\partial r}\right|$$

$$\mathbf{J} = \frac{\partial \mathbf{r}}{\partial \theta} \bigg/ \left|\frac{\partial \mathbf{r}}{\partial \theta}\right| = \frac{1}{h_\theta}\frac{\partial \mathbf{r}}{\partial \theta} \qquad \text{where } h_\theta = \left|\frac{\partial \mathbf{r}}{\partial \theta}\right|$$

$$\mathbf{K} = \frac{\partial \mathbf{r}}{\partial \phi} \bigg/ \left|\frac{\partial \mathbf{r}}{\partial \phi}\right| = \frac{1}{h_\phi}\frac{\partial \mathbf{r}}{\partial \phi} \qquad \text{where } h_\phi = \left|\frac{\partial \mathbf{r}}{\partial \phi}\right|$$

In each case, h is called the *scale factor*.

Move on

Scale factors for coordinate systems

1 *Rectangular coordinates* (x, y, z)
With rectangular coordinates, $h_x = h_y = h_z = 1$.

2 *Cylindrical coordinates* (ρ, ϕ, z)

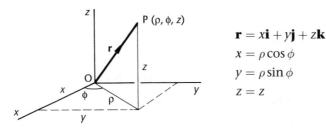

$$\mathbf{r} = x\mathbf{i} + y\mathbf{j} + z\mathbf{k}$$
$$x = \rho\cos\phi$$
$$y = \rho\sin\phi$$
$$z = z$$

$$\therefore \mathbf{r} = \rho\cos\phi\,\mathbf{i} + \rho\sin\phi\,\mathbf{j} + z\mathbf{k}$$

$$\mathbf{I} = \frac{\partial \mathbf{r}}{\partial \rho} \Big/ \left|\frac{\partial \mathbf{r}}{\partial \rho}\right| = \frac{1}{h_\rho}\frac{\partial \mathbf{r}}{\partial \rho} \qquad h_\rho = \left|\frac{\partial \mathbf{r}}{\partial \rho}\right| = |\cos\phi\,\mathbf{i} + \sin\phi\,\mathbf{j}|$$
$$= (\cos^2\phi + \sin^2\phi)^{1/2} = 1$$
$$\therefore h_\rho = 1$$

$$\mathbf{J} = \frac{\partial \mathbf{r}}{\partial \phi} \Big/ \left|\frac{\partial \mathbf{r}}{\partial \phi}\right| = \frac{1}{h_\phi}\frac{\partial \mathbf{r}}{\partial \phi} \qquad h_\phi = \left|\frac{\partial \mathbf{r}}{\partial \phi}\right| = |-\rho\sin\phi\,\mathbf{i} + \rho\cos\phi\,\mathbf{j}|$$
$$= (\rho^2\sin^2\phi + \rho^2\cos^2\phi)^{1/2} = \rho$$
$$\therefore h_\phi = \rho$$

$$\mathbf{K} = \frac{\partial \mathbf{r}}{\partial z} \Big/ \left|\frac{\partial \mathbf{r}}{\partial z}\right| = \frac{1}{h_z}\frac{\partial \mathbf{r}}{\partial z} \qquad h_z = \left|\frac{\partial \mathbf{r}}{\partial z}\right| = |\mathbf{k}| = 1$$
$$\therefore h_z = 1$$

$$\therefore h_\rho = 1;\ h_\phi = \rho;\ h_z = 1$$

3 *Spherical coordinates* (r, θ, ϕ)

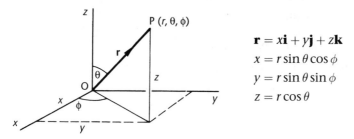

$$\mathbf{r} = x\mathbf{i} + y\mathbf{j} + z\mathbf{k}$$
$$x = r\sin\theta\cos\phi$$
$$y = r\sin\theta\sin\phi$$
$$z = r\cos\theta$$

$$\therefore \mathbf{r} = r\sin\theta\cos\phi\,\mathbf{i} + r\sin\theta\sin\phi\,\mathbf{j} + r\cos\theta\,\mathbf{k}$$

Then working as before

$$h_r = \ldots\ldots\ldots\ldots;\quad h_\theta = \ldots\ldots\ldots\ldots;\quad h_\phi = \ldots\ldots\ldots\ldots$$

20

$$h_r = 1; \quad h_\theta = r; \quad h_\phi = r\sin\theta$$

Because
$$\mathbf{r} = r\sin\theta\cos\phi\,\mathbf{i} + r\sin\theta\sin\phi\,\mathbf{j} + r\cos\theta\,\mathbf{k}$$

$$\mathbf{I} = \frac{\partial \mathbf{r}}{\partial r} \bigg/ \left|\frac{\partial \mathbf{r}}{\partial r}\right| = \frac{1}{h_r}\frac{\partial \mathbf{r}}{\partial r}$$

$$h_r = \left|\frac{\partial \mathbf{r}}{\partial r}\right| = |\sin\theta\cos\phi\,\mathbf{i} + \sin\theta\sin\phi\,\mathbf{j} + \cos\theta\,\mathbf{k}|$$
$$= (\sin^2\theta\cos^2\phi + \sin^2\theta\,\sin^2\phi + \cos^2\theta)^{1/2}$$
$$= (\sin^2\theta + \cos^2\theta)^{1/2} = 1$$
$$\therefore h_r = 1$$

$$\mathbf{J} = \frac{\partial \mathbf{r}}{\partial \theta} \bigg/ \left|\frac{\partial \mathbf{r}}{\partial \theta}\right| = \frac{1}{h_\theta}\frac{\partial \mathbf{r}}{\partial \theta}$$

$$h_\theta = \left|\frac{\partial \mathbf{r}}{\partial \theta}\right| = |r\cos\theta\cos\phi\,\mathbf{i} + r\cos\theta\sin\phi\,\mathbf{j} - r\sin\theta\,\mathbf{k}|$$
$$= (r^2\cos^2\theta\cos^2\phi + r^2\cos^2\theta\sin^2\phi + r^2\sin^2\theta)^{1/2}$$
$$= (r^2\cos^2\theta + r^2\sin^2\theta)^{1/2} = r$$
$$\therefore h_\theta = r$$

$$\mathbf{K} = \frac{\partial \mathbf{r}}{\partial \phi} \bigg/ \left|\frac{\partial \mathbf{r}}{\partial \phi}\right| = \frac{1}{h_\phi}\frac{\partial \mathbf{r}}{\partial \phi}$$

$$h_\phi = \left|\frac{\partial \mathbf{r}}{\partial \phi}\right| = |-r\sin\theta\sin\phi\,\mathbf{i} + r\sin\theta\cos\phi\,\mathbf{j}|$$
$$= (r^2\sin^2\theta\sin^2\phi + r^2\sin^2\theta\cos^2\phi)^{1/2}$$
$$= (r^2\sin^2\theta)^{1/2} = r\sin\theta$$
$$\therefore h_\phi = r\sin\theta$$

$$\therefore h_r = 1; \quad h_\theta = r; \quad h_\phi = r\sin\theta$$

So: (a) for cylindrical coordinates

$$\mathbf{I} = \frac{\partial \mathbf{r}}{\partial \rho}; \quad \mathbf{J} = \frac{1}{\rho}\frac{\partial \mathbf{r}}{\partial \phi}; \quad \mathbf{K} = \frac{\partial \mathbf{r}}{\partial z}$$

(b) for spherical coordinates

$$\mathbf{I} = \frac{\partial \mathbf{r}}{\partial r}; \quad \mathbf{J} = \frac{1}{r}\frac{\partial \mathbf{r}}{\partial \theta}; \quad \mathbf{K} = \frac{1}{r\sin\theta}\frac{\partial \mathbf{r}}{\partial \phi}$$

General curvilinear coordinate system (u, v, w)

Any system of coordinates can be treated in like manner to obtain expressions for the appropriate unit vectors **I**, **J**, **K**.

$$\mathbf{I} = \frac{\partial \mathbf{r}}{\partial u} \bigg/ \left|\frac{\partial \mathbf{r}}{\partial u}\right|; \quad \mathbf{J} = \frac{\partial \mathbf{r}}{\partial v} \bigg/ \left|\frac{\partial \mathbf{r}}{\partial v}\right|; \quad \mathbf{K} = \frac{\partial \mathbf{r}}{\partial w} \bigg/ \left|\frac{\partial \mathbf{r}}{\partial w}\right|$$

These unit vectors are not always at right angles to each other.

If they are mutually perpendicular, the coordinate system is ....

---

orthogonal

Unit vectors **I**, **J**, **K** are orthogonal if
$$\mathbf{I} \cdot \mathbf{J} = \mathbf{J} \cdot \mathbf{K} = \mathbf{K} \cdot \mathbf{I} = 0$$

Exercise

Determine the unit base vectors in the directions of the following vectors and determine whether the vectors are orthogonal.

1. $\mathbf{i} - 2\mathbf{j} + 4\mathbf{k}$
   $2\mathbf{i} + 3\mathbf{j} + \mathbf{k}$
   $-2\mathbf{i} + \mathbf{j} + \mathbf{k}$

2. $2\mathbf{i} - 3\mathbf{j} + 2\mathbf{k}$
   $\mathbf{i} + 2\mathbf{j} + 2\mathbf{k}$
   $-10\mathbf{i} - 2\mathbf{j} + 7\mathbf{k}$

3. $4\mathbf{i} + 2\mathbf{j} - \mathbf{k}$
   $3\mathbf{i} - 5\mathbf{j} + 2\mathbf{k}$
   $\mathbf{i} + 2\mathbf{j} + 6\mathbf{k}$

4. $3\mathbf{i} + 2\mathbf{j} + \mathbf{k}$
   $\mathbf{i} - 3\mathbf{j} + 3\mathbf{k}$
   $6\mathbf{i} + \mathbf{j} - \mathbf{k}$

The results are as follows:

1. $\mathbf{I} = \dfrac{1}{\sqrt{21}}(\mathbf{i} - 2\mathbf{j} + 4\mathbf{k}); \quad \mathbf{J} = \dfrac{1}{\sqrt{14}}(2\mathbf{i} + 3\mathbf{j} + \mathbf{k});$

   $\mathbf{K} = \dfrac{1}{\sqrt{6}}(-2\mathbf{i} + \mathbf{j} + \mathbf{k})$

   $\mathbf{I} \cdot \mathbf{J} = 0; \quad \mathbf{J} \cdot \mathbf{K} = 0; \quad \mathbf{K} \cdot \mathbf{I} = 0 \quad \therefore$ orthogonal

2. $\mathbf{I} = \dfrac{1}{\sqrt{17}}(2\mathbf{i} - 3\mathbf{j} + 2\mathbf{k}); \quad \mathbf{J} = \dfrac{1}{3}(\mathbf{i} + 2\mathbf{j} + 2\mathbf{k});$

   $\mathbf{K} = \dfrac{1}{\sqrt{153}}(-10\mathbf{i} + 2\mathbf{j} + 7\mathbf{k})$

   $\mathbf{I} \cdot \mathbf{J} = 0; \quad \mathbf{J} \cdot \mathbf{K} = 0; \quad \mathbf{K} \cdot \mathbf{I} = 0 \quad \therefore$ orthogonal

3    $I = \dfrac{1}{\sqrt{21}}(4\mathbf{i} + 2\mathbf{j} - \mathbf{k}); \quad J = \dfrac{1}{\sqrt{38}}(3\mathbf{i} - 5\mathbf{j} + 2\mathbf{k});$

$K = \dfrac{1}{\sqrt{41}}(\mathbf{i} + 2\mathbf{j} + 6\mathbf{k})$

$I \cdot J = 0; \quad J \cdot K \neq 0 \quad\quad \therefore$ not orthogonal

4    $I = \dfrac{1}{\sqrt{14}}(3\mathbf{i} + 2\mathbf{j} + \mathbf{k}); \quad J = \dfrac{1}{\sqrt{19}}(\mathbf{i} - 3\mathbf{j} + 3\mathbf{k});$

$K = \dfrac{1}{\sqrt{38}}(6\mathbf{i} + \mathbf{j} - \mathbf{k})$

$I \cdot J = 0; \quad J \cdot K = 0; \quad K \cdot I \neq 0 \;\; \therefore$ not orthogonal

Transformation equations

24

In general coordinates, the transformation equations are of the form

$x = f(u, v, w); \quad y = g(u, v, w); \quad z = h(u, v, w)$

where the functions f, g, h are continuous and single-valued, and whose partial derivatives are continuous.

Then

$\mathbf{r} = x\mathbf{i} + y\mathbf{j} + z\mathbf{k} = f(u, v, w)\mathbf{i} + g(u, v, w)\mathbf{j} + h(u, v, w)\mathbf{k}$

and coordinate curves can be formed by keeping two of the three variables constant.

Now $\mathbf{r} = x\mathbf{i} + y\mathbf{j} + z\mathbf{k} \quad \therefore d\mathbf{r} = \dfrac{\partial \mathbf{r}}{\partial u} du + \dfrac{\partial \mathbf{r}}{\partial v} dv + \dfrac{\partial \mathbf{r}}{\partial w} dw$ (1)

$\dfrac{\partial \mathbf{r}}{\partial u}$ is a tangent vector to the u-coordinate curve at P

$\dfrac{\partial \mathbf{r}}{\partial v}$ is a tangent vector to the v-coordinate curve at P

$\dfrac{\partial \mathbf{r}}{\partial w}$ is a tangent vector to the w-coordinate curve at P

$\mathbf{I} = \dfrac{\partial \mathbf{r}}{\partial u} \bigg/ \left|\dfrac{\partial \mathbf{r}}{\partial u}\right| \quad \therefore \dfrac{\partial \mathbf{r}}{\partial u} = h_u \mathbf{I} \;\; \text{where} \;\; h_u = \left|\dfrac{\partial \mathbf{r}}{\partial u}\right|$

$\mathbf{J} = \dfrac{\partial \mathbf{r}}{\partial v} \bigg/ \left|\dfrac{\partial \mathbf{r}}{\partial v}\right| \quad \therefore \dfrac{\partial \mathbf{r}}{\partial v} = h_v \mathbf{J} \;\; \text{where} \;\; h_v = \left|\dfrac{\partial \mathbf{r}}{\partial v}\right|$

$\mathbf{K} = \dfrac{\partial \mathbf{r}}{\partial w} \bigg/ \left|\dfrac{\partial \mathbf{r}}{\partial w}\right| \quad \therefore \dfrac{\partial \mathbf{r}}{\partial w} = h_w \mathbf{K} \;\; \text{where} \;\; h_w = \left|\dfrac{\partial \mathbf{r}}{\partial w}\right|$

Then (1) above becomes

$d\mathbf{r} = h_u du\, \mathbf{I} + h_v dv\, \mathbf{J} + h_w dw\, \mathbf{K}$

where, as before, h_u, h_v, h_w are the scale factors.

Element of arc ds and element of volume dV in orthogonal curvilinear coordinates

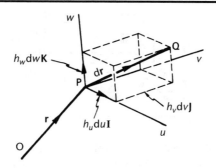

(a) *Element of arc ds*

Element of arc ds from P to Q is given by

$$d\mathbf{r} = h_u \, du \, \mathbf{I} + h_v \, dv \, \mathbf{J} + h_w \, dw \, \mathbf{K}$$
$$\therefore \; d\mathbf{r} \cdot d\mathbf{r} = (h_u du \, \mathbf{I} + h_v dv \, \mathbf{J} + h_w dw \, \mathbf{K}) \cdot (h_u du \, \mathbf{I}$$
$$+ h_v dv \, \mathbf{J} + h_w dw \, \mathbf{K})$$
$$\therefore \; ds^2 = h_u^2 du^2 + h_v^2 dv^2 + h_w^2 dw^2$$
$$\therefore \; ds = (h_u^2 du^2 + h_v^2 dv^2 + h_w^2 dw^2)^{1/2}$$

(b) *Element of volume dV*

$$dV = (h_u du \, \mathbf{I}) \cdot (h_v dv \, \mathbf{J} \times h_w dw \, \mathbf{K})$$
$$= (h_u du \, \mathbf{I}) \cdot (h_v dv \, h_w dw \, \mathbf{I}) = h_u du \, h_v dv \, h_w dw$$
$$\therefore \; dV = h_u \, h_v \, h_w \, du dv dw$$

Note also that

$$dV = \left| \frac{\partial \mathbf{r}}{\partial u} \cdot \left(\frac{\partial \mathbf{r}}{\partial v} \times \frac{\partial \mathbf{r}}{\partial w} \right) \right| du \, dv \, dw$$
$$= \frac{\partial(x, y, z)}{\partial(u, v, w)} du \, dv \, dw$$

where $\dfrac{\partial(x, y, z)}{\partial(u, v, w)}$ is the Jacobian of the transformation.

Grad, div and curl in orthogonal curvilinear coordinates

26

(a) *Grad V* (∇V)

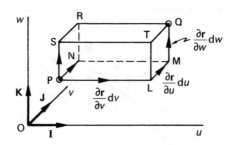

Let a scalar field V exist in space and let dV be the change in V from P to Q. If the position vector of P is $\mathbf{r}$ then that of Q is $\mathbf{r} + d\mathbf{r}$.

Then $dV = \dfrac{\partial V}{\partial u} du + \dfrac{\partial V}{\partial v} dv + \dfrac{\partial V}{\partial w} dw$

Let grad $V = \nabla V = (\nabla V)_u \mathbf{I} + (\nabla V)_v \mathbf{J} + (\nabla V)_w \mathbf{K}$

where $(\nabla V)_{u,v,w}$ are the components of grad V in the u, v, w directions.

Also $d\mathbf{r} = \dfrac{\partial \mathbf{r}}{\partial u} du + \dfrac{\partial \mathbf{r}}{\partial v} dv + \dfrac{\partial \mathbf{r}}{\partial w} dw$

But $\dfrac{\partial \mathbf{r}}{\partial u} = \left|\dfrac{\partial \mathbf{r}}{\partial u}\right| \mathbf{I} = h_u \mathbf{I}; \qquad \dfrac{\partial \mathbf{r}}{\partial v} = \left|\dfrac{\partial \mathbf{r}}{\partial v}\right| \mathbf{J} = h_v \mathbf{J};$

and $\dfrac{\partial \mathbf{r}}{\partial w} = \left|\dfrac{\partial \mathbf{r}}{\partial w}\right| \mathbf{K} = h_w \mathbf{K}.$

$\therefore d\mathbf{r} = h_u\, du\, \mathbf{I} + h_v\, dv\, \mathbf{J} + h_w\, dw\, \mathbf{K}$

We have previously established that $dV = $ grad $V \cdot d\mathbf{r}$

$\therefore dV = \{(\nabla V)_u \mathbf{I} + (\nabla V)_v \mathbf{J} + (\nabla V)_w \mathbf{K}\} \cdot$

$\{h_u du\mathbf{I} + h_v dv\mathbf{J} + h_w dw\mathbf{K}\}$

$= (\nabla V)_u h_u\, du + (\nabla V)_v h_v\, dv + (\nabla V)_w h_w\, dw$

But $dV = \dfrac{\partial V}{\partial u} du + \dfrac{\partial V}{\partial v} dv + \dfrac{\partial V}{\partial w} dw$

Curvilinear coordinates

∴ Equating coefficients, we then have

$$\frac{\partial V}{\partial u} = (\nabla V)_u h_u \quad \therefore (\nabla V)_u = \frac{1}{h_u}\frac{\partial V}{\partial u}$$

$$\frac{\partial V}{\partial v} = (\nabla V)_v h_v \quad \therefore (\nabla V)_v = \frac{1}{h_v}\frac{\partial V}{\partial v}$$

$$\frac{\partial V}{\partial w} = (\nabla V)_w h_w \quad \therefore (\nabla V)_w = \frac{1}{h_w}\frac{\partial V}{\partial w}$$

$$\therefore \text{grad } V = \nabla V = \frac{1}{h_u}\frac{\partial V}{\partial u}\mathbf{I} + \frac{1}{h_v}\frac{\partial V}{\partial v}\mathbf{J} + \frac{1}{h_w}\frac{\partial V}{\partial w}\mathbf{K}$$

i.e. grad operator $\nabla = \dfrac{\mathbf{I}}{h_u}\dfrac{\partial}{\partial u} + \dfrac{\mathbf{J}}{h_v}\dfrac{\partial}{\partial v} + \dfrac{\mathbf{K}}{h_w}\dfrac{\partial}{\partial w}$

Other results we state without proof.

27

(b) *Div* **F** ($\nabla \cdot \mathbf{F}$)

$$\text{div }\mathbf{F} = \nabla \cdot \mathbf{F}$$
$$= \frac{1}{h_u h_v h_w}\left\{\frac{\partial}{\partial u}(h_v h_w F_u) + \frac{\partial}{\partial v}(h_u h_w F_v) + \frac{\partial}{\partial w}(h_u h_v F_w)\right\}$$

Example 1

Show that the curvilinear expression for div **F** agrees with the earlier definition in Cartesian coordinates.

In Cartesian coordinates x, y, z we have $h_x = h_y = h_z = \ldots\ldots\ldots$ so that

$$\text{div }\mathbf{F} = \ldots\ldots\ldots$$

28

$$h_x = h_y = h_z = 1 \text{ so that}$$
$$\text{div }\mathbf{F} = \frac{\partial F_x}{\partial x} + \frac{\partial F_y}{\partial y} + \frac{\partial F_z}{\partial z}$$

(c) *Curl* **F** ($\nabla \times \mathbf{F}$)

$$\text{curl }\mathbf{F} = \nabla \times \mathbf{F} = \frac{1}{h_u h_v h_w}\begin{vmatrix} h_u\mathbf{I} & h_v\mathbf{J} & h_w\mathbf{K} \\ \dfrac{\partial}{\partial u} & \dfrac{\partial}{\partial v} & \dfrac{\partial}{\partial w} \\ h_u F_u & h_v F_v & h_w F_w \end{vmatrix}$$

Example 2

Show that the curvilinear expression for curl **F** agrees with the earlier definition in Cartesian coordinates.

In Cartesian coordinates x, y, z we have $h_x = h_y = h_z = \ldots\ldots\ldots$ and **I, J, K** = ..., ..., ... so that

$$\text{curl }\mathbf{F} = \ldots\ldots\ldots$$

29

$h_x = h_y = h_z = 1$ and $\mathbf{I, J, K} = \mathbf{i, j, k}$ so that

$$\text{curl } \mathbf{F} = \mathbf{i}\left(\frac{\partial F_z}{\partial y} - \frac{\partial F_y}{\partial z}\right) + \mathbf{j}\left(\frac{\partial F_x}{\partial z} - \frac{\partial F_z}{\partial x}\right) + \mathbf{k}\left(\frac{\partial F_y}{\partial x} - \frac{\partial F_x}{\partial y}\right)$$

Because in Cartesians
$h_x = h_y = h_z = 1$ and $\mathbf{I, j, K} = \mathbf{i, j, k}$ so that

$$\nabla \times \mathbf{F} = \frac{1}{h_u h_v h_w} \begin{vmatrix} h_u \mathbf{I} & h_v \mathbf{J} & h_w \mathbf{K} \\ \frac{\partial}{\partial u} & \frac{\partial}{\partial v} & \frac{\partial}{\partial w} \\ h_u F_u & h_v F_v & h_w F_w \end{vmatrix}$$

$$= \begin{vmatrix} \mathbf{i} & \mathbf{j} & \mathbf{k} \\ \frac{\partial}{\partial x} & \frac{\partial}{\partial y} & \frac{\partial}{\partial z} \\ F_x & F_y & F_z \end{vmatrix}$$

$$= \mathbf{i}\left(\frac{\partial F_z}{\partial y} - \frac{\partial F_y}{\partial z}\right) + \mathbf{j}\left(\frac{\partial F_x}{\partial z} - \frac{\partial F_z}{\partial x}\right) + \mathbf{k}\left(\frac{\partial F_y}{\partial x} - \frac{\partial F_x}{\partial y}\right)$$

(d) *Div grad V* $(\nabla^2 V)$

$\text{div grad } V = \nabla \cdot (\nabla V) = \nabla^2 V$

$$= \frac{1}{h_u h_v h_w}\left\{\frac{\partial}{\partial u}\left(\frac{h_v h_w}{h_u} \cdot \frac{\partial V}{\partial u}\right) + \frac{\partial}{\partial v}\left(\frac{h_u h_w}{h_v} \cdot \frac{\partial V}{\partial v}\right) + \frac{\partial}{\partial w}\left(\frac{h_u h_v}{h_w} \cdot \frac{\partial V}{\partial w}\right)\right\}$$

Example 3

Show that the curvilinear expression for $\nabla^2 V$ agrees with the earlier definition in Cartesian coordinates.

In Cartesian coordinates x, y, z we have $h_x = h_y = h_z = \ldots\ldots\ldots$ so that

$$\nabla^2 V = \ldots\ldots\ldots\ldots$$

30

$h_x = h_y = h_z = 1$ so that

$$\nabla^2 V = \frac{\partial^2 V}{\partial x^2} + \frac{\partial^2 V}{\partial y^2} + \frac{\partial^2 V}{\partial z^2}$$

Let's try another example, this time in coordinates other than Cartesians.

Example 4

If $V(u, v, w) = u + v^2 + w^3$ with scale factors $h_u = 2$, $h_v = 1$, $h_w = 1$, find $\nabla^2 V$ at the point (5, 3, 4).

There is very little to it. All we have to do is to determine the various partial derivatives and substitute in the expression above with relevant values.

$$\text{div grad } V = \ldots\ldots\ldots\ldots$$

Curvilinear coordinates

> 26

Because
$$\nabla^2 V = \frac{1}{h_u h_v h_w}\left\{\frac{\partial}{\partial u}\left(\frac{h_v h_w}{h_u}\cdot\frac{\partial V}{\partial u}\right)+\frac{\partial}{\partial v}\left(\frac{h_u h_w}{h_v}\cdot\frac{\partial V}{\partial v}\right)+\frac{\partial}{\partial w}\left(\frac{h_u h_v}{h_w}\cdot\frac{\partial V}{\partial w}\right)\right\}$$

In this case, $V = u + v^2 + w^3$ $\therefore \dfrac{\partial V}{\partial u} = 1$; $\dfrac{\partial V}{\partial v} = 2v$; $\dfrac{\partial V}{\partial w} = 3w^2$

Also $h_u = 2,\ h_v = 1,\ h_w = 1$

$$\therefore \nabla^2 V = \frac{1}{2}\left\{\frac{\partial}{\partial u}\left(\frac{1}{2}\right)+\frac{\partial}{\partial v}(4v)+\frac{\partial}{\partial w}(6w^2)\right\}$$
$$= \tfrac{1}{2}\{0 + 4 + 12w\}$$

$\therefore$ At $w = 4$, $\nabla^2 V = 26$

That is all there is to it. Here is another.

Example 5

If $V = (u^2 + v^2)w^3$ with $h_u = 3,\ h_v = 1,\ h_w = 2$, find div grad V at the point $(2,\ -2,\ 1)$.

$$\nabla^2 V = \dots\dots\dots$$

> $14\tfrac{2}{9}$

Because
$$V = (u^2 + v^2)w^3 \quad\therefore\quad \frac{\partial V}{\partial u} = 2uw^3;\quad \frac{\partial V}{\partial v} = 2vw^3;\quad \frac{\partial V}{\partial w} = 3(u^2+v^2)w^2$$

also $h_u = 3,\ h_v = 1,\ h_w = 2$

$$\therefore \nabla^2 V = \frac{1}{6}\left\{\frac{\partial}{\partial u}\left(\frac{2}{3}\frac{\partial V}{\partial u}\right)+\frac{\partial}{\partial v}\left(6\frac{\partial V}{\partial v}\right)+\frac{\partial}{\partial w}\left(\frac{3}{2}\frac{\partial V}{\partial w}\right)\right\}$$
$$= \frac{1}{6}\left\{\frac{\partial}{\partial u}\left(\frac{4}{3}uw^3\right)+\frac{\partial}{\partial v}(12vw^3)+\frac{\partial}{\partial w}\left(\frac{9}{2}(u^2+v^2)w^2\right)\right\}$$

$\therefore$ at $(2,\ -2,\ 1)$
$$\nabla^2 V = \tfrac{1}{6}\{(\tfrac{4}{3}w^3) + (12w^3) + 9(u^2+v^2)w\}$$
$$= \tfrac{1}{6}\{\tfrac{4}{3} + 12 + 72\} = \frac{256}{18} = 14\tfrac{2}{9}$$

Particular orthogonal systems

We can apply the general results for div, grad and curl to special coordinate systems by inserting the appropriate scale factors – as we shall now see.

▶

33

(a) *Cartesian rectangular coordinate system*

If we replace u, v, w by x, y, z and insert values of $h_x = h_y = h_z = 1$, we obtain expressions for grad, div and curl in rectangular coordinates, so that

$$\text{grad } V = \ldots\ldots\ldots ; \quad \text{div } \mathbf{F} = \ldots\ldots\ldots ; \quad \text{curl } \mathbf{F} = \ldots\ldots\ldots$$

34

$$\text{grad } V = \frac{\partial V}{\partial x}\mathbf{i} + \frac{\partial V}{\partial y}\mathbf{j} + \frac{\partial V}{\partial z}\mathbf{k}$$

$$\text{div } \mathbf{F} = \frac{\partial F_x}{\partial x} + \frac{\partial F_y}{\partial y} + \frac{\partial F_z}{\partial z}$$

$$\text{curl } \mathbf{F} = \begin{vmatrix} \mathbf{i} & \mathbf{j} & \mathbf{k} \\ \frac{\partial}{\partial x} & \frac{\partial}{\partial y} & \frac{\partial}{\partial z} \\ F_x & F_y & F_z \end{vmatrix}$$

$$\nabla^2 V = \frac{\partial^2 V}{\partial x^2} + \frac{\partial^2 V}{\partial y^2} + \frac{\partial^2 V}{\partial z^2}$$

all of which you will surely recognise.

(b) *Cylindrical polar coordinate system*

Here we simply replace u, v, w with ρ, ϕ, z and insert $h_u = h_\rho = 1$, $h_v = h_\phi = \rho$, $h_w = h_z = 1$ giving

$$\text{grad } V = \ldots\ldots\ldots ; \quad \text{div } \mathbf{F} = \ldots\ldots\ldots ;$$
$$\text{curl } \mathbf{F} = \ldots\ldots\ldots$$

35

$$\text{grad } V = \frac{\partial V}{\partial \rho}\mathbf{I} + \frac{1}{\rho}\frac{\partial V}{\partial \phi}\mathbf{J} + \frac{\partial V}{\partial z}\mathbf{K}$$

$$\text{div } \mathbf{F} = \frac{1}{\rho}\left\{\frac{\partial}{\partial \rho}(\rho F_\rho) + \frac{\partial}{\partial \phi}(F_\phi) + \frac{\partial}{\partial z}(\rho F_z)\right\}$$

$$\text{curl } \mathbf{F} = \frac{1}{\rho}\begin{vmatrix} \mathbf{I} & \rho\mathbf{J} & \mathbf{K} \\ \frac{\partial}{\partial \rho} & \frac{\partial}{\partial \phi} & \frac{\partial}{\partial z} \\ F_\rho & \rho F_\phi & F_z \end{vmatrix}$$

$$\nabla^2 V = \frac{\partial^2 V}{\partial \rho^2} + \frac{1}{\rho}\frac{\partial V}{\partial \rho} + \frac{1}{\rho^2}\frac{\partial^2 V}{\partial \phi^2} + \frac{\partial^2 V}{\partial z^2}$$

(c) *Spherical polar coordinate system*

Replacing u, v, w with r, θ, ϕ with $h_r = 1$, $h_\theta = r$, $h_\phi = r\sin\theta$,

$$\text{grad } V = \ldots\ldots\ldots ; \quad \text{div } \mathbf{F} = \ldots\ldots\ldots ;$$
$$\text{curl } \mathbf{F} = \ldots\ldots\ldots$$

Curvilinear coordinates

$$\text{grad } V = \frac{\partial V}{\partial r}\mathbf{I} + \frac{1}{r}\frac{\partial V}{\partial \theta}\mathbf{J} + \frac{1}{r\sin\theta}\frac{\partial V}{\partial \phi}\mathbf{K}$$

$$\text{div } \mathbf{F} = \frac{1}{r^2 \sin\theta}\left\{\frac{\partial}{\partial r}(r^2 \sin\theta\, F_r) + \frac{\partial}{\partial \theta}(r \sin\theta\, F_\theta) + \frac{\partial}{\partial \phi}(rF_\phi)\right\}$$

$$\text{curl } \mathbf{F} = \frac{1}{r^2 \sin\theta}\begin{vmatrix} \mathbf{I} & r\mathbf{J} & r\sin\theta\, \mathbf{K} \\ \dfrac{\partial}{\partial r} & \dfrac{\partial}{\partial \theta} & \dfrac{\partial}{\partial \phi} \\ F_r & rF_\theta & r\sin\theta\, F_\phi \end{vmatrix}$$

$$\nabla^2 V = \frac{\partial^2 V}{\partial r^2} + \frac{2}{r}\frac{\partial V}{\partial r} + \frac{1}{r^2}\frac{\partial^2 V}{\partial \theta^2} + \frac{\cot\theta}{r^2}\frac{\partial V}{\partial \theta} + \frac{1}{r^2 \sin\theta}\frac{\partial^2 V}{\partial \phi^2}$$

The results we have compiled are sometimes written in slightly different forms, but they are, of course, equivalent.

That brings us to the end of this Program which is designed as an introduction to the topic of curvilinear coordinates. It has considerable applications, but these are beyond the scope of this present course of study.

The **Review summary** follows as usual. Make any further notes as necessary: then you can work through the **Can You?** checklist and the **Test exercise** without difficulty. The Program ends with the usual **Further problems**.

Review summary

1 *Curvilinear coordinates in two dimensions*

$$u = f(x,y); \quad v = g(x,y)$$

2 *Orthogonal coordinate system in space*
  (a) *Cartesian rectangular coordinates* (x, y, z)

   $\mathbf{F} = F_x \mathbf{i} + F_y \mathbf{j} + F_z \mathbf{k}$ Scale factors $h_x = h_y = h_z = 1$

  (b) *Cylindrical polar coordinates* (ρ, ϕ, z)

   $\mathbf{r} = \rho \cos\phi\, \mathbf{i} + \rho \sin\phi\, \mathbf{j} + z \mathbf{k}$

   Base unit vectors: Scale factors:

   $\mathbf{I} = \dfrac{\partial \mathbf{r}}{\partial \rho} \Big/ \left|\dfrac{\partial \mathbf{r}}{\partial \rho}\right|$ $h_\rho = \left|\dfrac{\partial \mathbf{r}}{\partial \rho}\right| = 1$

   $\mathbf{J} = \dfrac{\partial \mathbf{r}}{\partial \phi} \Big/ \left|\dfrac{\partial \mathbf{r}}{\partial \phi}\right|$ $h_\phi = \left|\dfrac{\partial \mathbf{r}}{\partial \phi}\right| = \rho$

   $\mathbf{K} = \dfrac{\partial \mathbf{r}}{\partial z} \Big/ \left|\dfrac{\partial \mathbf{r}}{\partial z}\right|$ $h_z = \left|\dfrac{\partial \mathbf{r}}{\partial z}\right| = 1$

   $\mathbf{F} = F_\rho \mathbf{I} + F_\phi \mathbf{J} + F_z \mathbf{K}$

(c) *Spherical polar coordinates* (r, θ, ϕ)

$$\mathbf{r} = r\sin\theta\cos\phi\,\mathbf{i} + r\sin\theta\sin\phi\,\mathbf{j} + r\cos\theta\,\mathbf{k}$$

Base unit vectors: Scale factors:

$$\mathbf{I} = \frac{\partial \mathbf{r}}{\partial r}\bigg/\bigg|\frac{\partial \mathbf{r}}{\partial r}\bigg| \qquad h_r = \bigg|\frac{\partial \mathbf{r}}{\partial r}\bigg| = 1$$

$$\mathbf{J} = \frac{\partial \mathbf{r}}{\partial \theta}\bigg/\bigg|\frac{\partial \mathbf{r}}{\partial \theta}\bigg| \qquad h_\theta = \bigg|\frac{\partial \mathbf{r}}{\partial \theta}\bigg| = r$$

$$\mathbf{K} = \frac{\partial \mathbf{r}}{\partial \phi}\bigg/\bigg|\frac{\partial \mathbf{r}}{\partial \phi}\bigg| \qquad h_\phi = \bigg|\frac{\partial \mathbf{r}}{\partial \phi}\bigg| = r\sin\theta$$

$$\mathbf{F} = F_r\,\mathbf{I} + F_\theta\,\mathbf{J} + F_\phi\,\mathbf{K}$$

3 *General orthogonal curvilinear coordinates* (u, v, w)
$x = f(u, v, w); \quad y = g(u, v, w); \quad w = h(u, v, w)$

$$\mathbf{r} = x\mathbf{i} + y\mathbf{j} + z\mathbf{k}$$

$$\frac{\partial \mathbf{r}}{\partial u} = h_u\mathbf{I} \quad \text{where} \quad h_u = \bigg|\frac{\partial \mathbf{r}}{\partial u}\bigg|$$

$$\frac{\partial \mathbf{r}}{\partial v} = h_v\mathbf{J} \quad \text{where} \quad h_v = \bigg|\frac{\partial \mathbf{r}}{\partial v}\bigg|$$

$$\frac{\partial \mathbf{r}}{\partial w} = h_w\mathbf{K} \quad \text{where} \quad h_w = \bigg|\frac{\partial \mathbf{r}}{\partial w}\bigg|$$

Element of arc: $ds = (h_u^2\,du^2 + h_v^2\,dv^2 + h_w^2\,dw^2)^{1/2}$

Element of volume: $dV = h_u h_v h_w\,du\,dv\,dw$

$$= \frac{\partial(x, y, z)}{\partial(u, v, w)}\,du\,dv\,dw$$

4 *Grad, div and curl in orthogonal curvilinear coordinates*

(a) Grad $V = \nabla V = \dfrac{1}{h_u}\dfrac{\partial V}{\partial u}\mathbf{I} + \dfrac{1}{h_v}\dfrac{\partial V}{\partial v}\mathbf{J} + \dfrac{1}{h_w}\dfrac{\partial V}{\partial w}\mathbf{K}$

$$\text{grad operator} = \nabla = \frac{\mathbf{I}}{h_u}\frac{\partial}{\partial u} + \frac{\mathbf{J}}{h_v}\frac{\partial}{\partial v} + \frac{\mathbf{K}}{h_w}\frac{\partial}{\partial w}$$

(b) Div $\mathbf{F} = \dfrac{1}{h_u h_v h_w}\left\{\dfrac{\partial}{\partial u}(h_v h_w F_u) + \dfrac{\partial}{\partial v}(h_w h_u F_v) + \dfrac{\partial}{\partial w}(h_u h_v F_w)\right\}$

(c) Curl $\mathbf{F} = \dfrac{1}{h_u h_v h_w}\begin{vmatrix} h_u\mathbf{I} & h_v\mathbf{J} & h_w\mathbf{K} \\ \dfrac{\partial}{\partial u} & \dfrac{\partial}{\partial v} & \dfrac{\partial}{\partial w} \\ h_u F_u & h_v F_v & h_w F_w \end{vmatrix}$

(d) Div grad $V = \nabla\cdot\nabla V = \nabla^2 V$

$$= \frac{1}{h_u h_v h_w}\left\{\frac{\partial}{\partial u}\left(\frac{h_v h_w}{h_u}\cdot\frac{\partial V}{\partial u}\right) + \frac{\partial}{\partial v}\left(\frac{h_u h_w}{h_v}\cdot\frac{\partial V}{\partial v}\right) + \frac{\partial}{\partial w}\left(\frac{h_u h_v}{h_w}\cdot\frac{\partial V}{\partial w}\right)\right\}$$

5 Grad, div and curl in cylindrical and spherical coordinates

(a) Cylindrical coordinates (ρ, ϕ, z)

$$\text{grad } V = \frac{\partial V}{\partial \rho}\mathbf{I} + \frac{1}{\rho}\frac{\partial V}{\partial \phi}\mathbf{J} + \frac{\partial V}{\partial z}\mathbf{K}$$

$$\text{div } \mathbf{F} = \frac{1}{\rho}\left\{\frac{\partial(\rho F_\rho)}{\partial \rho}\right\} + \frac{1}{\rho}\left\{\frac{\partial F_\phi}{\partial \phi}\right\} + \frac{\partial F_z}{\partial z}$$

$$\text{curl } \mathbf{F} = \frac{1}{\rho}\begin{vmatrix} \mathbf{I} & \rho\mathbf{J} & \mathbf{K} \\ \frac{\partial}{\partial \rho} & \frac{\partial}{\partial \phi} & \frac{\partial}{\partial z} \\ F_\rho & \rho F_\phi & F_z \end{vmatrix}$$

$$\nabla^2 V = \frac{\partial^2 V}{\partial \rho^2} + \frac{1}{\rho}\frac{\partial V}{\partial \rho} + \frac{1}{\rho^2}\frac{\partial^2 V}{\partial \phi^2} + \frac{\partial^2 V}{\partial z^2}$$

(b) Spherical coordinates (r, θ, ϕ)

$$\text{grad } V = \frac{\partial V}{\partial r}\mathbf{I} + \frac{1}{r}\frac{\partial V}{\partial \theta}\mathbf{J} + \frac{1}{r\sin\theta}\frac{\partial V}{\partial \phi}\mathbf{K}$$

$$\text{div } \mathbf{F} = \frac{1}{r^2}\frac{\partial}{\partial r}(r^2 F_r) + \frac{1}{r\sin\theta}\frac{\partial}{\partial \theta}(\sin\theta\, F_\theta) + \frac{1}{r\sin\theta}\frac{\partial}{\partial \phi}(F_\phi)$$

$$\text{curl } \mathbf{F} = \frac{1}{r^2 \sin\theta}\begin{vmatrix} \mathbf{I} & r\mathbf{J} & r\sin\theta\,\mathbf{K} \\ \frac{\partial}{\partial r} & \frac{\partial}{\partial \theta} & \frac{\partial}{\partial \phi} \\ F_r & rF_\theta & r\sin\theta\, F_\phi \end{vmatrix}$$

$$\nabla^2 V = \frac{\partial^2 V}{\partial r^2} + \frac{2}{r}\frac{\partial V}{\partial r} + \frac{1}{r^2}\frac{\partial^2 V}{\partial \theta^2} + \frac{\cot\theta}{r^2}\frac{\partial V}{\partial \theta} + \frac{1}{r^2 \sin^2\theta}\frac{\partial^2 V}{\partial \phi^2}$$

✅ Can You?

Checklist 10

Check this list before and after you try the end of Program test

On a scale of 1 to 5 how confident are you that you can: Frames

- Derive the family of curves of constant coordinates for curvilinear coordinates?
  Yes ☐ ☐ ☐ ☐ ☐ No

- Derive unit base vectors and scale factors in orthogonal curvilinear coordinates? 12 to 24
  Yes ☐ ☐ ☐ ☐ ☐ No

Vector Analysis

- Obtain the element of arc ds and the element of volume dV in orthogonal curvilinear coordinates?
  Yes ☐ ☐ ☐ ☐ ☐ No

- Obtain expressions for the operators grad, div and curl in orthogonal curvilinear coordinates?
  Yes ☐ ☐ ☐ ☐ ☐ No

Test exercise 10

1. Determine the unit vectors in the directions of the following three vectors and test whether they form an orthogonal set.
   $3\mathbf{i} - 2\mathbf{j} + \mathbf{k}$
   $\mathbf{i} + 2\mathbf{j} + \mathbf{k}$
   $-2\mathbf{i} - \mathbf{j} + 4\mathbf{k}$.

2. If $\mathbf{r} = u\sin 2\theta\,\mathbf{i} + u\cos 2\theta\,\mathbf{j} + v^2\,\mathbf{k}$, determine the scale factors h_u, h_v, h_θ.

3. If P is a point $\mathbf{r} = \rho\cos\phi\,\mathbf{i} + \rho\sin\phi\,\mathbf{j} + z\,\mathbf{k}$ and a scalar field $V = \rho^2 z \sin 2\phi$ exists in space, using cylindrical polar coordinates (ρ, ϕ, z) determine grad V at the point at which $\rho = 1$, $\phi = \pi/4$, $z = 2$.

4. A vector field **F** is given in cylindrical coordinates by
   $\mathbf{F} = \rho\cos\phi\,\mathbf{I} + \rho\sin 2\phi\,\mathbf{J} + z\,\mathbf{K}$
   Determine (a) div **F**; (b) curl **F**.

5. Using spherical coordinates (r, θ, ϕ) determine expressions for (a) an element of arc ds; (b) an element of volume dV.

6. If V is a scalar field such that $V = u^2 v w^3$ and scale factors are $h_u = 1$, $h_v = 2$, $h_w = 4$, determine $\nabla^2 V$ at the point $(2, 3, -1)$.

Further problems 10

1. Determine whether the following sets of three vectors are orthogonal.
   (a) $4\mathbf{i} - 2\mathbf{j} - \mathbf{k}$
   $3\mathbf{i} + 5\mathbf{j} + 2\mathbf{k}$
   $\mathbf{i} - 11\mathbf{j} + 26\mathbf{k}$
   (b) $2\mathbf{i} + 3\mathbf{j} - \mathbf{k}$
   $4\mathbf{i} - 2\mathbf{j} + 2\mathbf{k}$
   $\mathbf{i} + 4\mathbf{j} + 2\mathbf{k}$

2. If $V(u, v, w) = v^3 w^2 \sin 2u$ with scale factors $h_u = 3$, $h_v = 1$, $h_w = 2$, determine div grad V at the point $(\pi/4, -1, 3)$.

3. A scalar field $V = \dfrac{u^2 e^{2w}}{v}$ exists in space. If the relevant scale factors are $h_u = 2$, $h_v = 3$, $h_w = 1$, determine the value of $\nabla^2 V$ at the point $(1, 2, 0)$.

4. If $\mathbf{r} = x\mathbf{i} + y\mathbf{j} + z\mathbf{k}$ and $x = r \sin\theta \cos\phi$, $y = r \sin\theta \sin\phi$, $z = r \cos\theta$ in spherical polar coordinates (r, θ, ϕ), prove that, for any vector field $\mathbf{F}$ where
$$\mathbf{F} = F_x \mathbf{i} + F_y \mathbf{j} + F_z \mathbf{k} = F_r \mathbf{I} + F_\theta \mathbf{J} + F_\phi \mathbf{K}$$
then $F_x = F_r \sin\theta \cos\phi + F_\theta \cos\theta \cos\phi - F_\phi \sin\phi$
$F_y = F_r \sin\theta \sin\phi + F_\theta \cos\theta \sin\phi + F_\phi \cos\phi$
$F_z = F_r \cos\theta - F_\theta \sin\theta$

5. If V is a scalar field, determine an expression for $\nabla^2 V$
   (a) in cylindrical polar coordinates
   (b) in spherical polar coordinates

6. Transformation equations from rectangular coordinates (x, y, z) to parabolic cylindrical coordinates (u, v, w) are
$$x = \dfrac{u^2 - v^2}{2}; \quad y = uv; \quad z = w$$
V is a scalar field and $\mathbf{F}$ a vector field.
   (a) Prove that the (u, v, w) system is orthogonal
   (b) Determine the scale factors
   (c) Find div $\mathbf{F}$
   (d) Obtain an expression for $\nabla^2 V$

Answers

Test exercise 1 (page 22)

1 (a) $\dfrac{\partial z}{\partial x} = 12x^2 - 5y^2 \quad \dfrac{\partial z}{\partial y} = -10xy + 9y^2 \quad \dfrac{\partial^2 z}{\partial x^2} = 24x \quad \dfrac{\partial^2 z}{\partial y^2} = -10x + 18y$

$\dfrac{\partial^2 z}{\partial y \cdot \partial x} = -10y \quad \dfrac{\partial^2 z}{\partial x \cdot \partial y} = -10y$ (b) $\dfrac{\partial z}{\partial x} = -2\sin(2x + 3y) \quad \dfrac{\partial z}{\partial y} = -3\sin(2x + 3y)$

$\dfrac{\partial^2 z}{\partial x^2} = -4\cos(2x + 3y) \quad \dfrac{\partial^2 z}{\partial y^2} = -9\cos(2x + 3y) \quad \dfrac{\partial^2 z}{\partial y \cdot \partial x} = -6\cos(2x + 3y)$

$\dfrac{\partial^2 z}{\partial x \cdot \partial y} = -6\cos(2x + 3y)$ (c) $\dfrac{\partial z}{\partial x} = 2xe^{x^2 - y^2} \quad \dfrac{\partial z}{\partial y} = -2ye^{x^2 - y^2}$

$\dfrac{\partial^2 z}{\partial x^2} = 2e^{x^2 - y^2}(2x^2 + 1) \quad \dfrac{\partial^2 z}{\partial y^2} = 2e^{x^2 - y^2}(2y^2 - 1) \quad \dfrac{\partial^2 z}{\partial y \cdot \partial x} = -4xye^{x^2 - y^2}$

$\dfrac{\partial^2 z}{\partial x \cdot \partial y} = -4xye^{x^2 - y^2}$ (d) $\dfrac{\partial z}{\partial x} = 2x^2 \cos(2x + 3y) + 2x \sin(2x + 3y)$

$\dfrac{\partial^2 z}{\partial x^2} = (2 - 4x^2)\sin(2x + 3y) + 8x\cos(2x + 3y)$

$\dfrac{\partial^2 z}{\partial y \cdot \partial x} = -6x^2 \sin(2x + 3y) + 6x \cos(2x + 3y) \quad \dfrac{\partial z}{\partial y} = 3x^2 \cos(2x + 3y)$

$\dfrac{\partial^2 z}{\partial y^2} = -9x^2 \sin(2x + 3y) \quad \dfrac{\partial^2 z}{\partial x \cdot \partial y} = -6x^2 \sin(2x + 3y) + 6x \cos(2x + 3y)$

2 (a) 2V   3 P decreases 375 W   4 ±2.5%

Further problems 1 (page 23)

10 ±0.67E × 10^{-5} approx.   12 ±(x + y + z)%   13 y decreases by 19% approx.
14 ±4.25%   16 19%   18 $\delta y = y\{\delta x \cdot p \cot(px + a) - \delta t \cdot q \tan(qt + b)\}$

Test exercise 2 (page 53)

1 (a) $\dfrac{4xy - 3x^2}{3y^2 - 2x^2}$ (b) $\dfrac{e^x \cos y - e^y \cos x}{e^x \sin y + e^y \sin x}$ (c) $\dfrac{5\cos x \cos y - 2\sin x \cos x}{5\sin x \sin y + \sec^2 y}$

2 V decreases at 0.419 cm^3/s   3 y decreases at 1.524 cm/s

4 $\dfrac{\partial x}{\partial r} = (4x^3 + 4xy)\cos\theta + (2x^2 + 3y^2)\sin\theta$

$\dfrac{\partial z}{\partial \theta} = r\{(2x^2 + 3y^2)\cos\theta - (4x^3 + 4xy)\sin\theta\}$

6 145.7 ± 2.6 mm   7 5.8 m/s   8 $\dfrac{-2(x + y)}{2x + 3y}; \dfrac{-2}{(2x + 3y)^3}$

9 $\dfrac{x}{2(x^2 - y^2)}; \dfrac{-y}{4(x^2 - y^2)}; \dfrac{-y}{2(x^2 - y^2)}; \dfrac{x}{4(x^2 - y^2)}$

Further problems 2 (page 54)

2 $3x^2 - 3xy$ **3** $\tan\theta = 17/6 = 2.8333$ **9** (a) $\dfrac{1-y}{x+2}$ (b) $\dfrac{8y - 3y^2 + 4xy - 3x^2y^2}{2x^3y - 2x^2 + 6xy - 8x}$

(c) $\dfrac{y}{x}$ **14** $a = -\dfrac{5}{2},\ b = -\dfrac{3}{2}$ **16** $\dfrac{\cos x(5\cos y - 2\sin x)}{5\sin x \sin y + \sec^2 y}$ **17** $\dfrac{y\cos x - \tan y}{x\sec^2 y - \sin x}$

20 (a) $-\left\{\dfrac{2xy + y\cos xy}{x^2 + x\cos xy}\right\}$ (b) $-\left\{\dfrac{xy + \tan xy}{x^2}\right\}$

21 $(8x\cos x - 6y\sin x)/J;\ -(4x^3\cos y + 6x\sin y)/J;$
$J = 4x\cos x \sin y + 2x^2 y \sin x \cos y$ **22** $e^{3y}/2(xe^{3y} + e^{-3y});\ e^{-3y}/2(xe^{3y} + e^{-3y});$
$-1/3(xe^{3y} + e^{-3y});\ x/3(xe^{3y} + e^{-3y})$

24 $(2e^{-x}\sinh 2x \sin 3y + 3ye^{-x}\cosh 2x \cos 3y)/(1 + 3y^2);$
$\{-4ye^x \sinh 2x \sin 3y + 3e^x(1 + y^2)\cosh 2x \cos 3y\}/2(1 + 3y^2)$

Test exercise 3 (page 77)

1 $\dfrac{56}{3\pi^3}$

2 (a) $r = 2\sin\theta$ \qquad (b) $r = 5\cos^2\theta$

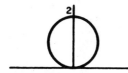

(c) $r = \sin 2\theta$ \qquad (d) $r = 1 + \cos\theta$

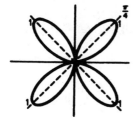
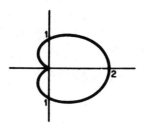

(e) $r = 1 + 3\cos\theta$ \qquad (f) $r = 3 + \cos\theta$

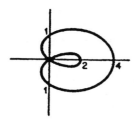
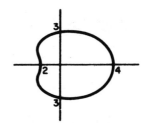

3 $\dfrac{40\pi}{3}$ **4** 8 **5** $\dfrac{32\pi a^2}{5}$

Further problems 3 (page 78)

1 (a) $A = \dfrac{3\pi}{16}$ (b) $V = \dfrac{2\pi}{21}$ **2** 3π **3** $\dfrac{4a^2}{3}$ **4** $\dfrac{13\pi}{8} + 3$ **5** $\dfrac{2}{3}$ **6** $\dfrac{20\pi}{3}$ **7** $\dfrac{8\pi a^3}{3}$

Vector Analysis

9 $\dfrac{3\pi a}{2}$ 10 $\dfrac{5\pi}{2}$ 11 $21.25a$ 12 $\dfrac{3\pi a}{2}$ 14 $\dfrac{a}{b}\{\sqrt{b^2+1}\}(e^{b\theta_1}-1);\ \dfrac{a^2}{4b}(e^{2b\theta_1}-1)$

15 $\pi a^2(2-\sqrt{2})$

Test exercise 4 (page 105)

1 (a) 0 (b) 0   2 (a) -1 (b) 120 (c) $\dfrac{17}{4}$ 3 3.67   4 170.67   5 $\dfrac{11\pi}{4}+6$

6 36

Further problems 4 (page 106)

1 $\dfrac{1}{3}$ 2 $\dfrac{243\pi}{2}$ 3 4.5 4 $\dfrac{abc}{3}(b^2+c^2)$ 5 $\dfrac{\pi r^3}{3}$ 6 4.5 7 $\pi+8$ 8 26 9 $\dfrac{22}{3}$

10 $\dfrac{1}{8}\left(\dfrac{\pi}{2}+1\right)$ 11 $\dfrac{1}{3}$ 12 $A=2\displaystyle\int_0^{\pi/6}\int_0^{2\cos 3\theta} r\,dr\,d\theta = \dfrac{\pi}{3}$ 13 $4\pi\left\{\dfrac{1}{\sqrt{2}}-\dfrac{1}{\sqrt{5}}\right\}$

14 $\dfrac{64}{9}(3\pi-4)$ 15 $M=\displaystyle\int_0^\pi\int_0^{a(1+\cos\theta)} r^2\sin\theta\,dr\,d\theta = \dfrac{4a^3}{3};\ h=\dfrac{16a}{9\pi}$ 16 (a) $\dfrac{1}{2}\pi ab$

(b) $\dfrac{1}{8}\pi ab^3$; centroid $\left(0,\dfrac{4b}{3\pi}\right)$ 17 19.56 18 $\dfrac{b^2}{4}(c^2-a^2)$ 19 $\dfrac{a^2}{6}(2\pi+3\sqrt{3})$

20 232

Test exercise 5 (page 148)

1 (a) $dz = 4x^3\cos 3y\,dx - 3x^4\sin 3y\,dy$ (b) $dz = 2e^{2y}\{2\cos 4x\,dx + \sin 4x\,dy\}$
(c) $dz = xw^2\{2yw\,dx + xw\,dy + 3xy\,dw\}$ 2 (a) $z = x^3y^4 + 4x^2 - 5y^3$
(b) $z = x^2\cos 4y + 2\cos 3x + 4y^2$ (c) not exact differential
3 9 square units   4 (a) 278.6   (b) $\pi/2$ (c) 22.5   (d) 48   (e) -21 (f) -54π
5 Area $= \dfrac{5}{12}$ square units   6 (a) 2   (b) 0

Further problems 5 (page 149)

1 14 2 1.6 3 $\dfrac{\pi}{36}\{9-4\sqrt{3}\}$ 4 $\dfrac{1}{2}\{\pi^4+4\}$ 5 $\dfrac{9\pi}{256}$ 6 $\dfrac{1}{2}\cdot\ln 2$ 7 $2-\pi/2$

8 $\dfrac{1}{8}$ 9 14 10 (a) 39.24 (b) 0 11 $\dfrac{2}{3}$

Test exercise 6 (page 192)

1 $4\sqrt{2}\pi$ 2 $a(\pi/2)^2$ 3 (a) (1) (4.47, 0.464, 3) (2) (5.92, 0.564, 0.322)
(b) (1) (3.54, 3.54, 3) (2) ($-0.832, 1.82, 3.46$) 4 12π 5 $a^3(8-3a)\pi/12$
6 (a) $I = \displaystyle\iint v(1+u)(1+u+v)\,dv\,dv$ (b) $I = \displaystyle\iiint \dfrac{(2u+v)(v-4w)}{vw}\,du\,dv\,dw$

Further problems 6 (page 192)

1 $4\sqrt{5}\pi$ 2 $\left(\dfrac{a}{2},\dfrac{a}{2},\dfrac{a}{2}\right)$ 3 $10\sqrt{61}$ 4 $\dfrac{4\sqrt{22}\pi}{3}$ 5 $\dfrac{\pi}{24}(5\sqrt{5}-1)$

6 $\pi\sqrt{5}$ 7 $16a^2$ 8 $2a^2(\pi-2)$ 9 $4\pi(a+b)\sqrt{a^2-b^2}$ 10 45π

11 $\dfrac{11}{30}$ 12 $\dfrac{\pi a^4}{2}$ 13 $2\left(\pi-\dfrac{4}{3}\right)$ 14 $\bar{x}=\bar{y}=\bar{z}=\dfrac{3a}{8}$

15 $\dfrac{\pi a^3}{3}\{4\sqrt{2}-3\}$ **16** $\dfrac{4\pi abc}{3}$ **17** $\dfrac{2a^3}{3}$ **18** $\dfrac{1}{4}\iint(u^2+v^2)\,du\,dv$

19 $u^2v\,du\,dv\,dw$ **20** $\bar{z}=-\dfrac{a}{5}$ **21** $\dfrac{7}{18}$ **22** $2-\dfrac{\pi}{2}$ **23** $\dfrac{1}{4}(\sqrt{2}-1)$

Test exercise 7 (page 228)

1 $\overline{AB}=2\mathbf{i}-5\mathbf{j}$, $\overline{BC}=-4\mathbf{i}+\mathbf{j}$, $\overline{CA}=2\mathbf{i}+4\mathbf{j}$, AB $=\sqrt{29}$, BC $=\sqrt{17}$, CA $=\sqrt{20}$
2 (a) -8 (b) $-2\mathbf{i}-7\mathbf{j}-18\mathbf{k}$ 3 (a) -8 (b) $-2\mathbf{i}-7\mathbf{j}-18\mathbf{k}$
4 (a) 6, $\theta=82°44'$ (b) 47.05, $\theta=19°31'$ 5 (a) -15 (b) $-16\mathbf{i}+10\mathbf{j}+17\mathbf{k}$
6 (a) 9 (b) $-(47\mathbf{i}+17\mathbf{j}+29\mathbf{k})$ 7 $\mathbf{A}\cdot(\mathbf{B}\times\mathbf{C})=0$ ∴ vectors coplanar

Further problems 7 (page 228)

1 $\overline{OG}=\dfrac{1}{3}(10\mathbf{i}+2\mathbf{j})$ 2 $\dfrac{1}{\sqrt{50}}(3,4,5)$; $\dfrac{1}{\sqrt{14}}(1,2,-3)$; $\theta=98°42'$

3 moduli: $\sqrt{74}$, $3\sqrt{10}$, $2\sqrt{46}$; direction cosines: $\dfrac{1}{\sqrt{74}}(3,7,-4)$, $\dfrac{1}{3\sqrt{10}}(1,-5,-8)$,

$\dfrac{1}{\sqrt{46}}(3,-1,6)$; sum $=10\mathbf{i}$ 4 8, $17\mathbf{i}-7\mathbf{j}+2\mathbf{k}$, $\theta=66°36'$

5 (a) -7, $7(\mathbf{i}-\mathbf{j}-\mathbf{k})$ (b) $\cos\theta=-0.5$ 6 $\cos\theta=-0.4768$
7 (a) 7, $5\mathbf{i}-3\mathbf{j}-\mathbf{k}$ (b) 8, $11\mathbf{i}+18\mathbf{j}-19\mathbf{k}$

8 $-\dfrac{3}{\sqrt{155}}\mathbf{i}+\dfrac{5}{\sqrt{155}}\mathbf{j}+\dfrac{11}{\sqrt{155}}\mathbf{k}$; $\sin\theta=0.997$ 9 $\dfrac{2}{\sqrt{13}},\dfrac{-3}{\sqrt{13}},0$; $\dfrac{5}{\sqrt{30}},\dfrac{1}{\sqrt{30}},\dfrac{-2}{\sqrt{30}}$

10 $6\sqrt{5}$; $\dfrac{-2}{3\sqrt{5}},\dfrac{4}{3\sqrt{5}},\dfrac{5}{3\sqrt{5}}$ 11 (a) 0, $\theta=90°$ (b) 68.53, $(-0.1459,-0.5982,-0.7879)$

12 $4\mathbf{i}-5\mathbf{j}+11\mathbf{k}$; $\dfrac{1}{9\sqrt{2}}(4,-5,11)$ 13 (a) $\mathbf{i}+3\mathbf{j}-7\mathbf{k}$ (b) $-4\mathbf{i}+\mathbf{j}+2\mathbf{k}$

(c) $13(\mathbf{i}+2\mathbf{j}+\mathbf{k})$ (d) $\dfrac{\sqrt{6}}{6}(\mathbf{i}+2\mathbf{j}+\mathbf{k})$ 14 61 15 $29\mathbf{i}-10\mathbf{j}+16\mathbf{k}$

16 (a) $22\mathbf{i}+14\mathbf{j}+2\mathbf{k}$ (b) $-2\mathbf{i}+14\mathbf{j}-22\mathbf{k}$
17 (a) 15 (b) -33 (c) 7 18 (a) $-6\mathbf{i}+4\mathbf{j}-7\mathbf{k}$ (b) $62\mathbf{i}+10\mathbf{j}-38\mathbf{k}$
(c) $18\mathbf{i}-21\mathbf{j}+10\mathbf{k}$ 19 $p=6$ 20 (a) (1) $p=15/4$ (2) $p=-33$
(b) $\dfrac{1}{7}(3\mathbf{i}-2\mathbf{j}+6\mathbf{k})$

Test exercise 8 (page 261)

1 (a) $4\mathbf{i}-4\mathbf{j}+24\mathbf{k}$ (b) $2\mathbf{i}-2\mathbf{j}+24\mathbf{k}$ (c) 24.66 2 $\mathbf{T}=\dfrac{1}{\sqrt{66}}(4\mathbf{i}+\mathbf{j}+7\mathbf{k})$
3 $\dfrac{8}{5}(25\mathbf{i}-6\mathbf{j}-15\mathbf{k})$ 4 5.08 5 $\dfrac{1}{\sqrt{101}}(2\mathbf{i}+4\mathbf{j}+9\mathbf{k})$
6 (a) $14\mathbf{i}-12\mathbf{j}-30\mathbf{k}$ (b) 8 (c) $5\mathbf{i}-2\mathbf{j}-4\mathbf{k}$ (d) $7\mathbf{i}+2\mathbf{j}+3\mathbf{k}$
(e) $3\mathbf{i}+2\mathbf{j}+\mathbf{k}$

Further problems 8 (page 261)

1 (a) $2x\mathbf{i}+3\mathbf{j}+\cos x\,\mathbf{k}$ (b) $2\mathbf{i}-\sin x\,\mathbf{k}$ (c) $(4x^2+9+\cos^2 x)^{1/2}$
(d) $34+\sin 2$ 2 (a) $2-2u-9u^2$
(b) $(3u^2+4u+3)\mathbf{i}+(3u^2+6)\mathbf{j}+(1-2u)\mathbf{k}$ (c) $\mathbf{i}-2\mathbf{j}+(3-2u)\mathbf{k}$
3 $\dfrac{1}{5\sqrt{21}}(2\mathbf{i}-20\mathbf{j}+11\mathbf{k})$ 5 $\dfrac{-1}{\sqrt{129}}(10\mathbf{i}+2\mathbf{j}-5\mathbf{k})$
5 $\dfrac{-1}{\sqrt{126}}(5\mathbf{i}-\mathbf{j}+10\mathbf{k})$ 6 $\dfrac{-1}{\sqrt{601}}(12\mathbf{i}+4\mathbf{j}-21\mathbf{k})$ 7 -8.285 8 -9.165

9 (a) $12\mathbf{i} - 4\mathbf{j} + 4\mathbf{k}$ (b) $24\mathbf{i} - 4\mathbf{j}$ (c) 144
10 (a) $(2\sin 2)\mathbf{i} + 2e^3\mathbf{j} + (\cos 2 + e^3)\mathbf{k}$ (b) $(4\sin^2 2 + \cos^2 2 + 2e^3\cos 2 + 5e^6)^{1/2}$
11 -5.014 **12** $p = \dfrac{1}{\sqrt{29}}(3\mathbf{i} + 2\mathbf{j} - 4\mathbf{k}); \ q = \dfrac{1}{\sqrt{38}}(6\mathbf{i} - \mathbf{j} + \mathbf{k}); \ \theta = 68°\ 48'$
13 (a) $(2t + 3)\mathbf{i} - (6\cos 3t)\mathbf{j} + 6e^{2t}\mathbf{k}$ (b) $2\mathbf{i} + (18\sin 3t)\mathbf{j} + 12e^{2t}\mathbf{k}$ (c) 12.17
15 $-4x\mathbf{i} + 4z\mathbf{k}$ **16** $(2\cos 5.5)\mathbf{i} - (6\sin 5.5)\mathbf{j} - (6\sin 5.5)\mathbf{k}$

Test exercise 9 (page 310)

1 $3\mathbf{i} + \dfrac{18}{7}\mathbf{j} - \dfrac{81}{8}\mathbf{k}$ **2** 12 **3** $18\pi(2\mathbf{i} + \mathbf{j})$ **4** $24(\mathbf{i} + \mathbf{j})$ **5** $8 + \dfrac{4\pi}{3}$
6 all conservative **7** $36\left(\dfrac{\pi}{4} + 1\right)$ **8** 0

Further problems 9 (page 311)

1 (a) $576\mathbf{k}$ (b) $\dfrac{576}{5}(3\mathbf{i} + \mathbf{j} + 2\mathbf{k})$ **2** $1771\mathbf{i} + 1107\mathbf{j} + 830.4\mathbf{k}$
3 $416.1\mathbf{i} + 718.5\mathbf{j} + 5679\mathbf{k}$ **4** 46.9 **5** -4.18 **6** 8π **7** $\dfrac{16\pi}{3}(\mathbf{i} + \mathbf{k})$
8 $\dfrac{1}{3}(48\mathbf{i} + 64\mathbf{j} - 24\mathbf{k})$ **9** $64\left(\dfrac{\pi}{4} - \dfrac{1}{3}\right)(6\mathbf{i} + 4\mathbf{j})$
10 $\dfrac{9}{2}\{(\pi + 2)\mathbf{i} + (\pi + 2)\mathbf{j} + 4\mathbf{k}\}$ **11** $\dfrac{12}{5}(32\mathbf{j} + 15\mathbf{k})$ **12** -1 **13** $\dfrac{250}{3}\pi$
14 $\dfrac{1}{6}(117\pi + 256 - 28\sqrt{7}) = 91.58$ **15** -80 **16** 96π **17** -2 **18** 12π
19 $-\dfrac{a^3}{3}$ **20** $\dfrac{81\pi}{4}$

Test exercise 10 (page 338)

1 yes, an orthogonal set **2** $h_u = 1, \ h_v = 2v, \ h_\theta = 2u$ **3** $4\mathbf{I} + \mathbf{K}$
4 (a) $(2\cos\phi + 2\cos 2\phi + 1)$ (b) $(2\sin 2\phi + \sin\phi)\mathbf{K}$
5 (a) $(ds)^2 = (dr)^2 + r^2(d\theta)^2 + r^2\sin^2\theta\,(d\phi)^2$ (b) $dV = r^2\sin\theta\,dr\,d\theta\,d\phi$
6 -10.5

Further problems 10 (page 339)

1 (a) yes (b) no **2** -50.5 **3** $2\dfrac{5}{18}$
5 (a) $\nabla^2 V = \dfrac{\partial^2 V}{\partial\rho^2} + \dfrac{1}{\rho}\cdot\dfrac{\partial V}{\partial\rho} + \dfrac{1}{\rho^2}\cdot\dfrac{\partial^2 V}{\partial\phi^2} + \dfrac{\partial^2 V}{\partial z^2}$
(b) $\nabla^2 V = \dfrac{1}{r^2}\cdot\dfrac{\partial}{\partial r}\left(r^2\dfrac{\partial V}{\partial r}\right) + \dfrac{1}{r^2\sin\theta}\cdot\dfrac{\partial}{\partial\theta}\left(\sin\theta\dfrac{\partial V}{\partial\theta}\right) + \dfrac{1}{r^2\sin^2\theta}\cdot\dfrac{\partial^2 V}{\partial\phi^2}$
6 (b) $h_u = h_v = \sqrt{u^2 + v^2}; \ h_w = 1$
(c) $\text{div}\,F = \dfrac{1}{u^2 + v^2}\left\{\dfrac{\partial}{\partial u}\left(\sqrt{u^2 + v^2}\cdot\dfrac{\partial F_u}{\partial u}\right) + \dfrac{\partial}{\partial v}\left(\sqrt{u^2 + v^2}\cdot\dfrac{\partial F_v}{\partial v}\right)\right\} + \dfrac{\partial F_w}{\partial w}$
(d) $\nabla^2 V = \dfrac{1}{u^2 + v^2}\left\{\dfrac{\partial^2 V}{\partial u^2} + \dfrac{\partial^2 V}{\partial v^2}\right\} + \dfrac{\partial^2 V}{\partial w^2}$

Index

addition of vectors 198–201
angle between two vectors 215–217, 226
applications of multiple integrals 88–93
applications of partial derivatives 25–55
arc length, line integral 130–131
area of a rectangle, double integrals 81–83
area, enclosed by closed curve 115–119
area, enclosed by polar curves 64–68, 88–98

Cartesian coordinates 57–59, 163–166, 189–190
change of variables 37–42, 177–178, 191
components of a vector 201–207
conservative vector fields 284–288, 308
coordinate systems 325–326
coplanar vectors 220–221, 226–227
curl 252–258, 331–333
curl of a vector function see *curl*
curvilinear coordinates 179–188, 314–339
cylindrical coordinates 164–166, 189–190

determination of volume 99–102
differentials 108–119, 137–138, 145–146
differentiation of sums 235–236
differentiation of vectors 231–238
differentiation of products 235–236
direction cosines 209, 225–226
direction of unit normal vectors 298–303
direction ratios 218
directional derivatives 245–248
div 251–252, 254–258, 331
divergence of a vector function see *div*
divergence theorem 289–295, 309
double integrals 84–85, 93–96, 152–167

element of arc ds 329
element of volume dV 329
equal vectors 197–198
equations of standard polar curves 61–64
errors 16–22
exact differentials 112–115, 137–138, 146

first partial derivatives 2–8

Gauss' theorem 289–295, 309
general curvilinear coordinate system 327–328
grad 242–244, 254–258, 330–331
grad of sums 250–251
gradient of a scalar function see *grad*
graphs of polar curves 59–61
Green's theorem 304–307, 309
Green's theorem in the plane 138–144, 146

integration of exact differentials 113–115
integration of vector functions 239–241
inverse functions 43–54

Jacobin 177–188

lengths of polar curves 70–72
line integral round a closed curve 126–129
line integrals 119–138, 146, 264–271, 307

multiple integrals 80–107, 177–178

orthogonal coordinate systems 320–324
orthogonal curvilinear coordinates 319–320, 329–333

parametric equations 131–132
partial derivatives, application 25–55
partial differentiation 1–24, 238–241
particular orthogonal systems 333–335
path of integration 132–136
polar coordinates 56–79, 308
polar curves 59–76
products of scalars 250–251
properties of line integrals 123–125
properties of scalar triple products 219–220

rate-of-change problems 29–36
regions enclosed by closed curve 125

scalar fields 241–254, 264–266, 276–279
scalar product of three vectors 218–219
scalar product of two vectors 210–212, 226

scalar quantities 196, 225
scalar triple products 219–220, 226
scale factors 324–326
second moments of area 88–98
second-order partial derivatives 8–14
small increments 16–22
solenoidal vector 252
space coordinate systems 163–167
spherical coordinates 165–166, 189–190
standard polar curves 61–76
Stokes' theorem 295–297, 309
summation in two directions 81–83
surface integrals 151, 157–167, 189, 275–283, 308
surface of revolution 73–76

three independent variables 137–138
transformation equations 328
transformation in three dimensions 187–188
triple integrals 86–88, 218–225

unit normal vectors 248–250, 298–303
unit tangent vectors 236–238
unit vectors 205–207, 226

vector differentiation 230–261
vector fields 241–254, 267–271, 279–283
vector integration 263–313
vector product of two vectors 212–215, 226
vector quantities 196
vector representation 197–201
vector triple products 222–225, 227
vector, components 201–207
vectors 195–229
vectors in space 207–208
vectors, types 198
volume in space 167
volume integrals 151, 168–178, 190, 271–275, 307
volume of revolution 68–70
volume, triple integrals 99–102